The
European
Mathematical
Awakening

The European Mathematical Awakening

A Journey Through the History of Mathematics, 1000–1800

Edited by Frank J. Swetz

Dover Publications, Inc.
Mineola, New York

Bibliographical Note

This Dover edition, first published in 2013, contains a selection of thirty-two corrected chapters from *Five Fingers to Infinity: A Journey Through the History of Mathematics*, originally published in 1994 by Open Court Publishing Company, Chicago. The Preface, Epilogue, Historical Exhibit 9, Editor's Notes, Bibliography, and introductory Perspectives have been specially prepared for this Dover edition by Frank J. Swetz.

Library of Congress Cataloging-in-Publication Data

From five fingers to infinity. Selections.
   The European mathematical awakening : a journey through the history of mathematics, 1000–1800 / edited by Frank J. Swetz. — Dover edition.
      p. cm.
   "This Dover edition, first published in 2013, contains a selection of thirty-two corrected chapters From Five Fingers to Infinity: A Journey Through the History of Mathematics, originally published in 1994 by Open Court Publishing Company, Chicago" —
   Summary: "A global survey of the history of mathematics, this newly corrected and updated collection of 32 highly readable essays features contributions by such distinguished educators as Carl Boyer and Morris Kline. Fascinating articles explore studies by Fibonacci, Descartes, Cardano, Kepler, Galileo, Pascal, Newton, Euler, and others. Suitable for readers with no background in math" — Provided by publisher.
   Summary: "A history of mathematics in Europe from 1000 to 1800" — Provided by publisher.
   Includes bibliographical references and index.
   ISBN-13: 978-0-486-49805-8 (pbk.)
   ISBN-10: 0-486-49805-0 (pbk.)
   1. Mathematics—Europe—History I. Swetz, Frank J., 1937– II. Title.

QA27.E85F76 2013
510.9—dc23

2012031301

Manufactured in the United States by Courier Corporation
49805001
www.doverpublications.com

Table of Contents

Preface

*T*he following journey through the history of mathematics spans eight centuries, from 1000 CE to 1800 CE. This is a long period of time, which marks a turning point in European mathematical thinking, An inflection point in the graphical display of the "Growth of Mathematical Knowledge" as shown in Historical Exhibit 1 (p.4). It signifies a true societal awakening to the scope and power of mathematics and its applications. Other societies also experienced such intellectual transitions but their experiences will not be encountered directly during this journey. One route of understanding will be circuitous and challenging enough.

The landmarks of mathematical achievement and their relevant impact are dispersed over time and location. As your principal tour guide, I have tried to mark a meaningful travel agenda focusing on the development of mathematics and the conditions supporting that development. At each stop during the journey other tour guides, the authors of the individual articles, provide more experienced guidance to for the mathematical encounter. The route I have selected is chronologically organized so that in traversing it a reader follows the path of historical influences, the circumstances and discoveries, which gave rise to the particular event. However, each event in itself maybe visited separately, and still conveys a meaningful encounter. Historical Exhibits, "snapshots," of some prominent features that reflect on the developments considered are included at relevant intervals. These Historical Exhibits are intended to provide further insights into the events visited. At each stop in our journey, further guidance is also provided by bibliographies and notes.

While the period of our intellectual journey covers 800 years, the developments in mathematics are not constrained or limited by particular intervals. Mathematical knowledge is, and has been, constantly growing. It is an ongoing process, an expression of human existence, one in which we all can participate.

FRANK J. SWETZ
Harrisburg, Pennsylvania
October, 2011

The
European
Mathematical
Awakening

Perspective: The European Mathematical Awakening

By the beginnings of the 11th century, Europe had survived a series of cataclysmic events: barbarian invasions; plagues and crop failures, to emerge from a period of intellectual and political stagnation known as the "Dark Ages." Advances in agriculture and animal husbandry provided a better and larger food supply. Improved nutrition stimulated population increase. Land reclamation projects were undertaken and new towns founded. The harnessing of wind and water power made many of life's daily chores easier. Opportunities in villages and towns attracted freemen, craftsmen and artisans. Road systems were improved. Trade and commerce increased. Cities grew and prospered. In particular, the Italian maritime city states of Pisa, Venice and Genoa benefited from the weakening Islamic domination of the Mediterranean to become trade entrepôts for the goods and commodities of the Levant. Merchants from these cities established trading houses abroad, conducting business at the sources of their imports. Interacting with local merchants, they learned their customs and habits and brought much of this new knowledge back to Europe. Civic authority and structure were resurrected and strengthened throughout Europe, giving rise to regional and local identities, beginnings of a sense of nationalism. Trade guilds provided a united voice for some skilled working classes. These new institutions of identity and wealth gave rise to political power, which would challenge and modify existing sources of authority. Since the fall of the Roman Empire, the dominant encompassing political, as well as spiritual, authority in Europe was the Catholic Church. The Church was primarily concerned with otherworldly matters, and initially viewed inherited Greek scientific and mathematical theories as suspect pagan knowledge. However, the study of mathematics was formally sanctioned by St. Augustine (ca.400) as worthy of Christian involvement; still, the Catholic Church's actual interest with mathematics was minimal, limited primarily to the determination of the Church calendar based on a correct dating of Easter. Handbooks called *computi* were written to assist in this task. A few churchmen pursued mathematics for its intrinsic and classical values. One such scholar was the French monk, Gerbert of Aurillac (ca. 950–1003), who eventually became Pope Sylvester II in the year 999. Gerbert sought out existing Arabic sources of Euclid's *Elements* to compile a practical geometry text, employed little known Hindu Arabic numerals and improved counting table computing techniques. The 11th century also witnessed a revitalization and reform of the Church's monastic movement, which resulted in a widespread establishment of new monastic centers. These centers also helped to foster a sense of regional coherence and also contributed to an intellectual revival by establishing libraries and schools. Monks labored in *scriptoria* to copy and preserve extant antique works. Cathedral and monastic schools became available for the education of youth.

The new sources of political power began to assert themselves and challenge the overriding power of the Catholic Church. The church in promoting other worldliness, discouraged intellectual curiosity of the physical world and disdained the accumulation and manipulation of wealth. The rising climates of commercial expansionism and humanistic inquiry found themselves in direct conflict with these beliefs. Scholars now began to take a closer look at the world around them and tried to understand the physical forces that controlled nature and human existence. Mathematics became a primary tool in these quests of understanding.

European merchants in their travels abroad were avid observers of foreign practices and customs that would improve their profits margins. Such practices would be brought back to Europe and adapted or refined to suit the local commercial milieu. Leonardo of Pisa (Fibonacci), (ca.1175-1250), a member of a prominent Pisan merchant family, worked in their trade colony in Bougie on the coast of North Africa. Leonardo studied mathematics under the tutelage of Arab instructors. He learned a new set of numerals, said to have originated in India. Accompanying these numerals were computing schemes, *algorisms* that could be carried out with pencil and paper; freeing problem solvers from the labors of a computing table. Leonardo published his findings on the Hindu Arabic numerals for a European reading audience in *Liber Abaci* (1202). Eventually, this new system of arithmetic became popular with the European merchant community replacing the use of the cumbersome counting table and the *figura imperiale*, Roman numerals. Soon, special teachers called reckoning masters: in Italian, *maestri d'abbaco;* in German, *Rechenmeister,* taught this new form of computation to the merchant community and paying students. Adam Riese (1492-1559) became a well respect member of this movement. Books and manuscripts called *abaci* and *practicae* appeared promoting the new arithmetic. The advent of printing with movable type greatly helped to disseminate this new knowledge. The first printed European arithmetic book appeared in 1478 in Treviso, a small commercial town, outside of Venice. It is called simply the *"Treviso Arithmetic."* Now, written or printed calculations allowed for retrospection, analysis and the possible perception of patterns and structure in mathematics. Also at this time, the classics of Greek scholarship, preserved in Islamic libraries were reintroduced into Europe, translated into local languages, read and studied. Printed copies of *Euclid's Elements* were particularly in demand. Soon, this new instruction on arithmetic moved from the limited tutelage of the reckoning masters to the *quadrivium* of the monastic schools and to the newly founded universities or *guilds of scholars*. The use of numbers and calculation became available to a wide segment of the population and with this knowledge came an increased awareness of the usefulness and power of mathematics; power to make a livelihood and power to better understand the world.

The new numerals assisted in communication involving mathematics. Repeated phrases or operations printed in arithmetic manuals lent themselves to standardization, and abbreviation. Now mathematical problems that had for so long been expressed in sometime obscure rhetorical form, could be understood as a form of relationships between numbers. Soon there were special symbols adopted to represent the basic operations of addition and subtraction, multiplication, division and "equals". Often, in written or printed statement, one word or phrase was repeated many times. For example, in Italian the word for "unknown" or "thing" was *"cosa";* on a single typeset page of arithmetic book, *cosa* could appear a dozen times during an explanation. Such repetition, prompted the mathematician Luca Pacioli (ca. 1445-1509) to use the abbreviation "co" for *cosa* in writing his *Summa de arithmetica, geometrica, proportioni et proportionalita* published in 1494. As a result of such innovations, rhetorical algebra evolved into symbolic algebra. Pacioli's *Summa* was believed to be the compendium of 15th century mathematical knowledge. In his closing comments, the author noted that obtaining an algebraic solution for the cubic equation was impossible. Within forty years of this statement, solution techniques for the cubic equation were developed and perfected by such mathematicians as Scipione del Ferro, Nicolo Tartaglia, Girolamo Cardano and Rafael Bombelli. In the work done to obtain the solution process for cubic equations, a foundation was laid for a development of a theory of equations. Algebraic manipulation now encompassed the use of imaginary numbers. Improved astronomical techniques and measurement prompted a maturing of trigonometry and high-lighted urgency for improved computational efficiency and mathematical accuracy. In part to meet these needs, a variety of concrete computing devices were invented and computational capacities were strengthened by Simon Stevin's systematization of decimal fractions (1585) and the appearance of John Napier's logarithms (1614). New scientific theories were proposed and mathematically investigated by a series of individuals. Nicolo Tartaglia (1499-1557) and Galileo Galilei (1564-1642) explored the forces acting on a cannonball in motion. René Descartes (1596-1650) sought to unravel the workings of a rainbow and the forces propelling the planets through space. Greek number mysticism was replaced by a more mathematically based theory of numbers. Communities of natural scientist/mathematicians working in unison or consecutively, sought to unravel the mysteries of nature. National academies and societies were founded to facilitate cooperation and the exchange of scientific information. The British Royal Society opened its doors in 1660 and the French Academy in 1666. The trajectories of the planets were explored by Nicolo Copernicus (1473-1543), Johannes Kepler (1571-1630) and Galileo, and finally explained in the *Principia* of Isaac Newton (1643-1727). In the process of examining motion in space, new geometric theories as well as geometries were established. Projective geometry and analytic geometry allowed an investigator to better follow a moving point in space. Considerations by Pierre de Fermat (1601-1665), Blaise Pascal (1623-1662), and Christiaan Huygens (1629-1695) of the phenomena of change and attempts to determine its predictability led to the new mathematics of probability. The world of the discrete and stationary became a world of dynamic motion and change. Mathematicians sought to understand the instantaneous, and the infinite. The differential calculus devised by Isaac Newton and Gottfried Wilhelm Leibniz (1646-1716) provided the powerful means for such explorations. The 17th century ushered in an era of scientific experimentation, skepticism and rationalism as advocated by such works as William Gilbert's *De Magnete* (1600), Galileo's *Dialogue* (1632) and Descartes' *Discourse on Method* (1637). It also begot an era of geographical exploration, conquest and European imperial expansion.

The mathematical momentum and spirit of intellectual adventurism that began in the 17th century extended into the following two centuries. Europe experienced the rise of the Industrial Revolution and with it an accompanying increase of complex technological, sociological and political problems. Enlightened rulers sanctioned and supported the use of science and mathematics in national development. Scientific accomplishments became a source of national pride. The era of the natural philosopher and generalist mathematician were replaced by the highly skilled specialist, and a new term, *"Applied Mathematics"* entered the lexicon of mankind.

As the horizons of mathematical challenges have broadened, able individuals have, and always will, come forth to meet those challenges.

HISTORICAL EXHIBIT 1

The Growth of Mathematical Knowledge

Mathematical knowledge is cumulative, that is, people use and build upon the mathematics developed before them. If this fact seems reasonable then a model for the growth of mathematical knowledge can easily be devised by counting the number of recorded mathematicians that have lived up to any period of time and then plotting that number against the chronological year. Such a count was undertaken by noting the mathematicians listed in the comprehensive *Dictionary of Scientific Biography*, 16 vols. (New York: Scribner's, 1970–80). Cumulative sums were computed for the various years 600 B.C. to A.D. 1850 and plotted against their respective years. The following graph was produced:

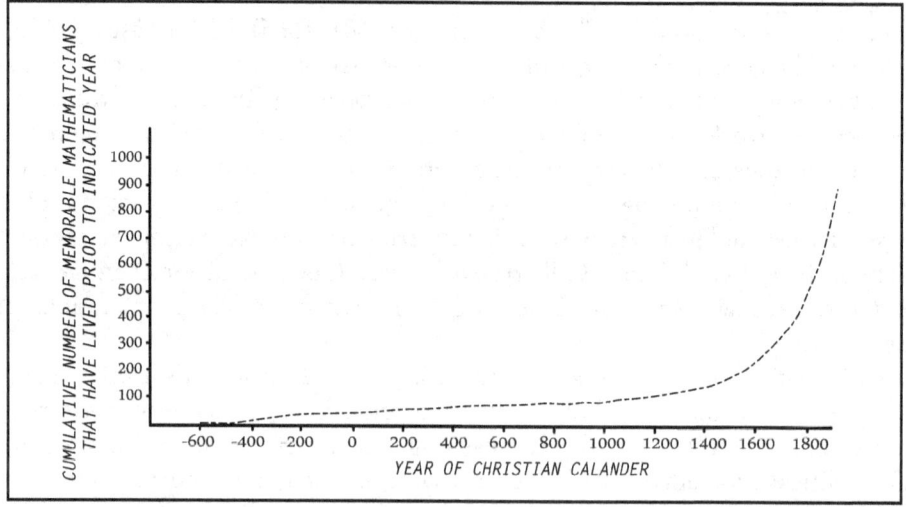

Several interesting trends can be noted from this graph, for example, the historical period A.D. 600 to 1200 is often termed the "Dark Ages," a time when little intellectual progress was taking place in Europe; yet the graph indicates a steady, albeit slow, growth rate of mathematical knowledge over this period. Around 1400, the rate of mathematical activity increased, due possibly to the introduction of printing and the writing of books in the common language. After A.D. 1500, the growth rate of mathematical knowledge increased exponentially.

Counters: Computing if You Can Count to Five

VERA SANFORD

ORDINARY computation can be accomplished with a minimum of learning by using the loose counter abacus and the counting board. The counting board is a flat surface marked with a series of parallel lines whose values are 1, 10, 100, 1000. The counters are small, easily handled objects,—pebbles, metal disks about the size of a penny, or even, for present-day experimentation, paper clips. The position of a counter shows its value. The number 1432 is indicated by placing one counter on the thousands line, four on the hundreds line, three on the tens, and two on the units line.

Reckoning with counters has left its mark in such words as "calculater" and "calculus" from the Latin *calculus* (a small stone), the "counter" in a store which originally was a counting board, to "cast up accounts" from throwing the counters on the board, and the terms "carry" and "borrow" which described the actual process.

The origin of the counting board and much of its history are not known. Herodotus (ca. 425 B.C.) notes that, "In writing and in reckoning with pebbles, the Greeks move the hand from left to right, but the Egyptians from right to left." This indicates that the reader was familiar with the process and would be interested in the difference between the Greek practice and that of the Egyptians and it also indicates that the lines of the

abacus were vertical. In both Greek and Latin literature, references to the counters appear in context such as the following,—"He (Solon) used to say that men who surrounded tyrants were like the pebbles used in calculations; for just as each pebble stood now for more, now for less, so the tyrants would treat each of their courtiers now as great and famous, now as of no account." We also know that the equipment of a Roman school boy included a bag of counters as well as a wax tablet. And there are still extant Roman abacuses of the type in which beads slide on rods or where studs move in grooves. Details as to the use of the abacus are lacking.

There are references to the use of counter casting in the fourteenth and fifteenth centuries, but there appears to be no actual description of its operation.

In the sixteenth century, a considerable number of arithmetics appearing in northern Europe, especially in Germany, included accounts of computation with the loose counter abacus which seems to have become a well established mercantile practice. German arithmetics were outstanding in this regard. In England, the earliest known book in English on arithmetic (1539) has the following descriptive title:

AN INTRODUCTION

for to lerne to recken with the pen, or with the counters accordynge to the trewe cast of Algorysme, in hole nombers or in broken/newly corrected. And certayne notable and goodlye rules of false posytions

Reprinted from *Mathematics Teacher* 43 (Nov., 1950): 368–70; with permission of the National Council of Teachers of Mathematics.

therevnto added, not before sene in oure Englyshe tonge, by whiche all maner of difficyle questionyons may easily be dissolved and assoylyd.

This was quickly followed by Robert Recorde's *Ground of Artes* (ca. 1542) which had a section on the use of counters. The subject seemed to have been neglected in Italy and in France, the textbooks keeping to the arithmetic with the pen. Strangely enough, although the subject was omitted from the earlier editions of a popular arithmetic in France, a book first printed in 1656, a section on counters was introduced in the edition of 1705 and appeared in at least three editions including that of 1781. It is a bit surprising that the later editors chose to put this topic in, but the explanation is as follows: "This arithmetic is quite as useful as that which is done with the pen since by counters, one can perform every calculation of which he has need in business. This way of computing is more practiced by women than by men; nevertheless many people who are employed in the Treasury and in all the government departments make use of this with great success."

The loose counter abacus of the sixteenth century differed from the Greek one in that the lines were marked from right to left instead of up and down. The line nearest to the computer had the lowest value. The counting board might be of stone with the lines cut on it. It might be of wood with lines drawn in chalk for temporary use or painted on permanently. It might be a table cloth with the lines embroidered on it. The lines had the values 1, 10, 100, etc., and the spaces between had the values 5, 50, 500. There is a close correspondence between these markings and Roman numerals. A number is indicated by placing counters on the lines and spaces as the case requires. The accompanying figure shows how the numbers 1285 and 431 would appear.

In addition, the addends are indicated on the counting board. Then whenever a line has five counters on it, as is the case with the tens line and the hundreds line in the case given here, the five

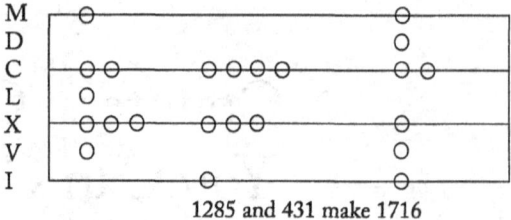

1285 and 431 make 1716

counters are picked up and one is "carried" to the next space. In the example under consideration, there now are two counters in the fifties space. But two fifties make one hundred, so these two counters are picked up and a counter is laid on the hundreds line. The process is repeated until no line has more than four counters and no space more than one.

In subtraction, minuend and subtrahend are entered on the counting board. Then the counters of the subtrahend are matched with those of the minuend and each pair is removed from the board. In some cases it is necessary to "borrow" a counter of higher value from the minuend and to replace it by the equivalent value of counters in the next lower space or line. The process continues until no counters are left in the subtrahend.

To multiply a number by 10, the counters are laid out as if the tens line were actually the units line. To multiply by 100, the hundreds line represents the units. To multiply by 200, you multiply by 100 twice. Since 50 is half of 100, multiplying by 50 is accomplished by taking half of the number of counters on each line or space in 100 times the number. In the following example the number 284 is multiplied by 153. (See solution below.)

Division is difficult and a number of different methods are used. The computer is expected to know the multiplication facts. He decides on the proper quotient figure, and subtracts the partial product from the dividend, using the various lines as the units line as was done in multiplication.

Except for the process of division, computation with the loose counter abacus made no demands on the learner beyond learning how to enter the counters, how to read a number represented by counters, and how to count to five.

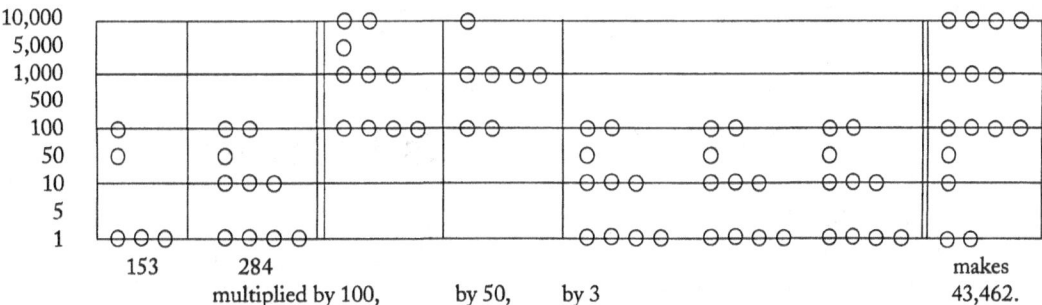

Multiplication was clumsy. Division demanded a knowledge of multiplication combinations unless the computer avoided this issue by using repeated subtraction.

The loose counter abacus is simpler and slower than is the Chinese or the Japanese abacus which requires more mental work. On the other hand, it is a simpler device and one which is easier to master.

REFERENCES

Barnard, F. P. *The Casting Counter and the Counting-Board*, Oxford, 1916. This is technical and exhaustive.

Smith, D. E. *History of Mathematics* II, Boston, 1925, pp. 156–192.

Yeldham, F. A., *The Story of Reckoning in the Middle Ages*, London, 1926.

Editor's Note: For more complete information on the medieval computing table, see: *The History of the Abacus.* J.M. Pullan (1969)

Bede's Finger Mathematics

Bede was a British monk who lived in the seventh century (ca. 673–735) and who was considered to be one of the greatest of the medieval Church scholars. Among his works were mathematical tracts on the Church calendar, i.e. computi, ancient number theory, and finger mathematics or numeration which was actually a method for designating numbers by the use of finger gestures. The following diagram, published a thousand years after Bede's death, illustrates some of Bede's medieval number postures.

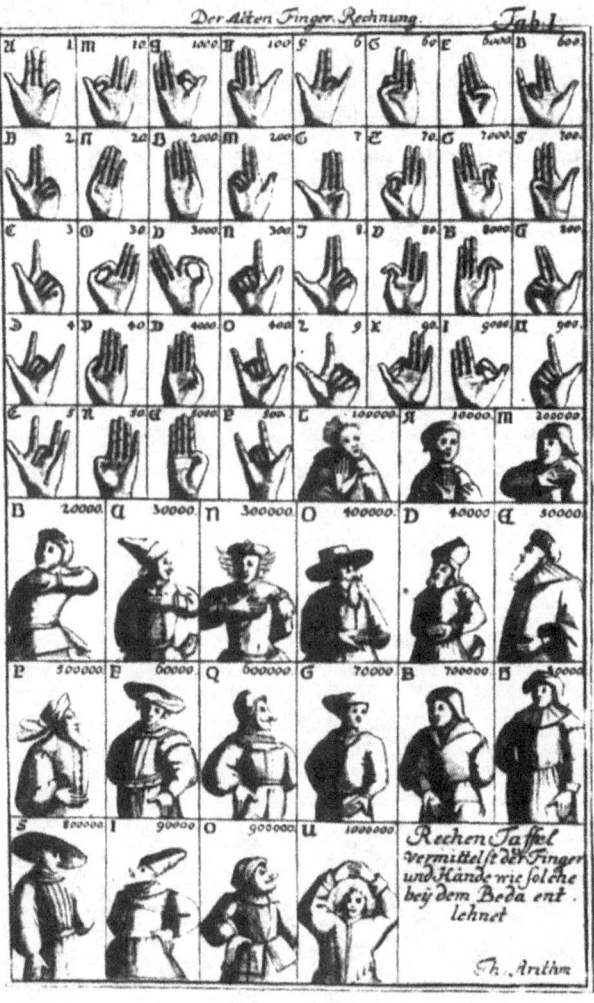

Gerbert's Letter to Adelbold

G. A. MILLER

ONE OF THE MOST astounding comedies of errors in the history of elementary mathematics relates to Gerbert's letter to Adelbold, as may be inferred from the brief articles by the present writer published in the May, 1921, number of the *Scientific Monthly* and in the June, 1921, number of the *American Mathematical Monthly*. Neither of these two articles furnishes the full data upon which the conclusions were based and hence they serve mainly to direct the attention of the careful reader to a few sources of interesting information and to present to the indifferent reader statements which he may accept or reject as his confidence, or lack of confidence, in the judgment of the present writer on such questions may dictate.

Teachers of mathematics should be especially interested in Gerbert, who died in 1003 as Pope Sylvester II, since he is the only man who rose to the position of Pope of the Catholic Church from that of being the most influential teacher of mathematics and other subjects in his generation. The mathematics which Gerbert taught would now be regarded as very elementary, and hence he should be classed with the teachers of secondary mathematics rather than with those of college grade. He lived at a time when even the leading mathematicians of the world used little of their mathematical heritage and added practically nothing thereto for the good of future generations.

Statements found in various well known modern histories of mathematics might lead one to infer that the letter under consideration had been a real contribution towards the increase of mathematical knowledge. Among these statements are the following: It is "the first mathematical paper of the Middle Ages which deserves this name." It explains "the reason why the area of a triangle, obtained by taking the product of the base by half its altitude, differs from the area calculated according to the formula $1/2\, a\, (a + 1)$." It "gives the correct explanation that in the latter formula all the small squares, into which the triangle is supposed to be divided, are counted in wholly, even though parts of them project beyond it." To provide a solid basis for further observations and on account of the intrinsic value of this letter for teachers we quote in full the extant part thereof, as follows[1]:

"In these geometric figures which you have taken from us there was an equilateral triangle whose side was 30 feet, altitude 26, and, according to multiplication of side and altitude, the area 390. If you measure the same triangle without paying attention to the altitude, viz., that one side be multiplied into itself and to this product the numerical measure of one side be added and from this sum half be taken, the area is 465. Do you not see how these two rules disagree? But also the geometric rule, which, by taking into consideration the altitude makes the area 390 feet, has been minutely discussed by me, and I grant that its altitude is only $25^5/7$ feet and its area $385^5/7$ feet. And let this be a universal rule for you for finding the altitude in every equilateral triangle; from the side always take away a seventh and assign the six remaining parts to the altitude."

Reprinted from *School Science and Mathematics* 21 (Oct., 1921): 649–53; with permission of the School Science and Mathematics Association.

"To make what has been said more intelligible permit me to exemplify with smaller numbers. I give you a triangle whose side is 7 feet long. This by the geometric rule I measure thus. Take away a seventh of the side and the six sevenths which are left I give to the altitude. I multiply the side by this and say 6 times 7, that makes 42, the half of this, 21, is the area of said triangle. If you measure the same triangle by the arithmetic rule and say: 7 times 7, that makes 49, and you add the side, making 56, and divide so that you may find the area, you will obtain 28. Behold thus in a triangle of one magnitude there are different areas, which is impossible."

"But that you may not wonder longer I shall explain to you the cause of the diversity. I believe it is known to you what feet are said to be linear, what square, and what cubic, and that in measuring areas we take only square feet. However small a part of these the triangle touches the arithmetic rule computes them as integral. Allow me to give a diagram[2] that what is said may be more clear."

"Behold in this little diagram there are included 28 feet, not all of which are integral. Whence the arithmetic rule, taking the part for the whole, counts the halves with the integers. But the skill of the geometric discipline throwing away the small parts extending beyond the sides and putting together the remaining halves cut in two on the inside of the lines considers that only which is enclosed by the lines. For in this little diagram of which the sides are each 7 feet long if you seek the altitude it is 6 feet. Multiplying this number 6 by 7 you complete, as it were, a rectangle of which the front is 6 feet, the side 7 and the area of it you thus determine to be 42 feet. If you take half of this you leave a triangle of 21 feet. To understand more clearly use your eyes and always remember me."

It may be assumed that all modern teachers of mathematics agree that if any of their students would present the vague arguments found in this letter to explain why the formula $\frac{1}{2}a\,(a + 1)$ does not give the same result for the area of an equilateral triangle of side a as the formula $(a/2)^2\,\sqrt{3}$, which is equivalent to the product of one-half the base into the altitude, they would reply that these arguments failed to explain the difference in question. Hence the statement that Gerbert gave a *correct* explanation of this difference, which appears in various well-known modern histories of mathematics, including that of M. Cantor, volume I (1907), page 865, is inaccurate.

It may be of interest to consider here briefly a possible explanation of the fact that the Roman surveyors assumed that the formula $\frac{1}{2}a\,(a + 1)$ represents the area of an equilateral triangle whose side is a. The ancient Greeks used the term *triangular numbers* for the positive integers of the form $\frac{1}{2}a\,(a + 1)$. From the facts that the number of equal tangent circles arranged

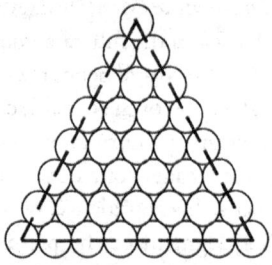

as in the adjoining figure is of this form and that the centers of the outside circles in this figure determine an equilateral triangle, it seems natural to assume that the term triangular numbers originated in connection with some such figure.

If, in the given figure, the number of circles is increased indefinitely without changing the centers of the corner circles, the number of these equal tangent circles will always be expressed by the formula $\frac{1}{2}a\,(a + 1)$ where a represents the number of those circles whose centers determine a side of the given equilateral triangle. As the lower limit of each of these circles is a point it would appear that the number of points in the surface bounded by the given equilateral triangle should be $\frac{1}{2}a\,(a + 1)$, where a represents the number of points in a side. On the other hand, if the area of the given triangle is expressed in smaller and smaller square units, such that the length of the base is equal to a of the corresponding linear units, the area of this triangle is always $(a/2)^2\sqrt{3}$. In fact, $^1/_2a$ $(a + 1)$ has for its limit as a is indefinitely increased the area of an isosceles triangle whose base is equal to the altitude as may be seen directly by means of figure below.

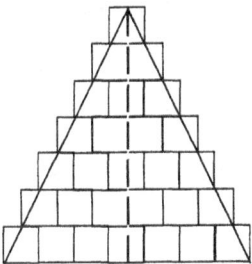

It would thus seem that the assumption that a surface is composed of points which are the limits of equal tangent circles leads to a contradiction. At any rate, if our hypothesis in regard to the origin of the formula $^1/_2a(a + 1)$ for the area of an equilateral triangle whose side is a is correct, it is interesting to note the dilemma into which the ancient Roman surveyors were led by assuming that a surface is made up of points which are the limits of circles. While these surveyors were not good mathematicians they had sufficient mathematical instinct to recognize that if two assumptions lead to contradictory results they cannot both be true—an instinct which is sometimes disregarded by eminent modern mathematicians.

Returning to Gerbert's letter it may first be noted that the rule for finding the altitude of an equilateral triangle which is found therein is inaccurate since this altitude is equal to a side multiplied by $^1/_2\sqrt{3}$ = .8660+, while $^6/_7$ = .8571+. It is true that the latter is a fairly close approximation but the language of the letter implies that the writer does not realize that his rule gives a result which is only approximately correct.

The main claims for the mathematical merits of the letter in question relate to the so-called explanation of the reasons why the "geometric rule" and the "arithmetic rule" give different re-

sults. Judging from his language the diagram to which Gerbert refers may have been like the adjoining figure which he probably assumed to represent an equilateral triangle. At any rate, if the number of squares on the base line had been 30, in accord with the first triangle to which Gerbert refers in this letter, the vertex of the equilateral triangle constructed on this base would have been below several of the topmost squares, and hence a number of the squares of the corresponding figure would not have been touched by this triangle, lying entirely outside of it.

The method of counting as wholes also those squares which lie only partly within the triangle would therefore not have yielded the number 465, required by the formula $^1/_2a$ $(a + 1)$. Even for the special case when a = 7 the explanation is too vague to be called correct. Hence it seems clear that the statement that Gerbert gives a *correct* explanation[3] of the fact that different results are obtained by using two different methods for finding the area of an equilateral triangle is very misleading.

The claim that the letter in question is "the first mathematical paper of the Middle Ages which deserves this name" is even more misleading. It is true that not much mathematical progress was made from the beginning of the Middle Ages to the time of Gerbert, but various papers which were produced during this period are much more meritorious than this letter by Gerbert. In support of this assertion it is only necessary to recall that much of the work of the Hindu and the Arabian mathematicians belongs to this period. In particular, the work from which our term for algebra is derived appeared therein, as well as the dissertation on amicable numbers by Tabit ibn Korra, which has been called "the first known specimen of original work in mathematics on Arabian soil."[4]

Teachers of mathematics should be interested in the letter of Gerbert quoted above not so much on account of the many historical errors relating thereto as on account of the fact that it exhibits a type of the best mathematical thinking among the

Romans during the tenth century. From this standpoint the letter is very instructive and presents to us a picture of weak mathematical thinking on the part of a gifted man. We are thus reminded of the need of carefully graded mathematical aids for our students in order that they may reach even at an early age a mathematical maturity which excels that of the most gifted among the Romans who had to find their way without such aids. Gerbert's letter can profitably be read repeatedly by teachers of mathematics, since it illustrates a type of mathematical thinking which is not only interesting

historically but is represented by many students at the present time.

NOTES

1. Gerberti, *Opera Mathematica,* by N. Bubnov, 1899, p. 43. Professors H. J. Barton and W. A. Oldfather assisted me in the translation of this letter.

2. Cf. Fig. 2.

3. This claim seems to have been made first by H. Hankel, *Zur Geschichte der Mathematik,* 1874, p. 314.

4. F. Cajori, *History of Mathematics,* 1919, p. 104.

Editor's Note: Adelbold [*sic*], i.e. Adalbold studied mathematics under Gerbert. Eventually, he went on to become Bishop of Utrech (1010 – 1026) and wrote a treatise on the volume of a sphere, which he dedicated to Gerbert when he became Pope Sylvester II. Adalbold and Pope Sylvester II remained mathematical correspondents. For more information on Pope Sylvester II and his influence on mathematics, see:
The Abacus and the Cross: The Story of the Pope Who brought the Light of Science to the Middle Ages. Nancy Brown (2010).

HISTORICAL EXHIBIT 3

The Geometry of Gothic Church Windows

One of the outstanding architectural features of European Gothic churches is tracery, ornamental stone work of interlacing or branching arcs. Tracery depends on the ingenious and creative use of circular arcs and attests to the skill of medieval stonemasons as both craftsmen and users of geometry. Individual designs were quite simple but combined they resulted in structures of striking symmetry and beauty. Two tracery patterns used in constructing vaulted windows are given below.

 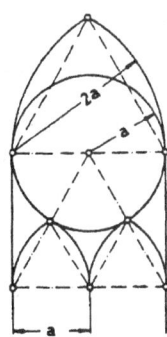  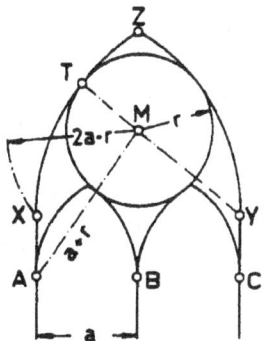

Some other designs are:

Illustrations adapted from Benno Artmann, "The Cloisters of Hauterine," *The Mathematical Intelligencer* 13 (Spring 1991): 44–49; with the permission of Springer-Verlag and the cooperation of the author.

3

The Arithmetic of
the Medieval Universities

DOROTHY V. SCHRADER

The Seven Liberal Arts in Antiquity

The seven liberal arts—the trivium composed of grammar, logic, and rhetoric and the quadrivium made up of arithmetic, music, geometry, and astronomy—were the basis of the curricula in the medieval universities. Whence came these liberal arts? They were not the creation of the Christian Middle Ages but rather a heritage from pagan antiquity. The very word "liberal" implies that these arts belonged to the education of free men, not to the technological training of slaves. The educational system of the medieval universities was an outgrowth, modification, and development of the ancient Greek and Roman educational patterns, adapted and oriented to the Christian ideal.

The Greeks were concerned with the education of free men as future citzens. Plato, whose plan was a theoretical one probably never put into actual practice but nevertheless reflecting the spirit and ideal of his period, conceived of such education as the sole occupation of the first thirty-five years of a man's life. He would have the first twenty years spent on gymnastics, music, and grammar, the next ten on arithmetic, geometry, astronomy, and harmony, and the next five on philosophy.[1] Only then would a man be equipped to take his rightful place as a useful member of society. Aristotle advanced a similar plan in which

the elementary training consisted of reading, writing, gymnastics, and music; and the advanced studies included arithmetic, geometry, and astronomy, with the emphasis on the natural sciences.[2] The Sophists, who asserted that rhetoric rather than natural science was the essential study for higher education, advocated gymastics, music, and drawing in the early years and arithmetic and geometry as advanced studies.[3] Philo Judaeus, about A.D. 30, suggested grammar, music, arithmetic, geometry, astronomy, dialectic, and rhetoric as elementary studies, and philosophy as the one higher study.[4] Sextus Empiricus, in the first half of the third century, mentioned grammar, rhetoric, geometry, astronomy, music, and arithmetic as elementary subjects, reserving dialectic for advanced work.[5] Thus we see that there was a tendency among the Greeks to consider six or eight different types of learning as essential to the education of free men, with some difference of opinion as to the order in which the various subjects should be studied.

The Roman writers have mentioned similar subjects of study in their educational plans. Varro (116–27 B.C.) tried to institute a Roman educational system based on that of the Greeks but with reference to Roman rather than Greek literature. He mentioned grammar, literature, arithmetic, geometry, and music as subjects of study.[6] Seneca (2 B.C.–A.D. 65) was less definite in his plans, sometimes placing medicine, rhetoric, and dialectic as basic liberal arts and sometimes considering them as higher studies.[7] Quintillian (A.D. 35–95) suggested that a boy study grammar, music, geometry,

Reprinted from *Mathematics Teacher* 60 (Mar., 1967): 264–75; with permission of the National Council of Teachers of Mathematics.

and astronomy until he was sixteen, after which age he might advance to higher studies.[8] Thus the Romans, like the Greeks, seem to have accepted a few basic subjects as worthy of the efforts of free men and to have expected the young citizens to concentrate on such subjects. However, there is some difference. The Greeks, with characteristic interest in the speculative, placed a greater emphasis on mathematical theory than did the Romans, who rejected much of the Greek mathematics as impractical and therefore unimportant. Moreover, the Romans expected a boy to be ready for advanced work by the time he was sixteen; the Greeks considered him a beginner until he was twenty or older.

It was Martianus Capella in his *De nuptiis philologiae et mercurii*, written about A.D. 330, who set the number of liberal arts at seven and named them: grammar, dialectic, rhetoric, arithmetic, music, geometry, and astronomy. Capella rejected medicine and architecture as purely technical subjects, pursued only for practical and not speculative ends and so unworthy of free men.[9] By the fourth century, this curriculum of the seven liberal arts, as Capella named them, was well established in the pagan schools.

Christianity and the Liberal Arts

In the first ages of the Church, when Christianity was struggling for its existence in a pagan world, the seven liberal arts were denounced by such Christian writers as Origen, Tertullian, and St. Jerome, not because the arts were evil in themselves, but because they were the basis of the pagan educational system which was a threat to the infant Christian Church. Later, when Christianity gained the ascendency over paganism and the pagan schools were no longer a danger, the pagan educational methods were reexamined and were eventually adopted by the Christians. In fact, the study of the seven liberal arts became prerequisite to the study of theology.[10] St. Augustine wrote on all the liberal arts except astronomy and "although not

the originator of the curriculum of the seven liberal arts, he, more than anyone else, made possible its general adoption by the Christian world of the west."[11]

St. Augustine was thoroughly educated in both pagan and Christian learning. He could see much that was good, true, and irrefutable in the work of the pagan scholars and accepted truth where he found it. "We may well say that St. Augustine defends . . . the principles of logic as the inviolable foundations of knowledge. . . . Side by side with logic we find the truths of mathematics . . . all these truths are necessarily and unconditionally true; they cannot be contested."[12] Augustine himself said (A.D. 386), in illustration of the reality of absolute truth: "However, if there are one world and six worlds, it is clear that there are seven worlds, no matter how I may be effected . . . for even if the whole human race were fast asleep, it would still be necessarily true that three times three are nine and that this is the square of intelligible numbers."[13]

Cassiodorus Senator (A.D. 490–585) was probably the first to use the term "seven liberal arts." With his writing, these particular seven subjects became fixed in Christian education and remained so for the next nine hundred years. Seven, as the number of subjects to be studied, was considered to be sanctioned by Holy Scripture itself (Proverbs 9:1), "See where wisdom has built herself a house, carved out for herself those seven pillars of hers,"[14] and continued to hold full sway in educational circles throughout the Middle Ages.

Arithmetic in the Quadrivium

Of the seven liberal arts, the quadrivium comprised what we might call the scientific studies of the day. It has been frequently stated and more frequently implied that the trivium-quadrivium division of studies existed in theory but that in actuality only the trivium was studied. This is not entirely true. It is true that there was, especially in the early years of the Middle Ages, little creative work in the quadrivium and so little of interest for later generations, but instruction was given and the knowledge possessed was imparted to those students who were interested. That there was much teaching in quadrivial subjects is attested by the many manuscripts extant today in European libraries. Most of these manuscripts have not been published because, being mere copies, translations, or commentaries, they contain little original work and so do not evoke scientific interest. It is probably true that there were fewer students of the quadrivium than of the trivium, just as there are fewer graduate students than undergraduates in the average modern university, and for the same reasons.

Considering specifically the field of mathematics, it is known that Church councils from the time of Charlemagne demanded that the clergy have a knowledge of music and be able to compute the date of Easter. To fulfill these two apparently simple requirements, a knowledge of three subjects of the quadrivium was involved: music, arithmetic, and astronomy. In England, from the eighth to the twelfth centuries it was forbidden to the bishop to ordain a priest who could not compute the date of Easter and teach the method to others.[15] Therefore, it is evident that the quadrivium must have been studied.

All the subjects of the quadrivium were based on the learning of antiquity, but perhaps none shows its Greek origins better than arithmetic. Among the Greeks computation or reckoning, the arithmetic of business, was called logistic and was considered to be entirely different from the study of number as such, which philosophical study was called arithmetic. "Logistic is the theory which deals with numerable objects and not with numbers. It does not consider number in the proper sense of the term . . . "[16] but rather the counting of flocks, addition, subtraction, multiplication, and division, always dealing with sensible objects. Arithmetic corresponded roughly to present-day number theory, being a philosophical approach to what is implied in number; it was a mathematical discussion of properties of numbers, proof, and for-

mal demonstration, a mixture of mathematical rigor and pseudo-scientific, semi-magical mysticism. Arithmetic was a study of the universities; logistic was not.

There seems to be some confusion among modern writers concerning the distinction between logic and logistic. Some have asserted or implied that logistic, not logic, was studied in the trivium and then was followed by the advanced mathematical study, arithmetic. However, it should be clear from the Aristotelian treatises used as texts that logic and not logistic was studied in the trivium of the medieval university. Logistic was practical and utilitarian, a study for children and slaves; logic was a liberal art, a study for free men.

The arithmetical knowledge of the Middle Ages can be divided roughly into three periods, and the type of teaching in the schools and universities falls into similar divisions. The first period extended to the end of the tenth century. Arithmetical knowledge was based on the writing of Nicomachus of Gerasa and his interpreters and commentators. The abacus was used for calculations and Roman letter-numerals for recording the results. The most complicated practical problems attempted had to do with determining the date of Easter, and theoretical arithmetic was confined to the classification of numbers by varied and sometimes fantastic properties.

The second period covered approximately the eleventh and twelfth centuries although the termination of this period varied in different localities. It was a period of little creative activity. However, these years did see an extension of the use of the abacus and the introduction of columnar computations.

The third period, from the end of the twelfth century to the end of the Middle Ages, was one of great activity and great change, one might almost say of intellectual revolution, in Europe. The Hindu-Arabic number system was introduced: zero became a familiar concept in a hitherto zeroless world; through translation into Latin from Arabic and Syriac, the ancient Greek mathematical writings were made available to the medieval world. Much of this intellectual activity was centered in the Islamic universities of Spain, and from there learning spread throughout Europe.

In each period, the arithmetic taught in the schools and universities paralleled the general development of arithmetical learning.

Stress on Number Mysticism and Calendric Reckoning

The first period of mathematical development in Europe closed at the end of the tenth century, before the rise of the great universities. However, arithmetic was considered to be an essential part of the curriculum in the cathedral schools and was also taught in all the monastery schools. The emphasis was on the art of computation, especially the method of establishing the date of Easter. In fact, "computus," which originally meant merely computation, soon came to be associated exclusively with the technical study of Easter reckoning.[17] The exclusive use of Roman numerals made computations with large numbers cumbersome, difficult, and close to impossible. Generally, the actual figuring was done on the abacus and then the results recorded in the Roman letter-numerals. There was much study of the mystical and symbolic number properties and relations, based on the work of Nicomachus.

Nicomachus of Gerasa (ca. A.D. 200) wrote much on number mysticism, a sort of theology of numbers. Numbers were identified with the various gods. He considered the odd numbers to be male and the even ones to be female. He made a strange distinction between the "divine number," a sort of general concept of number which existed only in the mind of the creator-god, and scientific numbers, which were the common numbers known to men on earth. Nicomachus' *Introduction to Arithmetic* is a restatement of commonly known facts, not an original treatise, and is largely Pythagorean. Theon of Smyrna wrote a book dealing with those mathematical matters which were essential for one

who was to read Plato; there are so many similarities between the work of Nicomachus and that of Theon that it is difficult to say which wrote first.[18] Nicomachus' book was the source for Martianus Capella, Boethius, Cassiodorus Senator, and Isidore of Seville.

Martianus Capella, who flourished from about 410 to 429, wrote *The Marriage of Philology and Mercury*, which is actually little more than a textbook, an allegorical setting for a book on the seven liberal arts. The first two chapters establish the setting; the rest of the book is devoted to instruction. It is "strictly instructive, as sapless as the rods of the mediaeval schoolmasters."[19] In it, "the most bizarrre imagination was allied with the most arid intellect."[20] The general level of the mathematical instruction contained in it is shown in the scene in which Philology, fearing to marry Mercury because he is a god and she cannot claim divinity, realizes, after much calculation, that the numbers forming her name and his indicate that this is a propitious match, and so she enters into the marriage happily and without fear.[21] Besides this numerology, Capella discusses proportions and multiples of proportions,[22] prime numbers, perfect numbers, perfect triads (cubes), and similar mystic properties of numbers, stressing the relationship of numbers with the planets.[23]

Martianus Capella's forty-seven pages of number mysticism pale to insignificance in the light of Boethius' hundred or more pages of mathematical work. In his *De institutione arithmetica libri duo*,[24] which was the direct or indirect source of arithmetical knowledge for close to a thousand years, Boethius (ca. 470–525) has not merely translated Nicomachus' treatise. True, he did write after the manner of the Greek, borrowing freely and at times obviously copying, but he also considerably augmented the other's work. In a sense he "baptized" Nicomachus, making his work acceptable to the Christian schoolmen. Boethius' work has the same sort of classification of number properties as does Nicomachus', with many extended or original interpretations and examples; he also includes a mystical interpretation of scriptural numbers. Like Nicomachus, Boethius gives no rules of operation.

The *De arithmetica* of Cassiodorus Senator (490–585) is almost a condensation of Boethius' arithmetical work. Cassiodorus quotes Scripture to show that God planned the universe on a basis of number, weight, and measure, and he advocated this as a reason for the study of arithmetic. At the end of his treatise on arithmetic, after a recapitulation of the importance of number, he pays tribute to Nicomachus, to one Apuleius of Madura, and to "Boethius, a man of distinction." He ends with a naively pious interpretation of the first seven digits: one God, two Testaments, three Persons in the Trinity, four Gospels, five books of Moses, six days of Creation, seven Gifts of the Holy Spirit.[25]

Isidore of Seville (560–636) also wrote a condensation of Boethius in his *Epistemologies*. There is little that is original in his encyclopedic work except his advanced number mysticism, which is almost ridiculous to the modern reader. He finds a mystic signification in all the numbers from one to 20 except 17; also 24, 30, 40, 46, 50, and 60.[26] He makes only the vaguest references to rules of operation, simply listing instances where like numbers occur and assuming from the similarity that there is an explanation and interrelation. To him, a pound is a perfect weight because it has the same number of ounces as there are months in a year.[27] He is deeply religious and finds mystic and scriptural significance in all numbers. The fact that God performed 22 works in creation is adequate explanation for Isidore as to why there are 22 generations from Adam to Jacob, 22 books of the Old Teatament to Esther, and 22 sextarii in a bushel.[28] He is Pythagorean in his stress on the vital importance of number. "Take number from all things and all things perish."[29] In spite of this phase of his work, Isidore has written a fairly good condensation of Boethius' *Arithmetica*.

Venerable Bede, in his *De tempore ratione*, wrote what was perhaps the first treatise on practical computation, a method of determining the date of Easter.[30] Rabanus Maurus, in *Liber de computo*, wrote a most complete textbook, giving a method of finding the date of Easter, instructions for finger reckoning, instructions for using Roman numerals, and some astronomical material.[31]

The method for fixing the date of Easter occupied a very important place in the arithmetic of the Christian era. Almost the entire liturgical year depends upon the date of Easter. The beginning of the Church year (the first Sunday of Advent) is determined by the fixed date of Christmas, but nearly all the other great feasts and important days—Septuagesima, Lent, Maundy Thursday, Good Friday, Ascension, Pentecost, the number of weeks after Epiphany and after Pentecost—all depend on the date of Easter. In the sixth century there was a serious schism threatened about this very topic; the Irish monks, calculating one way, arrived at a date a week different from the one determined by the continental Benedictines, who figured the date by another method. The controversy was settled by the Synod of Wisby.

The problem of setting the date of Easter is essentially a problem of determining the date on which will fall the first Sunday after the first full moon after the vernal equinox. To do this, one must know the vernal equinox, the day of the following full moon, and the method of correcting the error in the metonic cycle. (The metonic cycle is a period of nineteen years, after which the full moon falls on the same day of the year.) The astronomical problems were settled as early as 525. Tables of Dionysius, Isidore, and Bede were available: Bede's tables went as far as 1063. To use the tables, one had to be able to find the "Golden number" and the "Dominical letter" for the year involved. The "Golden number" was found by adding one to the number of the year, dividing by nineteen: the remainder is the "Golden number." The "Dominical letter" is found by dividing the number of the year by four, adding the quotient to the number of the year, adding four, dividing by seven, and subtracting the quotient from seven; the remainder determined the place of the "Dominical letter." By referring to the table, the date of Easter was easily determined, using these two answers.[32] After the Carolingian revival, every priest was required to understand the method and principles of computation.

This period was, then, one of slow growth and dissemination of arithmetical learning. At a time when many of the educated men were clerics, the major practical application of arithmetic was related to the Church. Nevertheless, there was an interest in every phase of arithmetic, which interest would grow during the second period and burst forth into the intellectual revolution of the third period.

The Rise of the Universities and the Influence of the Arabic Texts

The second period of medieval mathematical development extended from the end of the tenth to the end of the twelfth century. The outstanding mathematical genius of the period was Gerbert, who taught the quadrivium with marked success in the cathedral school at Rheims from 972 to 982. Gerbert improved the abacus by placing symbols at the top of each column, and so extended its use. He developed a method of division, making possible all four fundamental operations on the abacus. He introduced columnar computation but did not use zero. he wrote a treatise, *Regulae de abaci numerorum,* on the use of the abacus, perhaps the first of its kind.[33]

Towards the end of this period, the Hindu-Arabic number system was beginning to be known in Europe. Those who adopted the new system were known as algorists as distinct from the abacists, who used the older methods. The abacists used the abacus, the old Roman notation, no zero, and the Roman duodecimal fractions, and had no method for finding square root. The algorists used written calculations, the Hindu-Arabic numbers, zero, decimal fractions, and the Arabic method for finding square roots. Addition, subtraction, and multiplication were comparable in both systems, but division was very different, so much so that the complicated abacist division was called *divisio ferrea* while the simpler algorist division was called *divisio aurea.* Nevertheless, the algorists did not supplant the abacists at this time; in fact there was little controversy between the two groups.

Hermanus Contractus, in the first half of the eleventh century, wrote a *Liber abaci,* based on Gerbert's work, as did several other writers of the period.

With the rise of the universities, with their curricula based on the trivium and the quadrivium there came an opportunity for a more rapid spread of arithmetical knowledge. In the twelfth century and the first half of the thirteenth, the arithmetic taught at Oxford was largely derived from Boethius, Cassiodorus, and Isidore; it consisted of the study of the properties of numbers: ratio, proportion, fractions, and polygonal numbers. No practical calculations were done so no abacus was necessary.[34] An almost identical situation existed at Cambridge.[35] This was probably the situation in the continental universities, also. By the mid-thirteenth century, there was a wider range of mathematics available for those students who were interested but there seems to be no evidence that many took advantage of the opportunity.

The Arabic texts were being translated into Latin and so made available to the scholars of Europe. Gerard of Cremona is perhaps typical of the translators. Born in Cremona in 1114, he studied all the arts but was especially interested in astronomy. As Ptolemy's works were not available to him, he went to Toledo, where he learned Arabic and translated the *Almagest.*[36] In 1187 he translated Al-Khowarizmi's works on calculation and on algebra into Latin. Later he wrote a text on calculation, *Algorismus.* Various other translators did likewise: Adelard of Bath, Abraham ben Ezra, John of Seville, and others, whose names have not survived. These may not be considered typical works of this period but rather the foundation of the university texts which were to follow; these twelfth-century works formed a link between the second and third periods of intellectual development.

Arithmetic as a Science

The third period of mathematical development extended from the thirteenth century until the end of the Middle Ages. One might almost say that it was a period of intellectual revolution. Algorist arithmetic replaced the use of the abacus. There was a new emphasis on the practical and scientific aspects of arithmetic, although the mystical side was not neglected.

Leonard of Pisa, in his *Liber abaci* of 1202, exhibited an amazing theoretical and practical knowledge. He explained the Arabic numeration, using nine digits and zero, proved elementary formulae, and even solved some elementary algebraic equations.[37] Strangely, his work had little effect on the universities, although it seems to have had a great influence in the world of commerce. Until his time, the abacus and finger-reckoning had been the principal means of computation. Now the new symbols became popular, and algorism, as distinct from Boethian arithmetic, spread throughout Europe. By 1300, there were few almanacs which did not include an explanation of it. By 1350, the almanacs had added explanations of the four fundamental processes and the more important ones included rules of proportion, examples, and formulae for the more common commercial transactions.[38] In spite of the popularity of the Hindu-Arabic numbers for actual computation, the old Roman numerals were still used in keeping records by merchants until 1550, in colleges and monasteries until 1650, and for parish registers as late as 1700.

With the intellectual revolution of the twelfth and thirteenth centuries and the rise of the universities, brought about by a renewed interest in Roman and canon law, the systematization of theology, and a new interest in dialectic, there was a decided change in the status of the seven liberal arts. "Dialectic or logic became so important that it tended to obscure all the other arts."[39] There was a revival of interest in science, and for a while all the subjects of the quadrivium were carried along on the tide of the new intellectualism. This interest was short-lived, however, and the quadrivial subjects declined rapidly. In Paris and the other French universities, the quadrivium was almost entirely obscured during the latter part of the thirteenth and during the fourteenth centuries. "As for the quadrivium, as the sciences are called, since they have little to attract in themselves and produce only a meager profit, most of the students neglect them or else omit them entirely."[40] A utilitarian spirit grew up among the students. To obtain a prebend or a prelature, all that was necessary was to have studied the liberal arts. So, after the trivium, many of the students left the universities and received benefices. Those who did go on into law or medicine did so for profit. "Only in theology was the pure love of learning retained, and that was guarded jealously."[41]

In Oxford, the quadrivium was a little, but not much, more in vogue than in Paris. The quadrivium was neglected in Paris, Oxford, and Cambridge ostensibly because quadrivial studies had practical application and Paris, Oxford, and Cambridge "systematically discouraged all technical instruction, holding that a university education should be general and not technical."[42] The real reason, though, seems to have been that distinction could be more easily attained in theology and philosophy than in the sciences.

Nevertheless, by the mid-thirteenth century, arithmetic was being taught as a science in the universities. Algorism and arithmetic flourished together for a while, but gradually the mystical and over-refined classifications of Boethius gave place to the Arabic algorism and to algebra. Perhaps the most influential mathematician of this time was Jordanus Nemorarius. Manuscripts of Jordan's work have been found in Basle, Cambridge, Dresden, Erfurt, Munich, Oxford, Paris, Rome, Thorn, Venice, Vienna, and many places in southern Germany.[43] The three treatises with which we are most directly concerned are *Algorithmus demonstratus*, *Arithmetica demonstrata*, and *De numeris datis*, although Jordan is perhaps best known for his *De ponderibus*, a treatise on weights. *Algorithmus demonstratus* is an elementary treatment of practical computations, including Arabic notation, fundamental operations, and common and astronomical fractions. Jordan treats of nine operations: numeration, addition, subtraction, duplation, multiplication, mediation (finding the mean), division, progression, and extraction of roots.[44] About two-fifths of the work is devoted to fractions. *Arithmetica demonstrata*, in ten books, is in the Boethian tradition, treating of properties and relations of numbers, prime, perfect, polygonal, and solid numbers, and the Greek theory of ratio. Jordan made a notable advance over earlier mathematicians and imitators of Boethius in that he used letters as general symbols instead of only concrete numbers. That the *Arithmetica* was popular with generations of university students is shown by the number of printed editions which are extant. D. E. Smith has identified five authentic editions, those of 1496, 1503, 1507, 1510, and 1514, and one probable one, 1480.[45] *De numeris datis* is a treatise on arithmetic and algebra, four books of problems with linear and quadratic equations in one or more unknowns. All Jordan's books are purely academic, abstract and scientific, containing no business methods, and so they were eminently fitted for use as texts in the medieval universities.

The arithmetic of the schools did not receive the wholehearted approbation of all the people. This is readily seen in Henri d'Andeli's *Battle of the Seven Arts*. The medieval author, a cleric of the University of Paris, portrays a battle among the liberal arts, with logic and grammar striving for the ascendency. Arithmetic is portrayed as a maiden sitting under a tree, placidly counting and figuring, entirely unmoved by the practical problems of warfare surrounding her. In the lines, "And three times twenty by themselves make sixty," and "Five twenties make hundred . . . ," d'Andeli is ridiculing the affectation of the schoolmen who coin the terms "soixante" and "cent" instead of using the "trois-vingts" and "cinq-vingts" of the common people.

Arithmetic sat in the shade,

Where she says, and where she figures,

That ten and two and one make thirteen,

And three more make sixteen;

Four and three and nine to boot

Again make sixteen in their way;

Thirteen and twenty-seven make forty,

And three times twenty by themselves make sixty;

Five twenties make hundred and ten hundreds a thousand.

Does counting involve anything further? No.

One can easily count a thousand thousands

In the foregoing manner,

From the number which increases and diminishes

And which in counting goes from one to hundred.

The dame makes from this her tale,

That usurer, prince, and count

Today love the countress better

Than the chanting of High Mass.

Arithmetic then mounted

Her horse and proceeded to count

All the knights of the army; . . . [46]

Throughout the Middle Ages, university instruction was based on a lecture-disputation method. The students were obliged to "hear" certain books, that is, to attend lectures in which the text, glosses, and commentaries were discussed by the professors. Sometimes there were after-class discussions, reviews, and recapitulations of the lectures by the young bachelors. There were no examinations in the modern sense of the term. The student had simply to swear that he had read the books prescribed and attended the lectures. To qualify for a degree, he was required to participate in public disputations, either defending a proposition or opposing one defended by another student.

Detailed regulations and requirements varied in different universities, of course, and even at different periods in the same university, but the general practice was to follow this lecture-disputation method.

University Curricula

There was a rather wide range in the mathematical curricula offered by the various universities; this is to be expected, as some placed greater emphasis than others on arithmetical and scientific studies.

At Bologna, there was a chair of arithmetic in the faculty of arts, and a course on Jordan's Algorismus was prescribed for all students. Algorismi de minutes et integris, a text based on Jordan's book, was used and, in 1405, was the only arithmetical text prescribed. It was to be studied, along with the first book of Euclid's Elements, by the first-year students.[47]

In Paris, in 1215, no special books were prescribed for any subjects of the quadrivium. In 1366, in an effort to stimulate interest in mathematics, a university statute required Master of Arts candidates to attend lectures on "some mathematical books." This is generally interpreted to mean Sacrobosco's lectures on the sphere and one other book.[48]

Johannes Sacrobosco, also known as John of Holywood and John of Halifax, was a leading figure in the mathematical world of his day (d. 1244). Born in Yorkshire, he received his degree as Master at Oxford and then went to Paris, where he taught until his death. He was the first to give public lectures on algebra and algorism at any university. He wrote a work on algorism, based on Jordan's work, containing the rules of arithmetic, omitting fractions, and giving the rules of multiplication in verse. He also wrote on the sphere and the astrolabe. He was apparently a popular teacher and an influential member of the Paris faculty.

At Oxford, in the fourteenth century, mathematics flourished, especially at Merton College. Richard of Wallingford, Maudith, Simon Bredon,

John Ashender, William Rede all were capable mathematicians. Their works survive in manuscript but are little known. "Thus it is that we have no cognizance of work produced during a century in which Oxford could boast more mathematicians than any country in Europe."[49] Perhaps the outstanding Oxford mathematician of the century was Thomas Bradwardine, a Franciscan friar who was born near Chichester about 1290. Called "Doctor Profundis" because of his learning, he became proctor of Merton College in 1325. His mathematical writings include De tractatus de proportionibus, Arithmetica speculativa, Geometria seculativa, and De quadratu circuli. The tract on proportions was printed in Paris in 1495 and the arithmetic in 1502, some 160 years or more after they were written.[50] The Tractatus de proportionibus is inciuided in the 1389 list of required texts at the university at Vienna.[51] Bradwardine left Oxford in 1335 to become Archbishop of Canterbury, which post he held until his death in 1349.

In the fifteenth century, Oxford mathematics declined so that between 1449 and 1463, all the mathematics required for the Master's degree was the first two books of Euclid and Ptolemy's astronomy, either in the original or in commentary.[52]

The last of the great Oxford mathematicians was Robert Recorde (1510?–1558). He must have been a fine teacher as well as an eminent mathematician, for it is said that the students used to applaud his lectures and comment that they learned much of general interest in science as well as the rules of arithmetic. The earliest use of the word "algebra" in English occurs in his Pathway of Knowledge, in 1551. He wrote an arithmetic, The Grounde of Arts; a geometry, The Pathway of Knowledge; and an algebra, The Whetstone of Witte. Particularly interesting are the devices Recorde proposes for the painless use of Arabic numerals. For example, one need learn multiplication tables only as far as five times five if he chooses to multiply by Recorde's method. Using 8×7 as an example:

$$8 \diagdown 2 \qquad (10 - 8 = 2)$$
$$7 \diagup 3 \qquad (10 - 7 = 3)$$
$$\overline{5 \qquad 6} \qquad (2 \times 3 = 6)$$

$$(7 - 2 = 5)$$
$$\text{or } (8 - 3 = 5)$$

Therefore,

$$8 \times 7 = 56.^{53}$$

Might Recorde's little diagram be the origin of × as a symbol of multiplication? Recorde's works, written in dialogue form as a discussion between master and disciple, enjoyed great popularity in university and in social circles. Both *The Grounde of Arts* and *The Whetstone of Witte* went into several editions.

The state of mathematics at Cambridge was such that in the fourteenth century, only Sacrobosco's *De sphaera* was required for a bachelor; six books of Euclid and a book on optics were required for a master. There is record of lectures being given in arithmetic, finger reckoning, and the algorism of the integers, but none of these seems to have been required.[54]

There was a somewhat different situation at the University of Vienna, where there was a particular emphasis on science and mathematics. "Extraordinary" or "cursive"[55] lectures on mathematics were considered to be especially appropriate diversions for holiday afternoons. Also, disputations were held on mathematical subjects, a practice which was almost unknown elsewhere. In 1365, Vienna prescribed the first book of Euclid and some book on algorism for bachelors but required the first five books of Euclid, Bradwardine's *Tract on Proportion*, and two books on arithmetic for a license to teach. In 1389, five books of Euclid, treatises on proportional parts, perspective, and measurement of areas were required for a master.[56] In 1391–92, all of Jordan's works on integers, common fractions, astronomical fractions, proportions, and *arithmetica* were required for a master's degree. There was a distinction made between

algorism (practical) and arithmetic (theoretical); the fees for arithmetic were twice the fees for algorism.[57]

Most of the German universities seem to have followed the lead of Vienna in respect to their mathematical curricula. Prague in 1366 required algorism and six books of Euclid;[58] Erfurt in 1449 demanded algorism, computus, and six books of Euclid; Ingolstadt in 1472 prescribed the first book of Euclid, algorism, and common arithmetic.[59] Heidelberg in 1443 relegated algorism and proportion to the "brush up" category, charged extra fees, and gave no master's credit for such courses.[60]

A general view of the medieval universities would seem to indicate that, at any given time, whatever arithmetic was known was taught, being either required of degree candidates or made available to those students who were interested in reading beyond the narrow confines of course requirements. The teachers were often famous mathematicians, translators, commentators, and authors of texts.

Present-Day Judgments

Although it is true that much of what was, in the medieval university, course material for a master's degree is today common knowledge for third-grade school children, and although some of the more profound medieval processes of ratio and proportion are today taught in eighth-grade arithmetic classes, medieval arithmetic must not be regarded as superficial or merely elementary. Many of the concepts are as challenging to modern graduate students of number theory as they were to medieval students of *arithmeitca*. Moreover, the acceptance and spread of the Hindu-Arabic number system—a place system instead of an additive one, the use of zero as a number, number symbols instead of letter numerals, decimal instead of duodecimal fractions—with all of its implications and ramifications, required the reeducation of the entire population of Europe, no small task to accomplish, even over two centuries.

It is interesting to conjecture whether or not we are today facing a similar upheaval in theoretical mathematics. Is it not possible that someday a high school student may laugh condescendingly and say, "And they got *graduate* credit for that!"

NOTES

1. Paul Abelson, *The Seven Liberal Arts, A Study in Mediaeval Education* (New York: 1906), p. 2.

2. *Ibid.*, p. 2.

3. *Ibid.*, p. 3.

4. *Ibid.*, p. 3.

5. *Ibid.*, p. 4.

6. *Ibid.*, p. 4.

7. *Ibid.*, p. 5.

8. *Ibid.*, p. 5.

9. *Ibid.*, p., 7.

10. *Ibid.*, p. 8.

11. *Ibid.*, pp. 8–9.

12. Denis J. Davanagh, O.S.A., *Answer to Skeptics* (New York: 1943), Introduction, p. xv.

13. *Ibid.*, p. 183.

14. R. A. Knox translation.

15. Abelson, *op. cit.*, p. 91.

16. Nicomachus, *Introduction to Arithmetic*, tr., Martin L. D'Ooge (New York: 1926), pp. 3–4.

17. See later discussion in this article.

18. Nicomachus, *op. cit.*, Notes, pp. 37 ff.

19. Percival R. Cole, *Later Roman Education* (New York: 1909), p. 19.

20. *Ibid.*, p. 22.

21. *Ibid.*, p. 21.

22. E.g., 6:8::9:12—

$$6 \times 72 = 432$$

$$8 \times 72 = 576$$

$$9 \times 72 = 648$$

$$12 \times 72 = 864$$

Therefore,

$$432{:}576{::}649{:}864$$

23. Adolfus Dick (ed.), *Martianus capella* (Leipzig: 1925), VII, 735–802.

24. J. P. Migne, *Patrologiae cursus completus,* Tomus LXIII (Paris: 1847), columns 1079–1168.

25. Cassiodorus Senator, *An Introduction to Divine and Human Readings* (New York: 1946), pp. 180–89.

26. Ernest Brehaut, *An Encyclopedist of the Dark Ages* (London: 1912), p. 29.

27. *Ibid.*, p. 65.

28. *Ibid.*, p. 65.

29. *Ibid.*, p. 65.

30. Ableson, *op. cit.*, p. 97.

31. *Ibid.*, p. 97.

32. *Ibid.*, p. 98.

33. *Ibid.*, pp. 100–102.

34. R. T. Gunther, *Early Science in Oxford* (London: 1922), I, 94.

35. W. W. R. Ball, *A History of the Study of Mathematics at Cambridge* (Cambridge: 1889), p. 2.

36. Moritz Cantor, *Vorlesungen über Geschichte der Mathematik* (Leipzig: 1894), pp. 853–54.

37. Ball, *op. cit.*, p. 4.

38. *Ibid.*, p. 7.

39. Louis J. Paetow, *The Arts Course in Medieval Universities* (Champaign-Urbana: 1910), p. 7

40. Achille Luchaire, *L'Université de Paris sous Philippe-Auguste* (Paris: 1899), p. 25.

41. *Ibid.*, p. 25.

42. Ball, *op. cit.*, p. 152.

43. Abelson, *op. cit.*, p. 105.

44. *Ibid.*, p. 106.

45. David Eugene Smith, *Rara arithmetica* (Boston: 1908), p. 62.

46. Henri d'Andeli, *La Bataille des VII Arts* (Berkeley: 1927), pp. 48–49.

47. Hastings Rashdall, *The Universities of Europe in the Middle Ages* (Oxford: 1942), p. 248.

48. *Ibid.*, p. 249.

49. Gunther, *op. cit.*, p. 96.

50. *Ibid.*, p. 93.

51. Karl von Raumer, *German Universities* (New York: 1859), p. 159.

52. Gunther, *op. cit.*, p. 98.

53. The algebraic justification for Recorde's method is quite obvious:

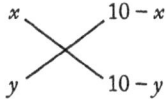

$$10[x - (10 - y)] + (10 - x)(10 - y)$$
$$= 10(x - 10 + y) + (10 - x)(10 - y)$$
$$= 10x - 100 + 10y + 100 - 10x - 10y + xy$$
$$= xy$$

54. Ball, *op. cit.*, p. 5 ff.

55. These lectures covered merely the text, without glosses and commentaries.

56. Ball, *op. cit.*, p. 8.

57. Abelson, *op. cit.*, p. 107.

58. von Raumer, *op. cit.*, p. 159.

59. *Ibid.*, p. 159.

60. Abelson, *op. cit.*, p. 111.

BIBLIOGRAPHY

Abelson, Paul. *The Seven Liberal Arts: A Study in Mediaeval Education.* New York: 1906.

Aron, Marguerite. *Saint Dominic's Successor.* St. Louis: 1955.

Augustine. *Answer to Skeptics.* A translation of St. Augustine's *Contra academicos* by Denis Kavanagh, O.S.A., S.T.M. Introduction by Rudolph Arbemann, O.S.A. New York: 1943.

Ball, W. W. Rouse. *Cambridge Papers.* London: 1918.

————. *A History of the Study of Mathematics at Cambridge.* Cambridge: 1889.

Boethius. *Patrologiae cursus completus* lxiii, Ser. 1 (*Opera omnia Boetii,* ed. J. P. Migne). Paris: 1847.

Brehaut, Ernest. *An Encyclopedist of the Dark Ages, Isidore of Seville.* London: 1912.

Cantor, Moritz. *Vorlesungen über Geschichte der Mathematik.* Leipzig: 1894.

Cassiodorus Senator. *An Introduction to Divine and Human Readings.* Translated by Leslie Webber Jones. New York: 1946.

Cole, Percival R. *Later Roman Education in Ausonius, Capella, and the Theodosian Code.* New York: 1909.

D'Andeli, Henri. *La Bataille des VII Arts,* ed. Louis J. Paetow, in *Memoirs of the University of California,* Vol. IV. Berkeley: 1927.

Dick, Adolphus (ed.). *Martianus capella.* Leipzig: 1925.

Gunther, R. T. *Early Science in Oxford,* Vol. VI, Part II. Oxford: 1922.

Isidore Hispalensis Episcopi. *Etymologiarum sive originium.* Oxford: 1911.

Luchaire, Achille. *L'Université de Paris sous Philippe-Auguste.* Paris: 1899.

Moody, Ernest A., and Clagett, Marshall. *The Medieval Science of Weights.* Madison: 1952.

Mullinger, James Bass. *The University of Cambridge from the Earliest Times to the Royal Injunctions of 1535.* Cambridge: 1873.

Nicomachus of Gerasa. *Introduction to Arithmetic.* Translated by Martin Luther D'Ooge ("Studies in Greek Arithmetic," ed. Frank E. Robbins and Louis C. Karpinski). New York: 1926.

4

The Craft of Nombrynge

E. R. SLEIGHT

$\mathcal{V}$ERY few arithmetics appeared in the English language before the sixteenth century. As late as the middle of the fifteenth century every person using even the simplest operations could read Latin, in which language such information was written. Until modern commerce was established, arithmetic was required only for addition and subtraction, and as late as the thirteenth century a student well advanced in science generally knew nothing of division.

The Earliest Arithmetics in English, edited by Robert Steele, and published for the English Text Society, lists only five arithmetics appearing in the English language before the beginning of the sixteenth century. All of these are of the fifteenth century as is shown by the style of English used. Three of them are mere fragments and have received very little attention. The other two, *The Crafte of Nombrynge* and *The Art of Nombrynge,* are more extensive in scope, and it is the purpose of this paper to review the first of these with special emphasis on the methods of operation.

The Crafte of Nombrynge is an interpretation of the *Canto de Algorismo* of Alexander de Villa Dei (1220), and is bound up, together with other scientific treatises, in the British Museum Library. It deals with the science of arithmetic rather than with the art. Each separate idea in the English translation is introduced by one or more lines of the *Canto*, the Latin form being retained. There are only thirty-two pages in which are discussed certain definitions,

notation, meaning and use of zero, how to read and write numbers, and the seven rules.

Algorism is the first definition introduced by

HEC ALGORISMUS ARS PRESENS DICITUR; IN QUA TALIBUS INDORUM FRUIMUR BIS QUINQUE FIGURIS.[1]

Then follows in fifteenth century English—*This book is called the book of Algorism, and the book treats the Craft of Numbering, the which Craft is called also Algorism. There was a King of India, the name of whom was Algor, and he made this his craft, and after his name he called it Algorism.*

Very frequently the question and answer method is used. *This present craft is called Algorism, in which we use ten signs of India. Questio. Why ten figures of India? Solucio. For as I have said before they were found first in India by a King of that country that was called Algor.*

It becomes necessary to define certain terms frequently used in this treatise. *Some numbers are digitus in Latin and digits in English. Some numbers are called articulus in Latin and articles in English. Some numbers are called composite in English.*

SUNT DIGITI NUMERI QUI CITRA DENARIUM SUNT.

Here he[2] tells what a digit is. A digit is a number that is within (or less than) ten, such as 9. 8. 7. 6. 5. 4. 3. 2. 1.[3]

ARTICUPLI DECUPLI DEGITORUM; COMPOSITI SUNT ILLI QUI CONSTANT EX ARTICULIS DEGITISQUE.

Here he tells what a composite number is, and what are articles. Articles are those which may be divided

---

Reprinted from *Mathematics Teacher* 32 (Oct., 1939): 243–48; with permission of the National Council of Teachers of Mathematics.

into numbers of ten and nothing left over, such as twenty, thirty, one hundred, one thousand, etc. A composite number is one that is composed of a digit and an article such as fourteen, fifteen, twenty-five, etc. And so every number that begins (at right) with a digit and ends with an article is a composite number. Thus twenty-five begins with the digit five and ends with the article twenty.

The meaning and place of these numbers is then given, followed by an explanation of the principles of notation with great emphasis on the use of the cypher.

NIL CIFRA SIGNIFICAT SED DAT SIGNARE SEQUENII

Explain this verse. A cypher means nothing but he[4] makes the figure that comes after him to mean more than he should if he were absent, as thus 10. Here the number means ten, and if the cypher were away and no figure in front of him it would mean only one, for then he would stand in the first place. And another cypher would mean nothing more unless it keeps the order of the place. A cypher is not a figure significant.

QUAM PRECEDENTES PLUS ULTIMA SIGNIFICABIT.

The last figure shall mean more than all the others though there were a hundred thousand before it. Thus 17689. The last figure, that is 1, means ten thousand, and all the other figures mean only seven thousand six hundred eighty and nine. And ten thousand is more than all that number. Ergo, the last figure means more than all the number before.

SEPTEM SUNT PARTES, NON PLURES ISTIUS ARTIS;
ADDERE, SUBTRAHERE, DUPLARE, DIMIDIARE,
SEXTAQUE, DIVIDERE, SED QUINTA MULTIPLICARE;
RADICEM EXTRAHERE PARS SEPTIMA DICITUR ESSE.

From this quotation the writer discovers that there are seven operations to be considered: addition, subtraction, duplication, mediation, multiplication, division, and extraction of roots. The remainder of the thirty-two pages of the treatise are devoted to the explanation and use of five of these seven operations, no mention being made of extraction of roots, and division is used only in mediation.

ADDERE SI NUMERO NUMERUM VIS, ORDINE TALI INCIPE; SCRIBE DUAS PRIMO SERIES NUMERORUM PRIMAM SUB PRIMA RECTE PONENDO FIGURAM, ET SIC DE RELIQUIS FACIAS, SI SINT TIBI PLURES.

Here begins the craft of addition. In this craft thou must know four things. First thou must know what is addition. Next thou must know how many rows of figures thou must have. Next thou must know how many diverse cases happen in this craft. And next what is the profit of this craft. As for the first thou must know that addition is a casting together of two numbers into one number. As for the second thou must know that thou shalt have two rows of figures, one under the other

1234

as here you may see: 2168. As for the third thou must know that there are four different cases. As for the fourth thou must know that the profit of this craft is to tell what is the whole number that comes of different numbers.

The four cases are:

1. No partial sum being greater than 9.

2. At least one partial sum being greater than 9.

3. The case in which at least one partial sum is 10 or a multiple of 10.

4. The case in which there is a cypher in the upper row.

The method of procedure is discussed at length, but it resolves itself into the method used today, the operations being performed from right to left as at present.

A NUMERO NUMERUM SI SIT TIBI DEMERE CURA SCRIBE FIGURARUM SERIES, VT IN ADDICIONE.

This is the chapter on subtraction in which thou must know four things.[5] These four things are identical in all operations, but the definitions are different. As for the first thou must know that subtraction is the withdrawing of one number from another. As for the second thou must know that there shall be two numbers. As for the third thou must know that there are four different cases. When all digits in the upper row are larger than the corresponding digit in the lower. When at least one digit in the lower row is larger than the corresponding digit of the upper. When the digit in the lower is larger than the corresponding digit in the upper, and the next figures to the left, both above and below are zeros.

Here again the method is entirely like the method of today, "borrowing" and all. A paragraph is given to *teaching the Craft how thou shalt know, when thou hast subtracted whether thou hast well done or no.* The method is the one now in use, adding the remainder to the subtrahend to give the minuend.

SEQUITUR DE MEDIACION. INCIPE SIC, SI VIS ALIQUEM NUMERUM MEDIARE SCRIBE FIGURARUM SERIEM SOLAM VELUT ANTE.

In this chapter is taught the Craft of mediation, in the which craft you must know four things. As for the first you shall understand that mediation is a taking out of a half of a number out of a whole number, as you would take 3 out of 6. As for the second thou shalt know that thou shalt have one row of figures and no more. As for the third thou must understand that five cases may happen in this craft, and as for the fourth, thou shalt know that the profit of this craft is when thou hast taken away a half of a number to tell what shall be left. If thou wouldst mediate, that is to say take a half of a number, thou must begin thus,—write one row of figures of what number you wish.

POSTEA PROCEDAS MEDIANS, SI PRIMA FIGURA SI PAR AUT IMPAR VIDEAS.

Here he says when thou hast written a row of figures, thou shalt take heed whether the first figure be even or odd in number, and understand that he speaks of the first figure in the right side. And in the right side thou shalt begin in this craft.

QUIA SI FUERIT PAR, DIMIDIABIS EAM, SCBIBENS QUIQUID REMANEBIT:

Here is the first case of this craft which is this, if the figure be even then thou shalt take away from the even figure a half and do away with that figure and set the

half in its place. Thus 4. Take a half out of four and that leaves 2. Do away with the 4 and set the 2 in its place.

IMPAR SI FVERIT VNUM DEMAS MEDIARE QUOD NON PRESUMAS, SED QUOD SUPEREST MEDIABIS INDE SUPER TRACTUM FAC DEMPTUM QUOD NOTAT VNUM.

Here is the second case of this craft, the which is this. If the first figure is a number that is odd, the odd number shall not be mediated, but thou shalt mediate that number less one, and write the result as in the first part of this craft. Where thou hast written that, then write such a mark as is here[w]. *Lo an example, 245. The 5 is odd. Then mediate 4 and replace the 5 by 2. That is to say replace 5 by 2 and make such a mark as w upon his head as thus 242*[w]. *Then mediate the 4 and the 2. The half is written 121*[w].

If the first figure is 1, *Thou shalt do away with the 1 and set there a cypher and a mark over his head.* The number 241 is used as an example which by mediation yields 120[w]. This is merely a special form of the second case.

The third case arises when the second digit is 1. But this again is merely a special form of the case in which the second digit is odd, which is listed as the fourth case and treated thus:

POSTEA PROCEDAS HAC CONDICIONE SECUNDA: IMPAR SI FUERIT HINC VNUM DEME PRIORI, INSCRIBENS QUINQUE, NAM DENOS SIGNIFICABIT MONOS PREDICTAM.

Here he puts forth the fourth case, the which is this. If it happens the second figure is an odd number, thou shalt take one away from the odd number, the which shall be reckoned as 10. The new number is then mediated, and a 5 is placed over the head of the second digit. As an example the mediated form of 4678 is 233[5]4.

SI MEDIACIO SIT BENE FACTA PROBARE VALEBIS DUPLANDO NUMERUM QUEM PRIMO DIMEDIASTI.

This couplet explains how the operation of mediation may be proved. *The second example was this, 245. When thou hast mediated this number, if thou hast done well, thou shalt have as the mediation this number, 122*[w]. *Now double this number, and begin with the left side. Double 1, that shall be 2. Do away with the 1 and set there the 2. Then double the 2 and set there the 4. Then double the other 2, that will be 4. Then double the mark (w) which stands for a half and that shall be 1. Cast that on to 4, and it shall be 5. Do away with the 2 and the mark and you shall have 5, then thou shalt have the number 245, and this was the number when thou began to mediate, as thou mayest see if thou takest heed.*

The same four things are to be known about multiplication as in the operations discussed. Multiplication is defined as *the bringing together of 2 numbers into one number.* The manner in which the two are united is illustrated by an example, thus: *Twice 4 is 8, and this number 8 contains as many times 4 as there are unities in the other number, the which is 2, for in 2 there are 2 unities and so 4 times 2 is 8, as thou knowest well.* The process is thus based on the number of "unities" involved.

In the "craft" of multiplication there are eight operations or types, all based on the product of a digit by a digit. After an elaborate description of this process of multiplying a digit by a digit the following consolation appears. *But nevertheless if thou hast haste to work thou shalt have here a table of figures whereby thou shalt see at once correctly what is the number that comes of the multiplication of two digits. Thus thou shalt work with this figure. As for example if 5 is to be multiplied by 3 look for the 5 in the left side of the triangle, look where the 3 sits in the lower-most row of the triangle. Now go from him upwards to the same row in which the 5 sits. And that number, the which you find here, is the number that comes from the multiplication of the 2 digits.*

1								
2	4							
3	6	9						
4	8	12	16					
5	10	15	20	25				
6	12	18	24	30	36			
7	14	21	28	35	42	49		
8	16	24	32	40	48	56	64	
9	18	27	36	45	54	63	72	81
1	2	3	4	5	6	7	8	9

A quotation of eight lines from the Latin shows how to work in this craft. An example will illustrate. It is desired to multiply 2465 by 232. The problem is thus written

$$2465$$
$$232$$

Then follows *Thou shalt begin to multiply from the left side. Multiply 2 by 2 and set the result over the head of 2, then multiply the same upper 2 by 3 of the lower number, thus thrice 2 shall be 6. Set the 6 over the head of 3, then multiply the same upper 2 by the 2 that stands under it, that will be four. Do away with the upper 2 and replace it by 4. The upper row will then be 464465, the 465 remaining unchanged as no operation was performed upon these digits.* Then follows a process called antery which means that the problem is now written thus: *464465*

$$232$$

The antery refers to the change produced on the lower line, the whole being moved one place to the right, the 2 which formerly appeared under the fourth digit from the right now appears under the third. Then as before the 4 of the upper line multiplies the digits of the lower line in succession, beginning with the left. The product is now placed above the lower digit, as in the case of the product of the 4 and 2, the 8 is placed above the 6. In case the product is a composite number, such as the product of 4 and 3 the units digit is placed above the corresponding digit of the lower line, while the tenths digit is placed above the topmost digit in the previous column; in this case above the 8. In the last multiplication in each antery, the units digit replaces the number above it in the upper line, while the tenths digit takes its position in the previous column. The result of this multiplication may thus be written:

1					
8	2				
4	6	4	8	6	5
	2	3	2		

Another antery and the problem becomes

1
82
464865, which on multiplication yields
232

> 11
>
> 121
>
> 828
>
> 464865
>
> 232

Another antery followed by the necessary multiplication and our problem takes this form, in which the multiplier is omitted.

> 11
>
> 110
>
> 1211
>
> 8285
>
> 464820

The sum of these numbers yields the product. The addition may be performed in the usual manner, but we are told to: *begin with the left,—now draw all these figures down together, as thus 6.8.1.[6] and 1; that whole is 16. Do away all this number save 6. Let him stand alone and set 1 over the head of 4 toward the left side, then draw onto 4, that will be 5. Do away with that 4 and 1 and set there the 5. Then draw 4. 2. 2. 1 and 1 together, that will be 10. Do away all that and write 0 and set the 1 over the next figure to the left side, the which is 6. Then draw that 6 and 1 together and that will be 7. Now do away with the 6 and set there the 7. Then draw that 8. 8. 1. and 1 together and that will be 18; do away all the figures that stand over the head of the 8, and let 8 stand still, and write the 1 over the next figures head to the left, which is 0. Then do away with the 0 and set there the 1, the which stands over the head of 0. Then draw the 2, 5, and 1[7] together, that will be 8. Then do away all that and write the 8. And then thou shalt have this number, 571880.*

The above process might be thus indicated:

> 4
>
> 16
>
> 10
>
> 18
>
> 80
> _____
>
> 571880

It will be noted that addition here is performed from left to right, and not from right to left as defined in the previous discussion concerning addition. It leads one to wonder why the two methods are used. It is quite probable that this second plan is introduced to be consistent with the order of multiplication as here used.

Eight cases in multiplication are recognized, depending upon the types of products, or upon the form of the problem.

1. When the product is an article.
2. When the product is a composite number.
3. When the product is a digit.
4. When the lower digit multiplies the upper digit directly above it.
5. When the lower digit multiplies the upper digit, and that upper digit is not directly above it.
6. If it happens that a zero stands right over the figure by which you multiply.
7. If it happens that the figure by which you multiply is a zero.
8. If there are several zeros in the upper row.

Each of these cases is treated at length, but all of them reduce to the method outlined in the given example.

Although division and extracting roots are listed, among the seven operations, no mention is made of the latter, and division is mentioned only in the process of mediation. One is led to wonder why this omission occurs, since the original poem discusses these operations in detail. It is highly probable that some of the original manuscript has been lost, which would account for the omission of these topics.

NOTES

1. Throughout the rest of the article, the Latin quotations will be in capitals and the translations from the fifteenth-century English in italic type.

2. "He" refers to the Latin author.

3. Note the use of the dot.

4. Note the personal pronoun.

5. See addition.

6. Here written without the dot. But I have used it as was the case in naming the digits.

7. Here the comma is used to separate the digits.

BIBLIOGRAPHY

Athenae Cantabrigienses.

Ball, W. W. R.: *History of the Study of Mathematics at Cambridge.*

De Morgan, Augustus: *List of Books in Arithmetic.*

Halliwell, J.: *Rara Mathematica.*

Leslie, John: *Philosophy of Arithmetic.*

Smith, David Eugene: *Rara Arithmetica.*

Steele, Robert (Editor): *The Earliest Arithmetics in English.*

Wilson, Duncan: *History of Mathematical Teaching in Scotland.*

HISTORICAL EXHIBIT 4

Algorist versus the Abacist

Although algorithmic computational procedures for the four basic operations of arithmetic were introduced into Europe in the ninth century, it was a long time before they were fully accepted or adopted. The controversy over which methods of computation were better, those performed with the counting board or those written out using algorithms, raged for centuries. Abacus techniques were deeply engrained in all commercial and financial enterprises. Business was conducted over the "counter," the table upon which abacus movements were made. Many early arithmetic books contained woodcut prints depicting this controversy. The following fanciful illustration showing an algorist in competition with an abacist comes from Gregor Reisch's, *Margarita philosophica* of 1508.

5

Leonardo Fibonacci

CHARLES KING

THE *Fibonacci Quarterly* receives its name from Leonardo of Pisa (or Leonardo Pisano), better known as Leonardo Fibonacci (Fibonacci is a contraction of *Filius Bonacci*, son of Bonacci). Leonardo was born about 1175 in the commercial center of Pisa. This was a time of great interest and importance in the history of Western civilization. One finds the influence of the crusades stirring and awakening the people of Europe by bringing them in contact with the more advanced intellect of the East. During this time the Universities of Naples, Padua, Paris, Oxford, and Cambridge were established, the Magna Carta signed in England, and the long struggle between the Papacy and the Empire was culminated. Commerce was flourishing in the Mediterranean world and adventurous travelers such as Marco Polo were penetrating far beyond the borders of the known world.

It is in this growing commercial activity that we find the young Leonardo at Bugia on the northern coast of Africa. Here the merchants of Pisa and other commercial cities of Italy had large warehouses for the storage of their goods. Actually very little is known about the life of this great mathematician. No contemporary historian makes mention of him, and one must look to his writings to find information about him. In the preface of his first and most important work, *Liber abaci* (I), Leonardo tells us that his father, the head of one of

the warehouses of Bugia, instructed him to study arithmetic. In Bugia, he received his early education from a Moorish schoolmaster.

Leonardo then traveled about the Mediterranean visiting Egypt, Syria, Greece, Sicily, southern France, and Constantinople. He met with scholars and studied the various systems of arithmetic then in use. Leonardo was persuaded that the Hindu-Arabic system was superior to the methods then adopted in the different countries he had visited and that it was even superior to the Algorithma and the method of Pythagoras. He busied himself with the subject and carried on his own research, intent upon bringing the Hindu-Arabic system to his Italian countrymen. The study and research in mathematics so absorbed him that he seems to have devoted his life to this pursuit and spent little time in commerce which was flourishing at that time and was the favorite occupation of his fellow citizens. Yet most of the applications Leonardo makes in his works are in the field of commerce. In one place, he gives a careful evaluation of the money systems of the countries of his travels.

Leonardo returned to Italy about 1200 and in 1202 wrote *Liber abaci* (I), in which he gave a thorough treatment of arithmetic and algebra, the first that had been written by a Christian. The work is divided into fifteen chapters. The chapter contents are given here to indicate the scope of the work: (1) Reading and writing numbers in the Hindu-Arabic system; (2) Multiplication of inte-

Reprinted from *Fibonacci Quarterly* 1 (Dec., 1963): 15–19; with permission of the Fibonacci Association.

gers; (3) Addition of integers; (4) Subtraction of integers; (5) Division of integers; (6) Multiplication of integers by fractions; (7) Additional work with fractions; (8) Prices of goods; (9) Barter; (10) Partnership; (11) Alligation; (12) Solutions of problems; (13) Rule of false position; (14) Square and cube roots; (15) Proportions, and Geometry and algebra.

The last and most important chapter is divided into three parts; the first relates to proportions, the second to geometry and the third, to algebra. each of the three parts begins with definitions and demonstrations credited to the Arabs, then Leonardo considers six questions, three simple and three complex, giving solutions for them.

Leonardo, in 1228, gave a second edition of the *Liber abaci* which he dedicated to Michel Scott, astrologer to the Emperor Frederick II and author of many scientific works. Copies of this edition exist today. Leonardo profusely illustrated and strongly advocated the Hindu-Arabic system in this work. He gave an extensive discussion of the Rule of False Position and the Rule of Three. Leonardo did not use a general method in problem solving; each problem was solved independently of the others. In the solution of a problem he not only considered the problem as it might occur, but considered all of the variations of the question, even those that were not reasonable.

In the *Liber abaci*, Leonardo states and gives the solution to the famous Rabbit Problem [1, Vol. 1, p. 285]. A pair of rabbits is placed in a pen to find out how many offspring will be produced by this pair in one year if each pair of rabbits gives birth to a new pair of rabbits each month starting with the second month of its life; it is assured that deaths do not occur.

Leonardo traces the progress of the rabbits: The first pair has offspring in the first month: thus two pair. The second month there are three pair, the first reproducing in this month. In the third month there are five pair. Continuing in this manner through the twelve months, Leonardo gives the following table:

0		Sixth Month	
Pairs		21	
1		Seventh Month	
First Month		34	
2		Eighth Month	
Second Month		55	
3		Ninth Month	
Third Month		89	
5		Tenth Month	
Fourth Month		144	
8		Eleventh Month	
Fifth Month		233	
13		Twelfth Month	
		377	

It is this sequence of numbers, 1, 2, 3, 5, 8, 13, . . ., that gives rise to the Fibonacci Sequence.

Of the many problems of elementary nature in the *Liber abaci*, the following are given as examples.

Seven old women are traveling to Rome and each has seven mules. On each mule there are seven sacks; in each sack there are seven loaves of bread: in each loaf there are seven knives; and each knife has seven sheaths. How many in all are going to Rome?

A man went into an orchard which had seven gates; and there took a certain number of apples. When he left the orchard he gave the first guard half the apples he had and one apple more. To the second he gave half the remaining apples and one apple more. He did the same in the case of each of the remaining five guards, and left the orchard with one apple. How many apples did he gather in the orchard?

A certain man puts one denarius at such a rate that in five years he has two denarii and in every five years thereafter the money doubles. How many denarii would he gain from this one denarius in 100 years?

A certain king sent thirty men into his orchard to plant trees. If they could set out a thousand trees in nine days, in how many days would thirty-six men set out four thousand four hundred trees?

Many readers will recognize these problems.

In 1220, Leonardo wrote *Practica geometriae*, which he dedicated to Master Dominique, a person of whom there is no record. In this work Leonardo systematized the subject matter of practical geometry with a specialization in measurements of bodies. He included some algebra and trigonometry, square and cube roots, proportions and indeterminate problems. The use of a surveying instrument called the quadrans is included. The work is skillfully done with Euclidean rigor and some originality.

Leonardo's reputation grew and from his writings it can be seen that he had a vast range of knowledge concerning Arabian mathematics and mathematics of antiquity, especially Greek. His treatment shows much originality, completeness, and rigor. It is especially noted that his writings did not contain the mysticism of numerology and astrology that were so prevalent in the writing of his day.

Because of Leonardo's great reputation, the Emperor Frederick II, when in Pisa (1225), held a sort of mathematical tournament to test Leonardo's skill. The competitors were informed beforehand of the questions to be asked, some or all of which were composed by Johannes of Palermo [1, Vol. II, p. 227], who was one of Frederick's staff. This is the first case in the history of mathematics that one meets with an instance of these challenges to solve particular problems which were so common in the sixteenth and seventeenth centuries.

The first question propounded was to find a number of which the square when decreased or increased by 5 would remain a square. The correct answer given by Leonardo was $41/12$. The next question was to find by the methods used in the tenth book of Euclid a line whose length x should satisfy the equation $x^3 + 2x^2 + 10x - 20 = 0$. Leonardo showed by geometry that the problem was impossible, but gave an approximation of the root $1.3688081075 \ldots$, which is correct to nine places.

The third question was:

Three men possess a certain sum of money, their shares in the ratio 3:2:1. While making the division, they were surprised by a thief and each took what he could and fled. Later the first man gave up half of what he had, the second gave up one-third, and the third, one-sixth. The money given up was divided equally among them and then each man had the share to which he was entitled. What was the total sum? Leonardo showed that the problem was indeterminate and gave as one solution 47 which is the smallest sum.

The other competitors failed to solve any of these questions. Through the consideration of these problems and others similar to them, Leonardo was led to write his *Liber quadratorum* (1225) [1, Vol. II, p. 253], a brilliant and original work containing a well arranged collection of theorems from indeterminate analysis involving equations of the second degree such as $x^2 + 5 = y^2$, $x^2 - 5 = z^2$. This work has marked him as the outstanding mathematician between Diophantus and Fermat in this field.

Two or three works of Leonardo that are known are the *Flos* [1, Vol. II, p. 227] (blossom or flower), which contains the last two problems of the tournament; the first problem is found in the *Liber quadratorum*, and a *Letter to Magister Theodoris* [1, Vol. II, p. 247], philosopher to Frederick II, relating to indeterminate analysis and to geometry. The last three works show clearly the genius and brilliance of Leonardo as a mathematician and were beyond the abilities of most contemporary scholars.

The works of Leonardo Fibonacci are available in some universities in the United States through B. Boncompagni, *Scritti di Leonardo Pisano*, Rome, (1857–1862). The first volume contains the *Liber abaci* and the second volume contains *Practica geometriae*, the *Flos*, *Letter to Magestrum Theodorum*, and *Liber quadratorum*. A treatment of square numbers composed by Leonardo and addressed to the Emperor Frederick II seems to have been lost.

REFERENCE

Boncompagni, Baldassarre. *Scritti di Leonardo Pisano: 2* vols. Rome, 1857.

6

Leonardo of Pisa
and His *Liber quadratorum*

R. B. McCLENON

*T*HE THIRTEENTH CENTURY is a period of great fascination for the historian, whether his chief interest is in political, social, or intellectual movements. During this century great and far-reaching changes were taking place in all lines of human activity. It was the century in which culminated the long struggle between the Papacy and the Empire; it brought the beginnings of civil liberty in England; it saw the building of the great Gothic cathedrals, and the establishment and rapid growth of universities in Paris, Bologna, Naples, Oxford, and many other centers. The crusades had awakened the European peoples out of their lethargy of previous centuries, and had brought them face to face with the more advanced intellectual development of the East. Countless travelers passed back and forth between Italy and Egypt, Asia Minor, Syria, and Bagdad; and not a few adventurous and enterprising spirits dared to penetrate as far as India and China. The name of Marco Polo will occur to everyone, and he is only the most famous among many who in those stirring days truly discovered new worlds.

Among the many valuable gifts which the Orient transmitted to the Occident at this time, undoubtedly the most precious was its scientific knowledge, and in particular the Arabian and Hindu mathematics. The transfer of knowledge and ideas from East to West is one of the most interesting phenomena of this interesting period, and accordingly it is worth while to consider the work of one of the pioneers in this movement.

Leonardo of Pisa, known also as Fibonacci,[1] in the last years of the twelfth century made a tour of the East, saw the great markets of Egypt and Asia Minor, went as far as Syria, and returned through Constantinople and Greece.[2] Unlike most travelers, Leonardo was not content with giving a mere glance at the strange and new sights that met him, but he studied carefully the customs of the people, and especially sought instruction in the arithmetic system that was being found so advantageous by the Oriental merchants. He recognized its superiority over the clumsy Roman numeral system which was used in the West, and accordingly decided to study the Hindu-Arabic system thoroughly and to write a book which should explain to the Italians its use and applications. Thus the result of Leonardo's travels was the monumental *Liber abaci* (1202), the greatest arithmetic of the middle ages, and the first one to show by examples from every field the great superiority of the Hindu-Arabic numeral system over the Roman system exemplified by Boethius.[3] It is true that Leonardo's *Liber abaci* was not the first book written in Italy in which the Hindu-Arabic numerals were used and explained,[4] but no work had been previously produced which in either the extent or the value of its contents could for a moment be compared with this. Even today it would be thoroughly worth while for any teacher of mathematics to become

Reprinted from *American Mathematical Monthly* 26 (Jan., 1919): 1–8; with permission of the Mathematical Association of America.

familiar with many portions of this great work. It is valuable reading both on account of the mathematical insight and originality of the author, which constantly awaken our admiration, and also on account of the concrete problems, which often give much interesting and significant information about commercial customs and economic conditions in the early thirteenth century.

Besides the *Liber abaci*, Leonardo of Pisa wrote an extensive work on geometry, which he called *Practica geometriae*. This contains a wide variety of interesting theorems, and while it shows no such originality as to enable us to rank Leonardo among the great geometers of history, it is excellently written, and the rigor and elegance of the proofs are deserving of high praise. A good idea of a small portion of the *Practica geometriae* can be obtained from Archibald's very successful restoration of Euclid's *Divisions of Figures*.[5]

The other works of Leonardo of Pisa that are known are *Flos*, a *Letter to Magister Theodorus*, and the *Liber quadratorum*. These three works are so original and instructive, and show so well the remarkable genius of this brilliant mathematician of the thirteenth century, that it is highly desirable that they be made available in English translation. It is my intention to publish such a translation when conditions are more favorable, but in the meantime a short account of the *Liber quadratorum* will bring to those whose attention has not yet been called to it some idea of the interesting and valuable character of the book.

The *Liber quadratorum* is dedicated to the Emperor Frederick II, who throughout his whole career showed a lively and intelligent interest in art and science, and who had taken favorable notice of Leonardo's *Liber abaci*. In the dedication, dated in 1225, Leonardo relates that he had been presented to the Emperor at court in Pisa, and that Magister Johannes of Palermo had there proposed a problem[6] as a test of Leonardo's mathematical power. The problem was, to find a square number which when either increased or diminished by 5 should still give a square number as result. Leonardo gave

a correct answer, $11^{97}/_{144}$. For $11^{97}/_{144} = (3^5/_{12})^2$, $6^{97}/_{144} = (2^7/_{12})^2$, and $16^{97}/_{144} = (4^1/_{12})^2$. Through considering this problem and others allied to it, Leonardo was led to write the *Liber quadratorum*.[7] It should be said that this problem had been considered by Arab writers with whose works Leonardo was unquestionably familiar; but his methods are original, and our admiration for them is not diminished by careful study of what had been done by his Arabian predecessors.[8]

In the *Liber quadratorum*, Leonardo has given us a well-arranged, brilliantly-written collection of theorems from indeterminate analysis involving equations of the second degree. Many of the theorems themselves are original, and in the case of many others the proofs are so. The usual method of proof employed is to reason upon general numbers, which Leonardo represents by line segments. He has, it is scarcely necessary to say, no algebraic symbolism, so that each result of a new operation (unless it be a simple addition or subtraction) has to be represented by a new line. But for one who had studied the "geometric algebra" of the Greeks, as Leonardo had, in the form in which the Arabs used it,[4] this method offered some of the advantages of our symbolism; and at any rate it is marvelous with what ease Leonardo keeps in his mind the relation between two lines and with what skill he chooses the right road to bring him to the goal he is seeking.

To give some idea of the contents of this remarkable work, there follows a list of the most important results it contains. The numbering of the propositions is not found in the original.

PROPOSITION I. THEOREM. Every square number[10] can be formed as a sum of successive odd numbers beginning with unity. That is,

$$1 + 3 + 5 + \cdots + (2n - 1) = n^2.$$

PROPOSITION II. PROBLEM. To find two square numbers whose sum is a square number. "I take any odd square I please, . . . and find the other from the sum of all the odd numbers from unity up to that odd square itself."[11] Thus, if $2n + 1$ is a square ($= x^2$) then

$$1 + 3 + 5 + \cdots + (2n - 1) + x^2 = n^2 + (2n + 1)$$
$$= \text{a sum of two squares} = (n + 1)^2$$

This is equivalent to Pythagoras's rule for obtaining rational right triangles, as stated by Proclus,[12] viz.,

$$\left(\frac{x^2 - 1}{2} \right)^2 + x^2 = \left(\frac{x^2 + 1}{2} \right)^2.$$

For, inasmuch as $2n + 1 = x^2$, we have

$$n = \frac{x^2 - 1}{2} \text{ and } n + 1 = \frac{x^2 + 1}{2}.$$

PROPOSITION III. THEOREM.

$$\left(\frac{n^2}{4} - 1 \right)^2 + n^2 = \left(\frac{n^2}{4} + 1 \right)^2.$$

This enables us to obtain rational right triangles in which the hypotenuse exceeds one of the legs by 2. It is attributed by Proclus to Plato. Leonardo also gives the rule in case the hypotenuse is to exceed one leg by 3, and indicates what the result would be if the hypotenuse exceeds one leg by any number whatever.

PROPOSITION IV. THEOREM. "Any square exceeds the square which immediately precedes it by the amount of the sum of their roots." That is, $n^2 - (n-1)^2 = n + (n-1)$. It follows from this that when the sum of two consecutive numbers is a square number, then the square of the greater will equal the sum of two squares. For, if $n + (n - 1) = u^2$, then $n^2 - (n-1)^2 = u^2$ or $n^2 = u^2 + (n-1)^2$.

PROPOSITION V. PROBLEM. Given $a^2 + b^2 = c^2$, to find two integral or fractional numbers x, y, such that $x^2 + y^2 = c^2$. Solution: Find two other numbers m and n such that[13] $m^2 + n^2 = q^2$. If $q^2 \neq c^2$, multiply the preceding equation by c^2/q^2, obtaining

$$\left(\frac{c}{q} \cdot m \right)^2 + \left(\frac{c}{q} \cdot n \right)^2 = c^2$$

so that $x = c/q \cdot m$, $y = c/q \cdot n$ is a solution.

PROPOSITION VI. THEOREM. "If four numbers not in proportion are given, the first being less than the second, and the third less than the fourth, and if the sum of the squares of the first and second is multiplied by the sum of the squares of the third and fourth, there will result a number which will be equal in two ways to the sum of two square numbers." That is,

$$(a^2 + b^2)(c^2 + d^2) = (ac + bd)^2 + (ad - bc)^2 =$$
$$(ad + bc)^2 + (ac - bd)^2.$$

This very important theorem should be called Leonardo's Theorem, for it is not found definitely stated, to say nothing of being proved, in any earlier work. Leonardo considers also the case where a, b, c, and d are in proportion, and shows that then $(a^2 + b^2) \cdot (c^2 + d^2)$ is equal to a square and the sum of two squares. This gives him still another way of finding rational right triangles.[14]

PROPOSITION VII. THEOREM. $(x^2 - y^2)^2 + (2xy)^2 = (x^2 + y^2)^2.$[15] Leonardo proves this very simply as a corollary of Proposition VI.

PROPOSITION VIII. PROBLEM. "To find two numbers the sum of whose squares is a number, not a square, formed from the addition of two given squares." That is, to find x and y such that $x^2 + y^2 = a^2 + b^2$. Choose any two numbers c and d, such that $c^2 + d^2$ is a square, and write $(a^2 + b^2) (c^2 + d^2)$ as a sum of two squares, let us say $p^2 + q^2$; this we can do by Proposition VI. Construct the right triangle whose legs are p and q; then the similar triangle whose hypotenuse is equal to $\sqrt{c^2 + d^2}$ will have as its legs the two required numbers x and y.

PROPOSITION IX. THEOREM.

$$6(1^2 + 2^2 + 3^2 + \cdots + n^2) = n(n + 1)(2n + 1).$$

The proof of this is strikingly original, and proceeds from the identity

$$n(n + 1)(2n + 1) = n(n - 1)(2n - 1) + 6n^2.$$

Hence

$$n(n - 1)(2n - 1) = (n - 1)(n - 2)(2n - 3)$$
$$+ 6(n - 1)^2,$$
$$\cdots\cdots\cdots\cdots\cdots\cdots\cdots\cdots\cdots\cdots\cdots\cdots\cdots\cdots$$
$$2 \cdot 3 \cdot (2 + 3) = 1 \cdot 2 \cdot (1 + 2) + 6 \cdot 2^2,$$
$$1 \cdot 2 \cdot (1 + 2) = \qquad\qquad 6 \cdot 1^2.$$

It follows by addition that

$$n(n + 1)(2n + 1) = 6(1^2 + 2^2 + 3^2 + \cdots + (n-1)^2 + n^2).$$

PROPOSITION X. THEOREM.

$$12[1^2 + 3^2 + 5^2 + \cdots + (2n-1)^2] = (2n-1)(2n+1)4n.$$

Leonardo gives a proof very similar to that of Proposition IX.

PROPOSITION XI. THEOREM.

$$12[2^2 + 4^2 + 6^2 + \cdots + (2n)^2] = 2n(2n+2)(4n+2),$$

and likewise

$$18[3^2 + 6^2 + 9^2 + \cdots + (3n)^2] = 3n(3n+3)(6n+3),$$

and

$$24[4^2 + 8^2 + 12^2 + \cdots + (4n)^2] = 4n(4n+4)(8n+4),$$

and in general

$$6a[a^2 + (2a)^2 + (3a)^2 + \cdots + (na)^2] = na(na+a)(2na+a).$$

Here Leonardo has almost discovered the general result

$$a^2 + (a+d)^2 + (a+2d)^2 + \cdots + [a+(n-1)d]^2$$
$$= \frac{6na^2 + 6n(n-1)ad + n(n-1)(2n-1)d^2}{6}.$$

His method needed no change at all, in fact.

PROPOSITION XII. THEOREM. If $x + y$ is even, $xy(x+y)(x-y)$ is divisible by 24; and in any case $4xy(x+y)(x-y)$ is divisible by 24. A number of this form is called by Leonardo a *congruum*, and he proceeds to show that it furnishes the solution to a problem proposed by Johannes of Palermo.

PROPOSITION XIII. PROBLEM. "To find a number which, being added to, or subtracted from, a square number, leaves in either case a square number." Leonardo's solution of this, the problem which had stimulated him to write the *Liber quadratorum*, is so very ingenious and original that it is a matter of regret that its length prevents its inclusion here. It is not too much to say that this is the finest piece of reasoning in number theory of which we have any record, before the time of Fermat. Leonardo obtains his solution by establishing the identities

$$(x^2 + y^2)^2 - 4xy(x^2 - y^2) = (y^2 + 2xy - x^2)^2$$

and

$$(x^2 + y^2)^2 + 4xy(x^2 - y^2) = (x^2 + 2xy - y^2)^2.$$

PROPOSITION XIV. PROBLEM. To find a number of the form $4xy(x+y)(x-y)$ which is divisible by 5, the quotient being a square. Take $x = 5$, and y equal to a square such that $x + y$ and $x - y$ are also squares. The least possible value for y is 4, in which case

$$4xy(x+y)(x-y) = 4 \cdot 5 \cdot 4 \cdot 9 \cdot 1 = 720.$$

PROPOSITION XV. PROBLEM. "To find a square number which, being increased or diminished by 5, gives a square number. Let a congruum be taken whose fifth part is a square, such as 720, whose fifth part is 144; divide by this the squares congruent to 720,[16] the first of which is 961, the second 1681, and the third 2401. The root of the first square is 31, of the second is 41, and of the third is 49. Thus there results for the first square $6^{97}/_{144}$, whose root is $2^7/_{12}$, which results from the division of 31 by the root of 144, that is, by 12; and for the second, that is, for the required square, there will result $11^{97}/_{144}$, whose root is $3^5/_{12}$, which results from the division of 41 by 12; and for the last square there will result $16^{97}/_{144}$, whose root is $4^1/_{12}$."

PROPOSITION XVI. THEOREM. When $x > y$, $(x+y)/(x-y) \neq x/y$. It follows that $x(x-y)$ is not equal to $y(x+y)$, and "from this," Leonardo says, "it may be shown that no square number can be a congruum." For if $xy(x+y)(x-y)$ could be a square, either $x(x-y)$ must be equal to $y(x+y)$, which this proposition proves to be impossible, or else the four factors must severally be squares, which is also impossible. Leonardo to be sure overlooked the necessity of proving this last assertion, which remained unproved until the time of Fermat.[17]

PROPOSITION XVII. PROBLEM. To solve in rational numbers the pair of equations

$$x^2 + x = u^2,$$

$$x^2 - x = v^2.$$

The solution is obtained by means of any set of three squares in arithmetic progression, that is, by means of Proposition XIII. Let us take x_1^2, x_2^2, and x_3^2 for the three squares, and let the common difference, that is, the congruum, be d. Leonardo says that the solution of the problem is obtained by giving x the value x_2^2/d. For then

$$x^2 + x = \frac{x_2^4}{d^2} + \frac{x_2^2}{d} = \frac{x_2^2(x_2^2 + d)}{d^2} = \frac{x_2^2 x_3^2}{d^2};$$

and

$$x^2 - x = \frac{x_2^4}{d^2} - \frac{x_2^2}{d} = \frac{x_2^2(x_2^2 - d)}{d^2} = \frac{x_2^2 x_1^2}{d^2}.^{18}$$

PROPOSITION XVIII. PROBLEM. To solve in rational numbers the pair of equations

$$x^2 + 2x = u^2,$$

$$x^2 - 2x = v^2.$$

The method is similar to that in Proposition XVII, the value of x being found to be $2x_2^2/d$. Leonardo adds, "You will understand how the result can be obtained in the same way if three or more times the root is to be added or subtracted."

PROPOSITION XIX. PROBLEM. To solve (in integers) the pair of equations

$$x^2 + y^2 = u^2,$$

$$x^2 + y^2 + z^2 = v^2.$$

Take for x and y any two numbers that are prime to each other and such that the sum of their squares is a square, let us say u^2. Adding all the odd numbers from unity to $u^2 - 2$,[19] the result is $((u^2 - 1)/2)^2$.

Now

$$\left(\frac{u^2 - 1}{2}\right)^2 + u^2 = \left(\frac{u^2 + 1}{2}\right)^2.$$

Thus

$$z^2 = \left(\frac{u^2 - 1}{2}\right)^2,$$

and

$$v^2 = \left(\frac{u^2 + 1}{2}\right)$$

PROPOSITION XX. PROBLEM. To solve in rational numbers the set of equations

$$x + y + z + x^2 = u^2,$$

$$x + y + z + x^2 + y^2 = v^2,$$

$$x + y + z + x^2 + y^2 + z^2 = w^2.$$

By an extension of the method used in Proposition XIX Leonardo obtains the results $x = 3^{1}/_5$, $y = 9^{3}/_5$, $z = 28^{4}/_5$. He even goes farther and obtains the integral solutions $x = 35$, $y = 144$, $z = 360$. He continues, "And not only can three numbers be found in many ways by this method but also four can be found by means of four square numbers, two of which in order, or three, or all four added together make a square number . . . . I found these four numbers, the first of which is 1295, the second $4566^{6}/_7$, the third $11417^{1}/_7$, and the fourth 79920." In the midst of the explanation of how these values were obtained, the manuscript of the *Liber quadratorum* breaks off abruptly. It is probable, however, that the original work included little more than what the one known manuscript gives. At all events, considering both the originality and power of his methods, and the importance of his results, we are abundantly justified in ranking Leonardo of Pisa as the greatest genius in the field of number theory who appeared between the time of Diophantus and that of Fermat.

NOTES

Editor's note: Notes have been renumbered. In the original, notes appeared as true footnotes and the numbering was restarted at 1 on each page.

1. This is probably a contraction for "Filiorum Bonacci," or possibly for "Filius Bonacci"; that is, "of the

family of Bonacci" or "Bonacci's son." See Boncompagni, *Della Vita e delle Opere di Leonardo Pisano, matematico del secolo decimoterzo*, Rome, 1852, pp. 8–12.

2. *Scritti di Leonardo Pisano*, 2 vols., Rome, 1857–61. Vol. I, p. 1.

3. Boethius, ed. Friedlein, Leipzig, 1867. The arithmetic occupies pages 1–173. This was the arithmetic that was very generally taught throughout Europe before the thirteenth century, and its use continued to be widespread long after better works were in the field.

4. Smith and Karpinski, *The Hindu-Arabic Numerals*, Boston and London, 1911. Chapter VII gives an account of the first European writings on these numerals.

5. Archibald, *Euclid's Book on Divisions of Figures; with a Restoration based on Woepcke's Text and on the Practica Geometriæ of Leonardo Pisano*, Cambridge, England, 1915.

6. In the introduction to *Flos* we are told that two other problems were propounded at the same time. *Scritti*, II, p. 227.

7. *Scritti*, II, p. 253.

8. See, for example, Woepcke, *Recherches sur plusieurs ouvrages de Leonard de Pise, et sur les rapports qui existent entre ces ouvrages et les travaux mathématiques des Arabes*, Rome, 1859.

9. Heath, T. L., *The Thirteen Books of Euclid's Elements*, Cambridge, 1908. Vol. I, pp. 372–374, 383–385, 386–388; Zeuthen, H. G., *Geschichte der Mathematik im Altertum und Mittelalter*, Copenhagen, 1896, pp. 44–53; Karpinski, L. C., *Robert of Chester's Latin translation of the Algebra of Al-Khowarizmi*, New York, 1915, pp. 77–89.

10. Throughout this article, unless otherwise stated, the word "number" is to be understood as meaning "positive integer."

11. The use of quotation marks indicates a literal translation of Leonardo's words; in other cases the exposition follows his thought without adhering closely to his form of expression.

12. Proclus, ed. Friedlein, Leipzig, 1873, p. 428.

13. This is possible by Proposition II or Proposition III.

14. For instance, letting $a = 6$, $b = 4$, $c = 3$, $d = 2$,

$$(36 + 16)(9 + 4) = 676 = (6 \cdot 3 + 4 \cdot 2)^2$$

$$= (6 \cdot 2 + 4 \cdot 3)^2 + (6 \cdot 3 - 4 \cdot 2)^2 = 26^2 = 24^2 + 10^2.$$

15. This is Euclid's general solution of the problem of finding rational right triangles. Heath, *op. cit.*, III, p. 63. (Euclid's Elements, X, Lemma to Theorem 29.)

16. That is, the three squares in arithmetic progression, whose common difference is the congruum 720. They are obtained by Proposition XIII, thus: Taking $x = 5$ and $y = 4$, $y^2 + 2xy - x^2 = 31$, the root of the first square; $x^2 + y^2 = 41$, the root of the second square; and $x^2 + 2xy - y^2 = 49$, the root of the third square.

17. *Fermat, Oeuvres*, Paris, 1891, vol. 1, p. 340; Heath, *Diophantus of Alexandria*, Cambridge, 1910, p. 293.

18. The simplest numerical example would be $x_1^2 = 1$, $x_2^2 = 25$, $x_3^2 = 49$, and this is the illustration given by Leonardo. It leads to $x = {}^{25}/_{24}$, from which we have $x^2 + x = {}^{1225}/_{576} = ({}^{35}/_{24})^2$ and $x^2 - x = {}^{25}/_{576} = ({}^{5}/_{24})^2$.

19. Here u^2 is odd, because it is the sum of the squares of two numbers x and y which are prime to each other. It is not possible that both x and y are odd, since $(2m + 1)^2 + (2n + 1)^2 = 4m^2 + 4m + 4n^2 + 2$, and this is divisible by 2 but not by 4, and hence cannot be a square. Thus, of the numbers x and y, one must be even and the other odd, hence $x^2 + y^2$ is odd.

Editor's Note: Since the appearance of this article, Fibonacci's major works have been translated into English:

Leonardo Pisano Fibonacci The Book of Squares. Laurence Sigler (1987).
Fibonacci's Liber Abaci. Laurence Sigler (2002).
Fibonacci's De Practica Geometrie. Barnabas Hughes (2008).

7

Some Uses of Graphing
before Descartes

THOMAS M. SMITH

During the first half of the seventeenth century at least four students of mathematics and natural philosophy were making use of a simple graphing technique.

In 1618, Isaac Beeckman, while writing in his journal on the subject of "a stone falling in a vacuum," employed this graphing technique to represent uniform acceleration analyzed in terms of what might be called "geometric infinitesimals." His approach suggests certain features of the infinitesimal calculus that Von Leibniz and Newton were to develop later in that century.

In 1637, René Descartes presented an application of the same graphing technique (without Beeckman's infinitesimal "individua," as he called them) in his newly published *La géométrie*. Analytical geometry, as it came to be called, properly owes its essential character to the simultaneous, independent discoveries of Descartes and Pierre Fermat, but like many another advance in scientific thought, its origins antedate the men who made it important.

In the following year, 1638, Galileo Galilei made use of the same graphing technique that Descartes, Fermat, and Beeckman had employed. He used it to provide a mathematical description of uniform acceleration in the physical example of a body in free fall.

All of these men were engaged, when using the same graphing technique, in describing two variables that are simple functions of each other. At that time the technique was neither new nor fully developed. Indeed, it appears to have been then not quite three hundred years old, according to present historical evidence, and it grew out of an even older tradition of purely rhetorical discussion that omitted geometry when discussing certain simple functional variables.

A common pair of functional variables that were being discussed by 1350 in many universities of Europe were "extension" and "intension." Richard Swineshead of Oxford—a logician known among later scholars simply as "The Calculator"—pointed out explicitly, for example, that the extension of a thing could be altered dimensionally without altering the *intensity* of the thing. One could, in the mind's eye, extend hotness or whiteness, for example, without altering the intensity of either the over-all hotness or the hotness at any particular point, if one wished. Or one could increase the intensity without increasing the extent.

Swineshead used a geometric analogy to clarify his point: geometrically speaking, one could place one rectangle beside another without increasing the over-all length, if one chose. Or he could place one rectangle next to another in such a way as to add to the over-all length but without affecting the width.

Some time between 1345 and 1365, these notions were systematically explored, developed, and sharpened, especially under the mind and the hand

Reprinted from *Mathematics Teacher* 54 (Nov., 1961): 565–67; with permission of the National Council of Teachers of Mathematics.

of one man, a Parisian scholar named Nicole Oresme. The essence of the technique that he systematized, however, was first employed by an Italian logician, Giovanni di Casali, while Oresme was still a student. Casali remarked in passing, when discussing the traditional topic of "the velocity of motion of alteration," that if one took the example of the quality hotness, one could conceive of a uniform hotness throughout "just as a rectangular parallelogram is formed between two equidistant lines, such that any part you wish is equally wide with another . . . . "

Again, Casali said, " . . . let there be throughout a uniformly difform hotness, such that it is a triangle . . . . "

We are quite unable to say whether or not Oresme read Casali's treatise. But about 1350 or 1360 Oresme wrote a lengthy work on "the configurations of qualities," and in this work he gave a detailed, systematic, and exhaustive explanation of how to use geometric figures to depict the extension and the intensity of any quality. "Although indivisible points or lines do not actually exist," he said in his introductory remarks, "yet it is necessary to picture them mathematically for the measure of things and for comprehending their proportions. Therefore, every intensity successively acquirable is to be imagined by a straight line perpendicularly erected upon some point of space or of the subject of that intensible thing."

Oresme also pointed out that his figures could be used to depict local motion. The perpendiculars represented speed, the base on which they were erected, time, and the area of the enclosed figure, distance. Uniform acceleration from rest would be portrayed by a right triangle, uniform speed by a rectangle.

These same geometric representations of motion were employed by Galileo and Beeckman nearly three hundred years later.

And in the meantime, what had happened to this simple graphing technique? Preliminary evidence indicates it persisted, often in the form of marginal illustrations, in a large number of documents. Thus, at the present time, thirteen extant manuscripts are known of Oresme's long work, *De configurationibus qualitatum,* and two copies are known of another treatise in which he described his graphing technique—his "Questions on the books of Euclid's Elements."

These treatises apparently were never printed after movable type and the printing press became available during the fifteenth century. However, another work, a primer, quite brief, was printed more than once. This little handbook was called "On the latitudes of forms." For a while, modern authorities, such as Pierre Duhem, H. Wieleitner, M. Curtze, Lynn Thorndike, Anneliese Meier, and Marshall Clagett, thought that Oresme had composed the primer. More recently, Miss Meier has suggested it was written by one Jacob of Florence (or Naples or St. Martin). Whoever wrote it, there is no question that it derives from Nicole Oresme's full-length treatise, "On the configurations of qualities."

Twenty-two versions of the short work that we know of are extant. Eighteen of these are manuscript copies; four are printed editions. The earliest dated manuscript is inscribed "1395." The last of the editions was printed in 1515.

Five of the manuscripts were written during the fourteenth century. Eight more copies survive from the fifteenth century; two of these are printed versions, the first dated 1482, the second 1486. Two more printed versions are known to be from the sixteenth century, one published in 1505, the other in 1515.

Seven manuscripts remain that cannot be sharply dated with any assurance at present. Tentatively, their provenance would seem to place them in the fourteenth century or the fifteenth century.

The primer was the most popular of all the treatises using or discussing the Casali-Oresme figures. However, scattered instances of the use of these simple graphs are to be found in other documents of the fourteenth, fifteenth, sixteenth, and seventeenth centuries. It is not always possible to say with assurance in every case who wrote a particular treatise that contains or refers to the figures embodying the crude graphing technique here under discussion, nor is it possible to assert that the figures in the margin were always placed there by the man who inscribed the text. Nevertheless, the fact remains that at present some fifty documents, written or printed, are known which reveal that this protographing technique was in use among European scholars of the fourteenth, fifteenth, sixteenth, and seventeenth centuries as an aid to their exploration and their understanding of certain abstract concepts of the uniform and the nonuniform.

REFERENCES

Clagett, M., *The Science of Mechanics in the Middle Ages*. Madison, Wisconsin: University of Wisconsin Press, 1959.

Duhem, P. *Études sur Leonard de Vinci*. 3 vols. Paris: F. De Nobelle, 1955.

Meier, A., *An der Grenze von Scholastik und Naturwissenschaft*. Rome: Edizioni di Storia e Letteratura, 1952.

Adam Riese

DOROTHY I. CARPENTER

*R*EFLECT, FOR A MOMENT, on the reeducation and records-changing entailed if we were to replace our illogical weights-and-measures systems by systems employing metric units. Such a Gargantuan task probably dooms us to the status quo. Yet a comparable change was effected in Europe during the period from the twelfth century to the sixteenth century as people gradually rejected Roman numerals in favor of the Hindu-Arabic. Many advocates of the new system have long been forgotten by even their own countrymen, but not Adam Riese. It was this arithmetic teacher and textbook writer who was largely responsible for the German acceptance of computation with numerals in lieu of the counting board. So influential was he that now, more than four centuries later, his name is still being perpetuated in the folk phrase *nach Adam Riese* used in connection with results in arithmetic computation, much as we might say that something is "according to Hoyle."

Riese was born in 1492 in Staffelstein, a town in Upper Franconia. Nothing is known of his early childhood. When he was fourteen his father died; and two years later he and a younger brother, Conrad, who had attended the famous Zwickauer Latin School, left home to make their own way. Although there is no record of formal education, Riese's arithmetics credit certain problems as being received from various individuals as early as 1509. This would indicate that by the time he

reached seventeen he was becoming proficient in the art of reckoning. His aptness in geometry is attested by this anecdote which has survived: A self-confident surveyor-engineer once wagered that he could construct more right angles in a given length of time than could Riese. Before the surveyor had completed the construction of his first perpendicular, Riese had drawn a great quantity of right angles in a semicircle.

FIGURE I
Portrait of Adam Riese from a woodcut on the title page of the 1550 edition of Riese's *Rechnung nach der lenge/auff den Linien vnd Feder.*

For a number of years Riese seems to have wandered about, returning once to Staffelstein for the contested settlement of his brother Conrad's estate. But there is no evidence that he received his just share, and he never went back again. During these years he visited arithmetic schools, examined their books, and was generally dissatisfied with the quality of instruction.

Reprinted from *Mathematics Teacher* 58 (Oct., 1965): 538–43; with permission of the National Council of Teachers of Mathematics.

His own first arithmetic appeared in 1518, a second in 1522. There is a record of his being comptroller of a mine in Epperstein in the spring of 1525. Later that year he married Anna Lewber, of Freiberg, bought a house, and took a freeman's oath. From 1529 to 1532 he was employed as mine comptroller in Marienberg. There is also evidence that he served as a *Rechenmeister*, or "master arithmetician," in Erfurt and Annaberg. In the smaller cities of the sixteenth century a *Rechenmeister* often assisted the town clerk in computational duties, serving as superintendent of measures, examiner of casks, etc. In 1536, at the request of city officials of Annaberg, Riese compiled and published a table of "bread regulations," whereby for a fixed price the weight of bread and rolls must vary with the price of grain. This booklet served as basis of the city's bread laws for over a century.

By 1550, Riese's reputation was so well established that he could risk asking the king for a five-year privilege for his arithmetic book. Along with this recognition, assistance came from Elector Moritz von Sachsen, who advanced the cost of printing.

To appreciate Riese's significance as "the arithmetic teacher of Germany" [2],* let us remind ourselves of several facts. First, printing had been known in Europe for fewer than seventy-five years when Riese's first book was published. Thus relatively few books on any subject were yet available to the common people, and this was especially true of textbooks in arithmetic. (Among the more important predecessors of Riese were Johannes Widman, who had published in 1489, and Jacob Koebel, whose text of 1514 ran through 22 editions.) Secondly, although Hindu-Arabic numerals had been introduced into Germany over a hundred years before, few but the monks and scholars were acquainted with them, and the common people were scarcely aware they existed. The latter used the Roman numerals and the counting board almost exclusively, and regarded the mysterious

new ciphers with considerable apprehension and superstition. Thirdly, some officials of Riese's day commanded that Roman numerals be used because alteration of records was less easily accomplished thereby. In fact, Riese himself had to keep his mining accounts in Roman numerals at the same time that he was advocating the Hindu-Arabic numerals in his books and using them in the bread regulations. Furthermore, it was the custom of the period to guard one's skills rather zealously. Since the thirteenth century, arithmetic masters had tended to settle in the large cities, where they often united in guilds, kept the city accounts, and were frequently well paid for lessons in arithmetic.

Many of the schools of Riese's day were conducted in Latin by the clergy and were only incidentally concerned with the practical art of reckoning. In a Wittenberg Church Order of 1533 one finds the statement: "After they can read, write and sing, one ought also in time teach them ciphers and something from arithmetic." While there were also "German schools" (for reading and writing, but usually not for arithmetic), and writing and arithmetic schools, one can see that, in general, the populace had little opportunity to learn the new symbols and how to compute with them. Typical of the situation of the day is this quotation from a French arithmetic published near the end of the fifteenth century: "... for there are many merchants who cannot read and write but yet must reckon." So as trade expanded, there was increased demand for those trained in the indispensable art of reckoning.

Adam Riese was one of the most successful of those who set up arithmetic schools to meet the above need. Then, as now, a natural consequence was to put courses of instruction into printed form. The oldest arithmetic book printed in Italian (1478) and the first one in German (1482) were both by a *Rechenmeister* of Nuremberg. Thus it is not surprising that Riese's books show direct evidence of teaching experience. In particular, in the 1529 edition, he explains that he has found that the

---

* Numerals in square brackets refer to the bibliography at the end of the article.

young people who begin with instruction "on the lines" (that is, with the equivalent of the use of a counting board) "obtain a better foundation" and "ought to execute their computation with the ciphers with less labor" [4]. His texts were designed to be used by the teachers in the arithmetic schools rather than by the pupils, but they were directed to the level of the people's education.

In his first arithmetic book, published in 1518, Riese himself used only "line reckoning." But his second book, *Rechnung auff der Linien vnd Federn,* published in 1522, began with a brief explanation of the familiar line reckoning and then led its readers into detailed and systematic use of the new "ciphers." He contended that they "will not be wearisome to learn, but ought to be comprehended with joy and gladness" [4]. This text was the most used of Riese's four arithmetics, and was published in eight different German towns. His 1550 edition is described as the "best exponent of the practical arithmetic of the middle of the century in Germany" [8]. In all, over forty editions of his textbooks appeared in the sixteenth century, and several more were published in the seventeenth century and were in use until the middle of that century. Their quality and popularity are attested to by a contemporary, Michael Stifel, who called them "elegant." He and others liked them so well that they "borrowed" some of Riese's illustrative examples—without due credit!

In 1524 Riese began an algebra text, *Die Coss,* completing it only through methods of solving first-degree equations. This book was never printed, and the manuscript was not found until 1855. Hence it exerted no influence. But it does serve now as representative of the state of algebra at the beginning of the sixteenth century.

Since the various editions differed primarily in the extent of their practical problems, one table of contents will suffice to show the organization and general scope of Riese's books: numeration, addition, subtraction, duplation, mediation, multiplication, division, progressions, rule of three, money exchange, profit, silver and gold computa-

tion, partnership, reduction (of fractions). Other editions included problems of inheritance, guardianship, coinage, alloys, proportion, and magic squares.

Compared with other texts, Riese's arithmetics were "head and shoulders above those of his predecessors and contemporaries in arrangement and preparation of subject matter, in the form of his presentation, and in the logical and superior execution of his pedagogical principles" [4]. These principles included: (1) proceeding from the concrete to the abstract (reckoning with counters to computation with ciphers), (2) passing from the simple to the complex, without and with short cuts, (3) passing from the particular to the general, and (4) offering numerous exercises to perfect a given technique.

While giving due credit to the above now commonly accepted teaching procedures, we must note that all of Riese's solutions omit development and proof, and instead are completely dogmatic. After the statement of each problem, he regularly continues, "Do it thus," or "Set it so," as in the following coin-exchange problem, literally translated from the 1559 edition:

Seven florin from Padua make 5 at Venice, and 10 at Venice make 6 at Nuremberg, and 100 from Nuremberg make 73 at Köln. How many make 1000 fl. from Padua to Köln. It makes 312 and 6/7. Set it so:

7 Padua	5 Venice
10 Venice	6 Nurenberg 1000 Pad.
100 Nuremberg	73 Köln

Multiply the first ones with one another, the same also the middle ones. Stand it so:

7000 . . . 2190 fl. . . . 1000

In modern notation we might write: Let

x = value of one florin from Padua,

y = value of one florin from Venice,

z = value of one florin from Nuremberg,

w = value of one florin from Köln.

Then we have the system of equations

$$7x = 5y,$$
$$10y = 6z,$$
$$100z = 73w,$$

from which we obtain

$$7000x = 2190w,$$

and

$$x = 312\frac{6}{7}w.$$

Likewise, from the 1559 edition comes this problem, solved by the method of double false position:

A son asks his father how old he is. The father answers him saying: When you were yet as much older, half as old, and ¼ as old, and a year older, so were you already 100 years old. [That is: When you shall be yet as much older, half your age still older, one fourth your age still older, and a year still older, you will be 100 years old.] The question is how old is the son?

Make thus. Take for yourself two numbers which encompass in themselves a half and a fourth [that is, are divisible by 2 and by 4], as 40 and 48. Examine those in the problem according as the 40 also. Say 40 but 40, half 20, the fourth are 10 and one year more make in sum 111 years. From that take the 100 years, leaving 11 years more. Likewise examine also the 48. Stand thus:

40	more	11
		22
48	more	33

Compute, so comes 36 years. So old is the son.

Using the 48 as one possible age, we obtain 48 + 48 + 24 + 12 + 1 = 133, which exceeds the 100 by 33. Previously, Riese had explained the remainder of such a solution as follows: "Take 11 from 33, leaving 22 the divisor. Thereafter cross-multiply, take one from the other, and divide." That is,

$$48 \times 11 = 528,$$
$$40 \times 33 = 1320,$$
$$1320 - 528 = 792,$$
$$792 \div 22 = 36.$$

For the most part, Riese's calculations and techniques are similar to ours. Among the exceptions may be noted that he taught subtraction by the equal additions method, that the divisor was written below the dividend and the remainders above it, and that the names million, billion, etc., were not used. Instead, he instructed that the number 86,789,325,178 be read: "86 thousand thousand times thousand, 700 thousand times thousand, 89 thousand times thousand, 300 thousand, 25 thousand, 178" [1].

The accompanying Figure 2 shows a page, from Riese's 1559 edition, devoted to division by what has come to be known as the "scratch method."** The division, at the top of the page, of 95,472 by 12, is accomplished by moving the divisor from left to right as the multiplication and subtraction are performed mentally and the remainders written above. The separate steps would appear as follows:

```
1
2 1
9 5 4 7 2 (7
1 2

1 2
2 1 6
9 5 4 7 2 (7 9
1 2 2
1

1 2 1
2 1 6 7
9 5 4 7 2 (7 9 5 6
1 2 2 2
1 1
```

** Sometimes called the "galley method" because of the pattern's resemblance, when viewed from the left, to a ship with a mast.

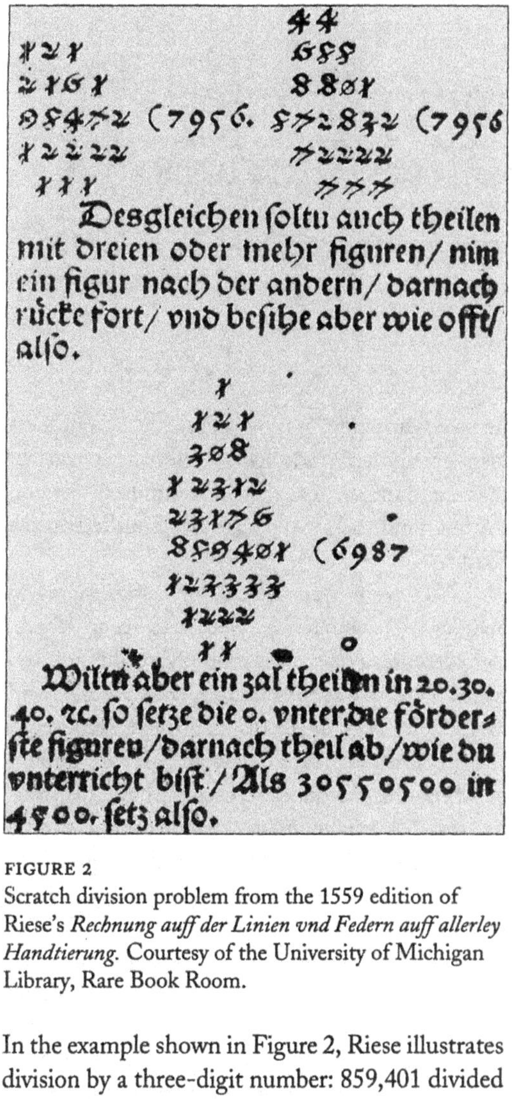
Desgleichen soltu auch theilen mit dreien oder mehr figuren/ nim ein figur nach der andern/ darnach rücke fort/ vnd besihe aber wie offt/ also.

Wiltu aber ein zal theilen in 20. 30. 40. ꝛc. so setze die o. vnter die förderste figuren/ darnach theil ab/ wie du vnterricht bist/ Als 30550500 in 4500. setz also.

FIGURE 2

Scratch division problem from the 1559 edition of Riese's *Rechnung auff der Linien vnd Federn auff allerley Handtierung.* Courtesy of the University of Michigan Library, Rare Book Room.

In the example shown in Figure 2, Riese illustrates division by a three-digit number: 859,401 divided by 123.

In the case of rather involved computations, Treutlein conjectured that Riese made use of his knowledge of algebra and then explained the solution in words, as in this example:

21 persons, men and women, have spent 81 pf. for drinking. A man ought to give 5 pf. and woman 3 pf. Now I ask how much have been each in particular. Set it so:

	man 5	
21 persons		81 pf.
	woman 3	

Take 3 pf. from 5 pf., leaving 2 the divisor. Now multiply 3 with 21, come 63, which take from 81 pf., leaving 18, which dividing with 2 some 9 men, which take from 21 persons, leaving 12. So much are the women [1].

We see that, if x denotes the number of men and $21 - x$ the number of women, the above explanation fits exactly the solution of the linear equation $5x + 3\,(21 - x) = 81$.

As Berlet assesses Rieses's greatness, "it consists not alone of his dexterity in arithmetic, but especially in that he recognized the adaptability of Hindu-Arabic numerals to calculation and demonstrated it to the German people by spoken and written word." Truly do his books "guard a part of their cultural heritage" [3].

BIBLIOGRAPHY

1. Berlet, Bruno. *Adam Riese, sein Leben, seine Rechenbücher und seine Art zu rechnen.* Leipzig: E. V. Mayer, 1892.

2. Cantor, Moritz. *Vorlesungen über Geschichte der Mathematik* (2nd ed.). Leipzig: B.G. Teubner, 1913. Vol. II, pp. 385-89.

3. Deubner, Fritz. "Adam Riese der Rechenmeister des deutschen Volkes," Sachsische Heimblatter (Dresden), V (1959), 602-8.

4. Falckenberg, Hans. "Adam Riese, ein deutsche Rechenmeister," *Deutsche Mathematik* (Leipzig), III (February 1938), 1–8.

5. Menninger, Karl. *Zahlwort und Ziffer.* Breslau: Vandenhoeck & Ruprecht, 1934. Pp. 252-54, 265, 340–45.

6. ———. *Zahlwort und Ziffer.* Göttingen: Vandenhoeck & Ruprecht, 1957. Vol. I, pp. 154–55.

7. Riese, Adam. *Rechnung auff der Linien vnd Federn auff allerley Handtierung.* Leipzig: Hans Rhambaw, 1559.

8. Smith, David Eugene. *Rara Arithmetica.* Boston: Ginn & Co., 1908. Pp. 138–40, 250.

9. ———. *History of Mathematics.* Boston: Ginn & Co., 1923. Vol. I, pp. 337–38.

10. Unger, Friedrich. *Die Methodik der praktischen Arithmetik in historischer Entwicklung vom ausgange des Mittelalters bis auf die Gegenwart 1888.* Leipzig: B. G. Teubner, 1888. Pp. 48–53.

Tangible Arithmetic: Finger Reckoning and Other Devices

PHILLIP S. JONES

THE FINGER SYMBOLS SHOWN in our Figure 1 were once used to communicate at international fairs and as an aid in remembering numbers when using an abacus. This picture is from Philippi Calandri's *Arithmetice* published in Florence in 1518. The first edition of this book is an incunabulum which means that it was printed during the "cradle" days of printing, which are taken to be its first fifty years, 1450–1499. The first edition was printed in 1491, but finger symbols existed in Greece as early as the fifth century B.C. and it has been stated that some also existed in China as early as the time of Confucius (551–478 B.C.)[1] Smith notes that finger symbols even appear in art and literature. For example, he cites a passage from Juvenal who said, "Happy is he who . . . numbers his years upon his right hand," referring to the fact that the numbers less than one hundred were on the left hand and greater than one hundred on the right.[2]

There were even devices for using fingers to operate with numbers as well as to record them. For example, a device for multiplying integers between 5 and 9 was to hold up fingers on one hand to show by how much the multiplier exceeded 5 and on the other hand to show similarly the amount by which the multiplicand exceeded 5. The product of the original numbers was the number whose tens digit was given by the sum of the fingers held up and whose units digit was the product of the fingers turned down on the two hands. For example 7×9 would lead to 2 and 4 fingers up and 3 and 1 fingers down. This would give the product 63. For this process one needed to know the multiplication tables only up to 5×5. The explanation of this device is enrichment material for algebra as well as for arithmetic since it hinges on the identity

$$ab = 10\,[(a-5)+(b-5)] + [(10-a)(10-b)].$$

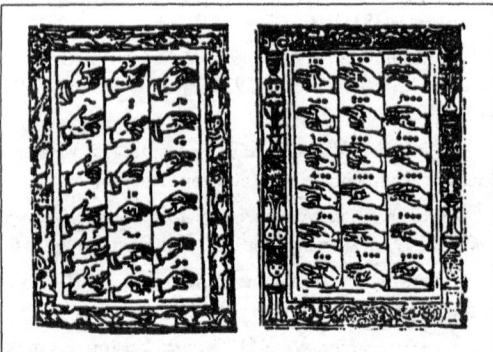

FIGURE I

Reprinted from *Mathematics Teacher* 48 (Mar., 1955): 153–59; with permission of the National Council of Teachers of Mathematics.

Other devices for recording numbers which should be included in an exhibit, display, report, or assembly program on tangible arithmetic are the *quipu*[3] used by Peruvian natives before the time of the Spanish conquerors, and *tally sticks*[4] used in

Eyn Newe geordent Rechēbüchlein vf den linien mit Rechē pfenigen /den Jungen angenden zū heüslichein gebzauch vnd hendeln leichtlich zū lernen /mit figuren vnnd Exempeln / volgt hernach cerlichen angetzeygt.

Gedzuckt zū Oppēheim.

FIGURE 2

FIGURE 3

medieval England. Encyclopedias will give some help on these items as well as the references cited in the notes.

Among computing devices the *abacus*[5] is so well known that we will only mention it here with a few references, and the *counting board* has been so recently discussed in this journal[6] that we will only add Figures 2 and 3 to the earlier data. Figure 2 is the title page of a German book on "Rechnenpfenigen" by J. Köbel published in 1514. It shows a merchant at his counting board with the lines drawn or cut into it. Note the title which reads "A new kind of little reckoning book (telling of reckoning) on lines with reckoning pennies." These

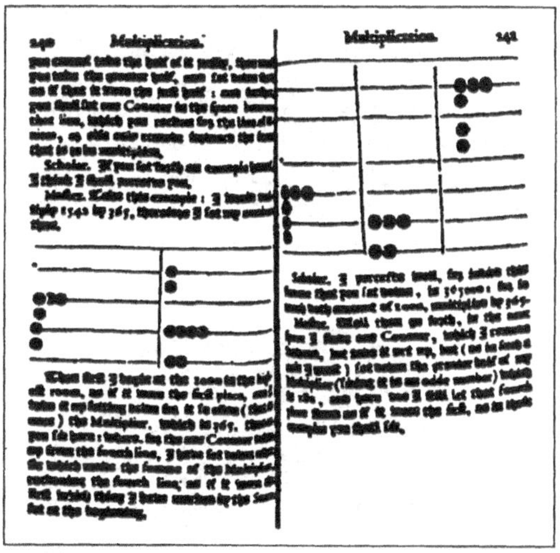

"pennies" were often little stones from whose Latin name, "calculi," we get our words "calculate" and "calculus" just as our word "counter" for a table or bench in a store comes from the medieval merchant's counting board.

Figure 3 shows an explanation of the multiplication of 1542×365 on a counting board from the most famous early English arithmetic. This is Robert Recorde's *The Grounde of Arts* first published about 1542. Our picture is from the 1654 edition and shows one of the many interesting features of Recorde's works; namely, they were written in the form of a dialogue between the "Master" and his "Student." The arithmetics we have pictured display three significant common and related characteristics: their period, the late fifteenth and early sixteenth century; their use of the common or "vulgar" languages rather than Latin, the scholarly international language; and their concern with the computations of commerce.

Figures 4 and 5 relate to two other well-known and recently discussed computing devices, *Gunter's scale* and the *slide rule*.[7] Figure 4 shows the scale as it appeared in the sixth (1698) edition of William Leybourn's *The Line of Proportion or Numbers, Commonly Called Gunters Line, Made Easie. . . .* Although Edmund Gunter was an Englishman, he

FIGURE 4

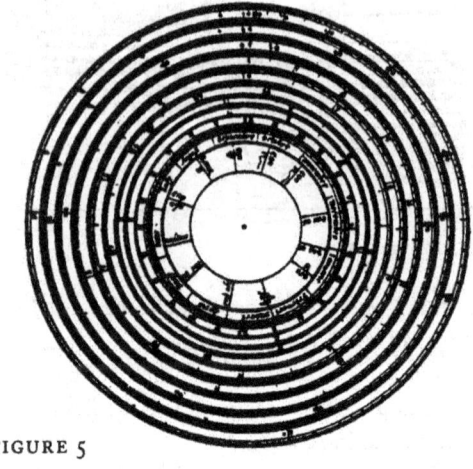

FIGURE 5

published his first discussion of his scale in French in Paris in 1624. The first discussion of the straight slide rule, which would seem the natural next step after Gunter's scale, was published by William Oughtred in 1633.

However, the first slide rule invented was a circular rule described by Richard Delamain in 1630. Oughtred, Delamain's teacher at one time, claimed to have invented a circular rule earlier, but he did not publish a description of it until 1632. Our Figure 5 is a picture of this rule from the 1660 edition of his book, *The Circles of Proportion*.

The *nomogram*[8] is a recently developed computing chart which uses logarithmic scales and could well be included in a tangible arithmetic project. Frenchman Maurice d'Ocagne began its modern phase in 1891.

Computing machines are so in the public eye and press today that it is not necessary to do more than mention them in the connection. Clippings, pictures, demonstations of desk calculators and an

outline of their historical development[9] would be part of a "tangible arithmetic" project.

Calandri's Arithmetic:

The book from which our Figure 1 came is such a fascinating little work that we include a few more pictures from it. Figure 6 is its title page. This illustrates a false notion of that day, that Pythagoras introduced the Hindu-Arabic numerals and their arithmetic into Europe. The square form of the multiplication table was also often

Pictagoras arithmetrice introductor

FIGURE 6

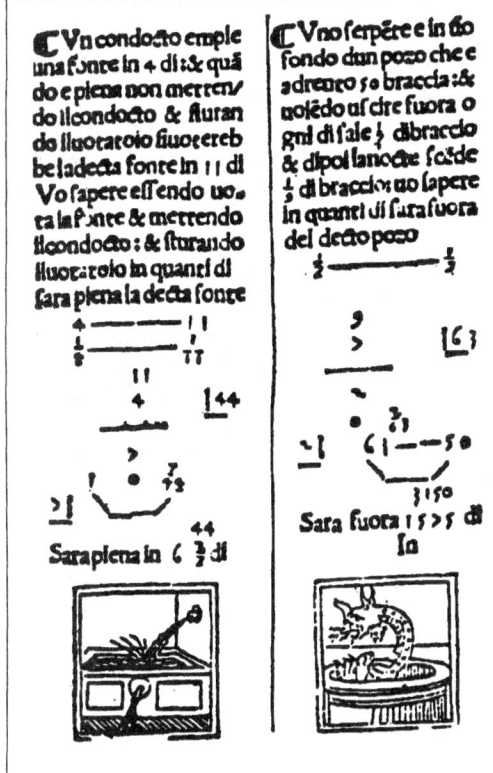

FIGURE 7

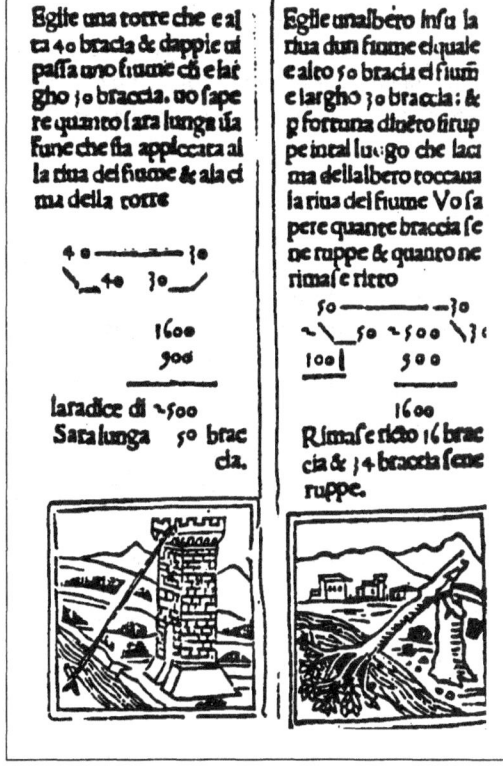

FIGURE 8

called the table of Pythagoras. Actually, the Greek "arithmetica" which owes much of its origins to the early Pythagoreans was the beginnings of what today we call number theory. "Arithmetica" neither used Hindu numerals nor concerned itself with "logistica," the Greek counterpart of modern elementary arithmetic.

Figures 7 and 8 show some of the illustrated pages from Calandri. These are of interest not only because Calandri's was the first printed Italian arithmetic with illustrations accompanying the problems, but also because the problems themselves are so similar to current ones that one can figure them out from the pictures, the numbers, and a few hints. Try them on some of your students. Figure 7 shows a serpent who each day climbs 1/7 "bracchia" (a unit of measure) out of a well only to slip back 1/9 bracchia. If the well is 50

bracchia deep how long does it take him to climb out?

The other picture in Figure 7 asks how long does it take to fill a trough or tub if water enters at a rate such that it would be full in 4 days if it weren't leaking at a rate which would empty it in 11 days?

Figure 8 asks at what height above the ground the 50 bracchia tree should be cut in order that its tip should touch the bank of the river at a distance of 30 bracchia as shown. Clearly this involves the Pythagorean theorem, but one might challenge students to set up algebraic expressions which would explain the process by which Calandri arrived at the correct answer, 16.

Also to be found in this book are "work problems," a lion, leopard, and wolf who eat at different rates, and ants who travel toward each other at different rates. Problems on the altitude of an equilateral triangle and circumference and area of a circle are also included in this book which was the first to give our modern form of long division and to omit the old "galley" type of division.[10]

NOTES

1. D. E. Smith, *History of Mathematics*, Vol. II (Boston: Ginn and Co.), p. 197. Chapter III, pages 156-207, titled "mechanical aids to computation" discusses many tangible computing devices.

2. Smith, *loc. cit.*

3. Smith, *op. cit.*, p. 195.

4. Smith, *op. cit.*, p. 192. Vera Sanford, "Art of Reckoning: I. Tallies," *The Mathematics Teacher*, Vol. XLIII (Oct. 1950), p. 292.

5. Smith, *op. cit.*, pp. 156-192. Vera Sanford and Yen Yi-Yun, "The Chinese Abacus," *The Mathematics Teacher*, Vol. XLIII (Dec. 1950), p. 402 ff. T. Kojima, *The Japanese Abacus, Its Use and Theory* (Rutland, Vt.: 1954). Jerry Adler, "So You Think You Can Count!" *Mathematics Magazine*, Vol. 28 (Nov.–Dec. 1954), pp. 83–86.

6. Smith, *op. cit.*, p. 181. Vera Sanford, "Counters: Computing If You Can Count to Five," *The Mathematics Teacher*, Vol. XLIII (Nov. 1950), p. 368.

7. Phillip Jones, "The Oldest American Slide Rule," *The Mathematics Teacher*, Vol. XLVI (Nov. 1953), p. 501. Florian Cajori, "On the History of Gunter's Scale and the Slide Rule During the Seventeenth Century," *University of California Publications in Mathematics*, Vol. 1 (Feb. 17, 1920), pp. 187–209. Smith, *op. cit.*

8. Florian Cajori, *A History of Mathematics* (New York: Macmillan, 1919), p. 481. Douglas P. Adams, "The Preparation and Use of Nomographic Charts in High School Mathematics," *Eighteenth Yearbook*, National Council of Teachers of Mathematics, pp. 164-181.

9. Smith, *op. cit.*, p. 202. Phillip S. Jones, "The Binary System," *The Mathematics Teacher*, Vol. XLVI (Dec. 1953), p. 575. Philip and Emily Morrison, "The Strange Life of Babbage," *Scientific American*, Vol. 186 (April 1952), pp. 66–73.

10. D. E. Smith, *Rara Arithmetica* (Boston: Ginn and Co., 1908), p. 47 ff.

The Cardano-Tartaglia Dispute

RICHARD W. FELDMANN

Introduction

As the Middle Ages came to a close, a rebirth of scientific inquiry occurred. Most large cities had universities, but these consisted primarily of lecturers, as there were few books beyond those of the classical authors. Hence the reputation of a university depended upon its ability to provide the best lecturers. But who was the best lecturer? One way to settle this was in a public debate with the winner gaining prestige and academic positions while the loser was ignored.

Since a scholar had to outperform all challengers to maintain his position, he needed every trick he knew. If someone knew something that no one else knew, his fame was insured. One assurance of success would be a knowledge of the general solution to the previously unsolved cubic equation. It was the search for this solution that led to the dispute between Gerolamo Cardano and Niccolo Tartaglia. But it is impossible to discuss the debate without first giving brief biographies of the antagonists to show why they acted as they did.

The Life of Tartaglia

Niccolo Tartaglia—his actual family name was Fontana—was born in Brescia, Italy, about the turn of the sixteenth century.[1] When the French sacked Brescia in 1512, the youthful Fontana suf-

Reprinted from *Mathematics Teacher* 54 (Mar., 1961): 160-63; with permission of the National Council of Teachers of Mathematics.

fered severe saber cuts about the face and mouth which caused a speech impediment. Because of this he was nicknamed Tartaglia, "the stutterer." He later wore a long beard to hide the scars, but he could never overcome the stuttering.

His education was very meager. In fact he tells, in one of his books, that his mother had accumulated a small amount of money so that he might be tutored by a writing master. The money ran out, so he says, when the instruction reached the letter K and his education stopped before he could write his own initials. But being an enterprising youth, he stole the master's lecture notes and completed the course on his own. In times of extreme poverty he even used tombstones in place of writing slates.

His mathematical knowledge was completely self-taught, but his attempts to teach himself Latin resulted in failure and he was forced to write his treatises in Italian instead of the accepted Latin. He taught primarily in Venice, and, once established, lived quite comfortably on his wages and the wagers he won in public disputes and challenges.

The Life of Cardano

The other participant, Gerolamo Cardano,[2] was born in Milan in 1501. His father, a lawyer and lecturer in geometry, married Cardano's mother, who came from a "socially unacceptable" family, a few years after Cardano's birth. This led to later complications.

He started his studies at the University of Pavia, but transferred to the University of Padua when war broke out. At college he supplemented his meager finances by constant gambling, at which he became very proficient. He wrote a treatise, *Liber de ludo aleae*,[3] which not only introduced the idea of probability as we use it today, but also included ways to cheat in these games.

At twenty-five, he graduated as a doctor of his chosen profession, medicine. Due to his illegitimate birth, he was not allowed to practice in Milan. Later he was recognized as a doctor and proved his ability by becoming the second most renowned medical expert in Europe at that time. Most of his 412 written works[4] were in the field of medicine, popular science, and astrology. One of these was an autobiography in which he describes himself as follows: "living from day to day, outspoken, despising religion, mindful of injuries caused by others, envious, melancholy, a spy, a betrayer of trusts, a deviner, a caster of charms, subject to frequent failures, hateful, given to shameful lust, solitary, unpleasant, strict, foretelling the future willingly, jealous, obscene, lascivious, abusive in speech, inconsistent, two-faced, dishonest, a slanderer, entirely unparalleled in vices, and by nature incompatible even with those with whom he converses daily."[5]

His sons were "chips off the old block," as one robbed his father and the other married and later murdered an immoral girl.

In 1570, Cardano was arrested and jailed on a charge of heresy. The charge could be substantiated by a horoscope of Christ, and by a book which Cardano had published praising Nero, the Roman emperor well known for his persecution of the early Christians. After he was convicted, he was forced to admit and renounce his heresies. As a punishment he was denied the right to lecture publicly and was ordered to refrain from writing and publishing books. Heartbroken, he accompanied one of his students to Rome, where he received an invitation to become a consultant to the College of Physicians. A pension was received from the Pope and soon, possibly insane,[6] he died.

A student of Cardano's who received fame as a mathematician was Ludovico Ferrari. He entered Cardano's house as a servant, soon was elevated to secretary, and became a public lecturer before he was twenty years old. His major contribution to mathematics was a general solution to the biquadratic equation $x^4 + ax^2 + b = cx$. Unfortunately his career was cut short by premature death when in his early forties.

The Dispute[7]

About 1510, Scipione del Ferro found a general solution to $x^3 + ax = b$, but he died before he could publish his discovery. His student, Antonio Maria Fiore, knew the solution and attempted to gain a reputation by exploiting his master's discovery. He challenged Tartaglia with thirty questions, all of which reduced to the solution of $x^3 + ax = b$. Tartaglia had the general solution to $x^3 + ax^2 = b$, so he responded with thirty questions of a more general theoretical nature, although some resolved to this equation. Besides the prestige to be gained, the winner and his friends were to receive thirty banquets from the loser. Just before the time limit elapsed, Tartaglia found general solutions to both $x^3 + ax = b$ and $x^3 = ax + b$.[8] With these, Tartaglia solved all of Fiore's problems, but Fiore was unable to solve any of the questions proposed by Tartaglia and so was vanquished. The banquets were not collected.

At this time Cardano was writing the *Practica arithmeticae generalis*, which would encompass arithmetic, geometry, and algebra. Inasmuch as Fra Luca Pacioli had earlier stated that there could not be a general solution to the cubic, Cardano had ignored this topic. Upon hearing that Tartaglia had a solution for $x^3 + ax = b$, he tried to find one. Failing, he asked Tartaglia for the solution so that he might publish it in a special section of the *Practica arithmeticae generalis* under Tartaglia's name. Tartaglia refused, stating he would publish the solution himself at a later date. This prompted Cardano to label Tartaglia as greedy and unwilling to help mankind.

Because of these insults, a correspondence developed between the two mathematicians which resulted in Tartaglia visiting Cardano in Milan. During this visit Tartaglia relented and offered Cardano a cryptic poem containing a solution to $x^3 + ax = b$, provided Cardano swore an oath that he would never reveal the solution. Cardano accepted the terms, but was unable to decipher the code, so he asked for and received the necessary clue from Tartaglia. Using the solution to $x^3 + ax = b$, Cardano and Ferrari found solutions for $x^3 + ax^2 = b$, $x^3 = ax^2 + b$, and $x^3 + b = ax^2$ by employing substitutions which reduced them to the known case.

In 1545, Cardano published the *Ars magna*, which contained Tartaglia's solution of the cubic with a statement that del Ferro and Tartaglia had each found solutions by independent research. Cardano also included some of his own discoveries, including the idea that every cubic should have three roots. Cardano also published Ferrari's solution to the biquadratic equation here, with due credit to Ferrari.

When he had seen the *Ars magna*, Tartaglia publicly denounced Cardano for breaking an oath sworn on the Gospels, and he ridiculed Cardano's mathematical ability.

Cardano disdained to refute the slur, but Ferrari attacked Tartaglia, charging that Tartaglia had built up his reputation by defaming others, had stolen one proof in his new book without giving credit, and in addition had at least one thousand errors in the text. Ferrari ended his published response by challenging Tartaglia to a public debate on mathematics and all related subjects.

Tartaglia answered with further insults and refused the debate on the grounds that Cardano knew the men who would be the judges. Perhaps he really feared a public debate because he stammered.

After a further exchange of insults, each proposed thirty-one questions which were exchanged, answered, and returned. However, no decision was reached, because each tore the other's answers to shreds.

Then for no given reason, Tartaglia accepted a debate to be held in Milan, Cardano's stronghold.[9]

On August 10th, 1548, Tartaglia and Ferrari met in combat, Cardano having left town.

Very little is known about the actual debate, but it appears to have degenerated into an invective match, with Tartaglia doing most of the shouting. Tartaglia left after the first day, claiming to have won, although it seems Ferrari won by default.

An indication of Ferrari's triumph is that Tartaglia lost his teaching post in Brescia, and Ferrari was invited to lecture in Venice, Tartaglia's stronghold.

Tartaglia died in 1557 without publishing his solution to the cubic, and when an attempt was made to publish his unpublished papers, none could be found which even mentioned the solutions to the cubic.

With Cardano's death in 1576, one of the most interesting and colorful episodes in the history of mathematics ended.

NOTES

1. The actual year is in doubt; Oystein Ore lists it as 1499 in *Cardano, the Gambling Scholar*, and D. E. Smith gives 1506 in his *History of Mathematics*.

2. Hieronymus Cardanus in Latin; also translated into English as Jerome Cardan.

3. *A Book on Games of Chance*.

4. At the time of his death, 131 of his works had been published, 111 existed in manuscript, and he claimed to have burned 170 which he found unsatisfactory.

5. A quote taken from the introduction by Gabriel Naude in Cardano's autobiography, *Liber de propria vita*, Amsterdam, 1654.

6. Charles W. Burr, *Jerome Cardan as Seen by an Alienist*, in University of Pennsylvania: University Lectures Delivered by Members of the Faculty in the Free Public Lecture Course, 1916–1917, vol. 4.

7. The two main sources for the dispute are Oystein Ore, *Cardano, the Gambling Scholar*, Princeton:

(Princeton University Press, 1953) and M. A. Nordgaard, "Sidelights on the Cardan-Tartaglia Dispute," *National Mathematics Magazine*, XII (1937), 327–346.

8. Due to the limited use of symbols and a lack of understanding of negative quantities, these equations had to be treated separately. For example, Cardano wrote $x^3 + 6x = 20$ as cub^s p: 6 reb^s aeqlis 20. Other methods of writing equations during this period can be seen in D. E. Smith's *History of Mathematics*, II, pp. 427–431.

9. Perhaps Brescia, Tartaglia's city, demanded it because of civic pride.

HISTORICAL EXHIBIT 6

Cardano's Technique for the Solution of a Reduced Cubic Equation

A solution is sought for an equation of the form

$$x^3 + ax = b, \; a > 0, \; b > 0.$$

It is known that

$$(p - q)^3 + 3pq(p - q) = p^3 - q^3;$$

therefore, if we let

$$x = (p - q) \text{ then } a = 3pq \text{ and } b = p^3 - q^3.$$

It follows then that:

$$p = \frac{a}{3q} \text{ and } b = \left(\frac{a}{3q}\right)^3 - q^3 \text{ or}$$

$$27b \; q^3 + a^3 - 27\left(q^3\right)^2 \text{ which can be rewritten as}$$

$$27\left(q^3\right)^2 + 27b \; q^3 - a^3 = 0.$$

This last equation is a biquadratic for which use of the existing quadratic solution scheme could supply a solution for q^3, that is,

$$q^3 = \frac{-b \pm \sqrt{b^2 + \dfrac{4a^3}{27}}}{2} \qquad \text{and}$$

q is found to be

$$q = \sqrt[3]{-\frac{b}{2} + \sqrt{\frac{b^2}{4} + \frac{a^3}{27}}} \; .$$

In a similar manner, the value for p is also obtained:

$$p = \sqrt[3]{\frac{b}{2} + \sqrt{\frac{b^2}{4} + \frac{a^3}{27}}} \; ,$$

and finally

$$x = p - q = \sqrt[3]{\frac{b}{2} + \sqrt{\frac{b^2}{4} + \frac{a^3}{27}}} - \sqrt[3]{\frac{-b}{2} + \sqrt{\frac{b^2}{4} + \frac{a^3}{27}}} \; .$$

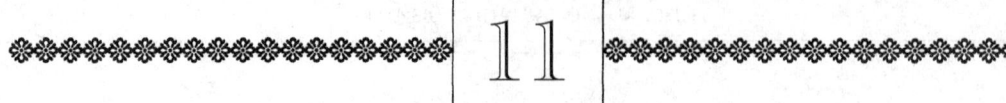

11

Complex Numbers: An Example of Recurring Themes in the Development of Mathematics—I

PHILLIP S. JONES

*T*HE STORY OF THE DEVELOPMENT of the concept and uses of complex numbers provides a fine example of how the history of mathematics may shed light on the meaning of terminology, the relative roles of "practical" needs and intellectual curiosity in the motivation of mathematicians, the utility of pure mathematics, and the development not only of mathematics itself but also of the concepts of rigor and proof. The story also involves intrigue and illustrates the international nature of mathematical scholarship. Hence, properly told, this story does much more than merely provide interest and motivation; in any event, it surely does no less.

Some historians[1] trace, in a negative sort of way, the beginnings of the concept of an imaginary number to Heron's *Sterometria* (ca. A.D. 75). In this Heron states a problem about a pyramid which properly computed would have led to an imaginary number for the length of a line. Heron (or a later copyist) reversed the order of subtraction, and thereby missed both the imaginary number and the correct answer. To the writer this hardly seems to represent a phase in the development of complex numbers, but it does illustrate that in Heron's day, as in our own textbooks, not all apparently

"real" problems were actually based upon real-life situations or measurements. Further this illustrates a situation which was later emphasized by René Descartes, that imaginary numbers or roots may be the algebraic counterpart of non-existent intersections in geometry, or of impossible physical conditions. In these situations the imaginary numbers obtained may however tell a real story since their mere occurrence may be significant in itself.

Still negative in its nature, but of real significance in the story of the growth of the idea because of its explicit recognition of the problem, is the statement of the Hindu Mahavir (ca. 850), "as, in the nature of things, a negative is not a square, it has no square root."[2]

The first printed book to contain algebra, Luca Pacioli's *Summa de Arithmetica Geometria* (Venice: 1494), in discussing quadratic equations, explained in words that a solution was possible only if the constant term was less than or equal to the square of one-half the coefficient of the first degree term (the coefficient of the second degree term being unity).

The real beginning of complex numbers is to be found in the work of the Italian, Jerome Cardan. His *Ars magna* was first printed in Nuremberg in 1545. Our Figure 1 showing the title page of its 1570 edition printed in Basel suggests many other things, however. Not only does it suggest the

Reprinted from *Mathematics Teacher* 61 (Feb., 1954): 106–14; with permission of the National Council of Teachers of Mathematics.

internationalism of printing revealed by the sources of these two editions, but also it points out the cosmopolitan nature of Cardan who is here titled *Mediolanensis* (of Milan) as well as *civisque Bononiensis* (citizen of Bologna). This recalls the heyday of the city-states of Italy, and the fact that Cardan was a contemporary of Machiavelli (1469–1527). This latter fact suggests that the famous "steal" of the solution of the general cubic by Cardan from Nicholas of Brescia (Tartaglia), using as a ruse the story of a noble sponsor, a possible patron, who was interested in Tartaglia's discovery, was quite in keeping with the political morals portrayed in Machiavelli's *The Prince,* as well as displaying something of the relationships between the persons of wealth, the nobility, and the scholars of the day.

Of further interest on the title page is the characterization of Cardan as *philosophi, medici, et mathematici clarissimi.* For not only was Cardan himself a remarkable person, famous in the history of medicine, cryptography, and gambling,[3] but also this suggests the parallel growth of philosophy and science with mathematics, and the continuing relationships which exist today among these disciplines.

Lastly, the words *Ars magnae, sive de Regulis Algebraicis* followed by a reference to *Arithmeticae* in contrast to which algebra was the "great art," with the added comment "*fecundum geometricas quantitates inquirentis*" (fruitful in inquiries concerning geometric quantities) shows that the interrelatedness of these three parts of mathematics was recognized then as it should be emphasized in our teaching now.

Figure 2 shows page 131 of the book whose title page we have just discussed. The first two lines state Cardan's famous problem "divide 10 into two parts such that the product of one times the remainder is (30 or) 40." Cardan immediately says *manifestum est—impossibilis.* He does not explain why this is manifestly impossible, but our students can be led to do good "functional" thinking by suggesting that they consider the products of pairs of numbers whose sum is 10, or on a higher level, that they graph $y = x(10 - x)$ and observe or even prove that its maximum is $y = 25$ for $x = 5$.

However, the fascinating thing is to note in the next step one way in which a mathematician, motivated by intellectual curiosity in this case, may move on to discover something new. Cardan says in line 3 (Fig. 2) *sic tamen operabimur,* "nevertheless we will operate," i.e., he says let's follow the procedures we used in other cases and see what happens. By completing the square he obtains $5 + \sqrt{-15}$ and $5 - \sqrt{-15}$. This is stated in line 8 using the abbreviations which characterized the "syncopated" algebra that was intermediary between the earliest period (sometimes called "rhetorical"), when all algebra was written out in words in full, and the modern "symbolic" period.

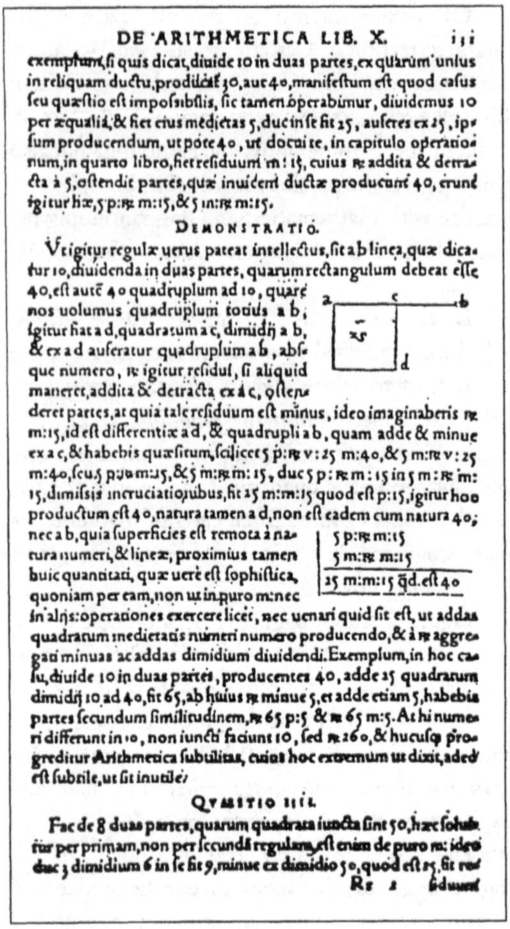

Thus we see a man proceeding to obtain by analogy and formalism results which he finds interesting but which he does not fully comprehend nor accept himself. In more recent times Oliver Heaviside had a similar experience when he developed and used his operational calculus in spite of the criticism of some of his mathematical contemporaries.

The first statement of the rules for operating with square roots of negatives is shown in our Figure 3 taken from the 1579 edition of Rafael Bombelli's *L'Algebra parte maggiore del arithmetica* published in Bologna (1st edition, 1572). Cardan's formulas for the roots of cubic equations involve complex numbers, oddly enough, in the so-called

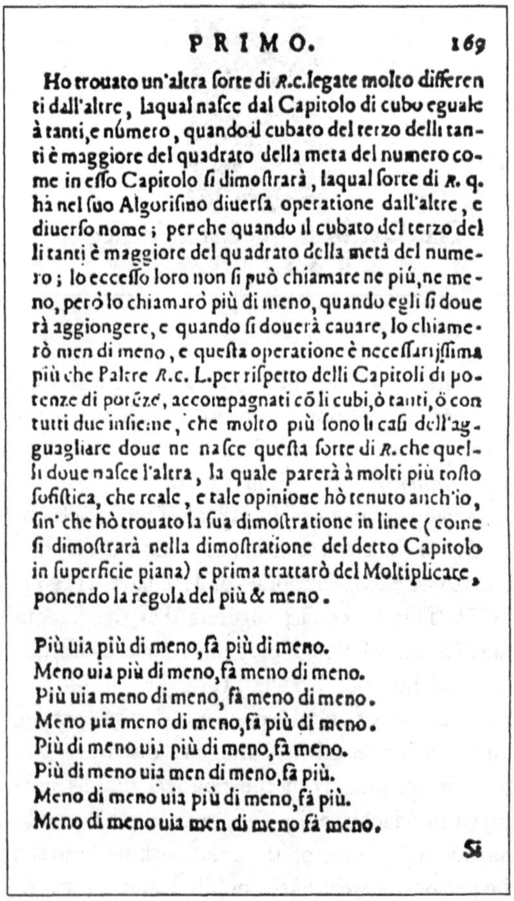

FIGURE 2

(Figure 2 reproduces a page headed **DE ARITHMETICA LIB. X.** *with a* **DEMONSTRATIO** *and* **QVÆSTIO IIII**, *in archaic Latin with a geometric diagram.)*

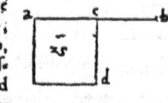

FIGURE 3

(Figure 3 reproduces a page headed **PRIMO. 169**, *in archaic Italian. The rules at the bottom read:)*

Più via più di meno, fa più di meno.
Meno via più di meno, fa meno di meno.
Più via meno di meno, fa meno di meno.
Meno via meno di meno, fa più di meno.
Più di meno via più di meno, fa meno.
Più di meno via men di meno, fa più.
Meno di meno via più di meno, fa più.
Meno di meno via men di meno fa meno.

From Arabic times the proof of algebraic procedures had been given using geometric diagrams such as one sees in the *"Demonstratio"* in our Figure 2. In this case, however, Cardan had to resort in line 9 of this *"Demonstratio"* to *imaginaberis* ("You will imagine"). Although he gives below the diagram a computational check of his results by showing that $5 + \sqrt{-15}$ times $5 - \sqrt{-15}$ is $25 - (-15)$, which is 40, in the discussion he notes that these quantities are *uere sophistica* ("truly sophisticated"), and in the last line of the *"Demonstratio"* he states that continuing to work with these numbers would involve one in an *Arithmetica subtilitas*— "as subtle as it would be useless."

irreducible case in which all three roots are real. He did not solve this case.

If $y^3 + py + q = 0$ Cardan's formula in modern form and symbols is $y = A + B$, where

$$A = \sqrt[3]{-\frac{q}{2} + \sqrt{\left(\frac{q}{2}\right)^2 + \left(\frac{p}{3}\right)^3}},$$

$$B = \sqrt[3]{-\frac{q}{2} - \sqrt{\left(\frac{q}{2}\right)^2 + \left(\frac{p}{3}\right)^3}}.$$

The other roots are $\omega A + \omega^2 B$ and $\omega^2 A + \omega B$ where $\omega = (-1 + \sqrt{3}i)/2$. The "irreducible case" is that in which

$$\Delta = \left(\frac{q}{2}\right)^2 + \left(\frac{p}{3}\right)^3 < 0.$$

Bombelli showed how to combine the two complex numbers given by Cardan's formula in this latter case to obtain finally the correct real root. To do this he had to develop rules for operating with square roots of negatives. One can read his statement of these rules at the bottom of his page 169 by use of the following vocabulary:

piu: a positive (quantity)
uia: times
di meno: a square root of a negative number
meno: a negative (quantity)

Some perception of the difficulties involved in the "rhetorical" stage of algebraic symbolism may be obtained by looking at the end of the second and third lines of the page. Here Bombelli writes that he is treating "cubes equal to quantities and numbers" (i.e., $x^3 = px + q$). Note that he uses a different word for each power of the unknown, and that he deliberately writes his equations so that there are no negative signs. He goes on to say, beginning in line 7, that he will treat the case in which *il cubato del terzo del li tanti è maggiore del quadrato della metà del numero* ("the cube of one third of the [coefficient of] x is greater that the square of one half the constant"). Since Bombelli's

D I S C O U R S
DE LA METHODE
Pour bien conduire fa raifon, & chercher
la verité dans les fciences.
P l u s
LA DIOPTRIQVE.
LES METEORES.
e t
LA GEOMETRIE.
Qui font des effais de cete METHODE.

A L e y d e
De l'Imprimerie de I a n M a i r e.
c I ɔ I ɔ c x x x v i i.
Auec Priuilege.

FIGURE 4

original equation is set up such that the p of our modern symbolic form of Cardan's formulas would be negative, this condition amounts to requiring that $\Delta < 0$, the irreducible case noted above.

Figure 4 is the title page of René Descartes's most famous work. It emphasizes the connections between mathematics, philosophy, and logic, and the interesting fact that his *La Geometrie* was merely the third illustrative appendix to his *Discourse on the Method of Reasoning and Seeking Truth in Science*. (Note the old use of cIɔ and Iɔ for M and D in writing Roman numerals.)

Figure 5 shows page 380 of this book and points out further that *La Geometrie* was far from being anything like our modern analytic geometry texts and that, in fact, it contained much that was purely algebraic. There are three further things to be remarked about this page. The middle para-

380   LA GÉOMÉTRIE.

eftoient ⅓, 1, & ⅖, & que celles de la premiere eftoient ½√3, ⅓√3, & ½√3.

Coment on read li quantité connuë de l'vn des ter-   Cete operation peut auffy feruir pour rendre la quantité connuë de quelqu'vn des termes de l'Equatió efgale a quelque autre donnée, comme fi ayant
$$x^3 \ .. - bbx + c^3 \infty o.$$

mes d'vne Equation efgale a vne autre qu'on veut.   On veut auoir en fa place vne autre Equation, en laquelle la quantité connuë, du terme qui occupe la troifiefme place, a fçauoir celle qui eft icy bb, foit $3aa$, il faut fuppofer $\infty x \sqrt{\tfrac{1aa}{11}}$; puis eftrire $y^1 \infty - 3aay + \tfrac{14^3c1}{11} \sqrt{3} \infty o$.

Que les racines, tant vrayes que fauffes peuuent eftre reelles ou imaginaires.   Au refte tant les vrayes racines que les fauffes ne font pas toufiours reelles; mais quelquefois feulement imaginaires; c'eft a dire qu'on peut bien toufiours en imaginer autant que iay dit en chafque Equation; mais qu'il n'y a quelquefois aucune quantité, qui correfponde a celles qu'on imagine. comme encore qu'on en puiffe imaginer trois en celle cy, $x^3 - 6xx + 13x - 10 \infty o$, il n'y en a toutefois qu'vne reelle, qui eft 2, & pour les deux autres, quoy qu'on les augmente, ou diminue, ou multiplie en la façon que ie viens d'expliquer, on ne fçauroit les rendre autres qu'imaginaires..

La reduction des Equatiós cubiques lorfque le probleme eft plan.   Or quand pour trouuer la conftruction de quelque problefme, on vient a vne Equation, en laquelle la quantité incónnuë a trois dimenfions; premierement fi les quantités connuës, qui y font , contienent quelques nombres rompus, il les faut reduire a d'autres entiers, par la multiplication tantoft expliquée , Et s'ils en contienent de fours , il faut auffy les reduire a d'autres rationaux, autant qu'il fera poffible, tant par cete mefme multiplication,

FIGURE 5

graph and its marginal note contain the first use of "imaginary" as the name of these new numbers. They had earlier been described variously. (We have seen them called "sophisticated" and "subtle.") Unfortunately the term "imaginary" stuck and has come down to us as an historical hangover—and as a headache to teachers, since students naturally think of "imaginary" as a descriptive adjective rather than as merely an arbitrary name for a mathematical object which is as real in its existence as a mathematical point or line. Note further that Descartes's *fausses racines* ("false roots") denoted negative, not imaginary, numbers. He apparently was not entirely willing to accept negatives as being as completely valid as positive numbers and did most of his geometrical drawing in what we would call the first quadrant. This illustrates how each extension of our number system from integers to fractions, to irrationals, to complex, to quaternions, to transcendentals has been first viewed with scepticism as an unreal or inapplicable abstraction. Acceptance and, eventually, fairly general understanding of such extensions have grown with familiarity, improved logical rigor and organization, graphical representation, and applications in both the mathematical and physical worlds. But these concepts, in their earliest days, bothered such geniuses as Descartes and Gauss.

All of these aspects appear in the growth of the acceptance of complex numbers. Their use in extending mathematics appears implicitly in the middle paragraph of our Figure 5 where Descartes is discussing the number of roots which an equation may have. Although, even before Descartes, Albert Girard and others had suggested that an *n*th degree equation may have *n* roots, it required a further expanded theory of complex numbers and their graphical representation before C. F. Gauss could prove the fundamental theorem of algebra, that every rational integral polynomial with real or complex coefficients has at least one root. From this the *n* root theorem follows directly.

Figure 6 serves to illustrate several phases in the development of ideas, as well as to point up again the continuity and internationalism so common in the history of mathematics. This picture is from the 1693 Latin edition of John Wallis's *Algebra*, which first appeared in English in 1685. The lower diagrams speak for continuity and internationalism because they appeared first in the work of Descartes. They show a geometrical construction for the roots of a quadratic equation of the type $x^2 \mp bx - c = 0$. (However, Wallis did not use *x* as Descartes did, but wrote $aa \mp ba - æ = 0$, using vowels for the unknown as Viete had done earlier. Wallis regarded the arbitrary constants as always representing positive numbers.)

If $CP = {}^1/_2 b$, and $PB = \sqrt{c}$ is perpendicular to CP, then the circle on CP determines the secant αB on which AB and $B\alpha$ are equal in length to the roots of the equation. (Note the connection between the geometric theorem that the product of the segments of a secant equals the square of the

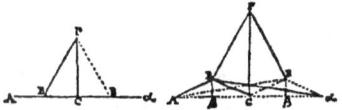

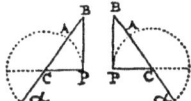

FIGURE 6

tangent from an external point, and the algebraic theorem that the product of the roots of this quadratic is c.)

The left-hand diagram at the top of the page gives Wallis' construction for the roots of $x^2 \pm bx + c = 0$ if $(b/2) \geq \sqrt{c}$. Take $AC = \frac{1}{2}b = C\alpha$ and $PC = \sqrt{c}$ perpendicular to $A\alpha$ at C. The radius $PB = \frac{1}{2}b$ with center at P determines the points B. The roots are then the numerical values of the two lengths AB.

If $(b/2) < \sqrt{c}$, the above construction fails, in which case Wallis used PC and PB to construct a right triangle with the right angle at B, as shown in the upper right hand diagram, rather than at C. He then says that the points B which are *above the line* may be regarded as representing the solution in

this case just as the B's *on the line* determined the solution when the roots were real. This, we see, was a step toward the modern graphical representation of complex numbers. Incidentally, recall that Descartes, whom Wallis had studied and admired, had interpreted imaginary roots for algebraic equations as indicating that the related geometric construction was impossible. Wallis also suggested representing $\sqrt{-1}$ by applying the Euclidean construction for a mean proportional to directed line segments representing $+1$ and -1, and he had argued that, just as people had become reconciled to representing both positive and negative numbers on a line and associating them with distances forward and backward, so one could imagine negative areas as land covered by rising water,[4] and use such a visualization to encourage the acceptance of negative squares.

Not long after Wallis, Lambert used complex variables in the development of map projections, and D'Alembert used them in the study of hydrodynamics. The stories of these and other applications, the story of the geometric representation of complex numbers, and the story of a modern rigorous formulation of an arithmetic for these numbers are interesting tales to which we will return later; but we hope that the story so far offers a way to an understanding of the numbers, their name, how and why they developed and the relative roles in their growth of application and abstraction, representation and generalization, practical necessity, and intellectual curiosity.

NOTES

1. D. E. Smith, *History of Mathematics* (Boston: Ginn and Co., 1925), p. 261. Smith also gives several references to other discussions of Heron's problem.

2. M. Rangacarya, *The Ganita-Sara-Sangraha of Mahaviracarya with English Translation and Notes* (Madras: 1912), p. 7 of the English translation. The sentence before the one quoted above was, "The square of a positive as well as of a negative (quantity) is positive; and the square roots of those (square quantities) are positive and negative in order."

3. There have been several biographies of Cardan. The most recent is by Oystein Öre, *The Gambling Scholar* (Princeton University Press, 1953). Cardan's own autobiography has been translated as *The Book of My Life* (New York: 1930, translated by Jean Stover; and London: J. M. Dent and Son, 1931), while his medical achievements are emphasized in Henry Morley's *Jerome Cardan: The Life of Girolamo Cardano of Milan, Physician* (London: Chapman and Hall, 1854). In addition to W.G. Walters, *Jerome Cardan a Biographical Study* (London: 1898), there have been many journal articles about him and his work as well as short biographies in books and in *Portraits of Eminent Mathematicians, Portfolio II* published by *Scripta Mathematica*.

4. A reprint of these discussions may be found in *A Source Book in Mathematics* edited by D. E. Smith (New York: McGraw-Hill and Co., 1929), pp. 46 ff.

12

Robert Recorde's
Whetstone of witte, 1557

VERA SANFORD

THE YEAR 1957 is the four hundredth anniversary of the publication of Robert Recorde's *Whetstone of witte*, the first algebra printed in English, the book in which the equality sign (=) was used for the first time. It is appropriate to mark this anniversary by quoting Recorde's explanation of his invention and by outlining its later history.

Contrary to the natural assumption that the value of this symbol would have been recognized at once, it develops that fully sixty years elapsed before its second appearance, although in less than a century after that time, it had general acceptance. This situation suggests that we examine the state of mathematics in England in the 1550s, that we review what we know about the author, and study the contents of the book itself.

I. The Equality Sign

At the beginning of his work with equations, Recorde says

> And to auoide the tediouse repetition of these woordes: is equalle to: I will sette as I doe often in woorke vse, a pair of paralleles, or Gemowe (twin) lines of one lengthe, thus: ===== bicause noe .2. thynges can be moare equalle.

The equality sign is Recorde's single contribution to the symbolism of algebra. Although the signs + and − were already in use, having letters represent

unknown quantities was in the future. Algebra looked very different from the way it looks today.

In writing equations, Recorde's contemporaries used the word *equal* in its different forms as "aequales," "faciunt," "gleich." Sometimes they abbreviated these words. There was considerable experimentation with symbols, governed in part by the type the printers had on hand. In the years following the publication of the *Whetstone of witte*, different writers used a number of symbols to represent equality and the symbol (=) stood for a variety of meanings.[1] In view of this competition and in consideration of the confusion of meanings assigned to the twin parallels, it is all the more remarkable that Recorde's invention ultimately won universal adoption. The process was slow.

The second appearance of the equality sign was in an appendix to a translation (1618) of Napier's first work on logarithms. It is supposed that this appendix was the work of William Oughtred, whose *Clavis mathematicae* (1631), a closely-packed work on arithmetic and algebra, was influential in winning for Recorde's symbol a general acceptance in England in the seventeenth century. By the eighteenth century, it had come into use on the continent.

II. Robert Recorde and His Times

Robert Recorde (ca. 1510–1558) was born shortly after Henry VIII came to the throne of England. He died just before the accession of Elizabeth I.

Reprinted from *Mathematics Teacher* 50 (Apr., 1957): 258–66; with permission of the National Council of Teachers of Mathematics.

67

His contemporaries on the continent included Cardan and Tartaglia who were doing pioneer work in mathematics in Milan and Venice. There were the German algebraists Rudolff, Stifel, and Scheubel. Copernicus, from Poland, published his great work in 1543. In 1538, Mercator, in Flanders, produced a map of the world on a projection used earlier by a Spaniard, but now called by his name. Methods of navigation were being improved. Almanacs of a sort were being published. The variation and the dip of the magnetic needle were being studied. The problem of the determination of longitude baffled mathematicians and other scientists. Sebastian Cabot maintained that he had solved it, but he died in 1557 without divulging his secret, and it was supposed that his boasts were the imaginings of an octagenarian.

England lagged behind in this activity. In fact the one important publication in mathematics prior to Recorde was Cuthbert Tunstall's *De arte Svppvtandi* (1522), a scholarly arithmetic in Latin based largely on continental models. So far as navigation and exploration were concerned, it is true that in 1497 the king had financed the voyage of the Genoese mariners John and Sebastian Cabot to the New World. John Cabot died the next year. Royal support stopped at this point, and Sebastian Cabot entered the employ of the King of Spain.

It was in the first half of the sixteenth century, however, that the center of world trade began to shift. Prior to this time, goods from China, India, and the East Indies had been brought by caravan to the ports on the eastern shores of the Mediterranean where they were loaded on ships from Genoa and Venice and other Italian cities for distribution to different parts of Europe. The Mediterranean was still the great highway it had been in the days of Greece and Rome.[2] But in the period under consideration, sea routes were being developed. England no longer lay at the periphery of the world but was shortly to assume an important position as a maritime power.

So far as mathematics was concerned, Englishmen were not isolated from scholars on the continent. Tartaglia had dedicated a book to Henry VIII in 1546. John Dee (1527–1608) studied on the continent and knew some of the outstanding mathematicians personally. Cardan visited London on his way to Scotland in 1552. Robert Recorde himself seems to have read the works of a number of these men, and we know that he made considerable use of Scheubel's algebra which was published in Paris in 1551.

But England had no mathematician to match the best of those in Europe, and so far as applied mathematics was concerned, the situation was deplorable. In England, navigation had not become a science. Map making was primitive. Surveying was casual. Little mathematics was taught. Arithmetic was neglected. Roger Ascham, tutor to the Princess Elizabeth, said that unless it was taught in moderation, mathematics overcharged the memory. It was a time when mathematics was needed to develop new techniques in surveying, navigation, and gunnery, but the workers knew no mathematics and the mathematicians, such as they were, lacked practical experience. The universities were of no help in the matter. The gap was bridged by a group, many of them amateurs, whom Dr. E. G. R. Taylor calls "Mathematical Practitioners."[3] Among them was Robert Recorde.

Recorde, a native of Pembrokeshire in Wales, studied at Oxford, received the degree of Doctor of Medicine from Cambridge, and taught mathematics privately at both places, making the subject "clear to all capacities to an extent wholly unprecedented." He wrote a medical treatise of considerable importance, for it had at least ten editions from 1547 to 1665. He is reported to have practiced medicine in London, and it is claimed that he was physician to Edward VI and to Queen Mary. About 1551 he was appointed Surveyor of the Mines and Monies of Ireland and was made Comptroller of the Mint at Bristol. The *Whetstone of witte* ends abruptly with the entrance of a messenger sent to fetch the author to answer charges presumably of mismanagement of one of these political jobs. Within a year he died in the King's Bench Prison.

III. Recorde's Mathematical Books

Recorde was an opportunist. In his mathematical books, his aim was to provide the reader with needed information in palatable form. Recorde's four published books in mathematics were:

> The *Grovnd of Artes*, an arithmetic, 1542 or perhaps 1540
> The *Pathewaie to Knowledge: First Principles of Geometry*, 1551
> The *Castle of Knowledge*, an astronomy, 1551
> The *Whetstone of witte*, an algebra, 1557

In the prefaces of these volumes he refers to others he intends to write, in one case speaking of "other sundrye woorkes partly ended and partly to bee ended."

He took a realistic view of his qualifications as a writer. In the Preface of the *Grovnd of Artes*, he says

> ... I know that no man can satisfie euery man, and therefore like as many do esteeme greatly other bookes, so I doubt not but some will like this my booke aboue any other English Arithmeticke hitherto written, & namely such as shal lacke instructers, for whose sake I haue plainly set forth the exa[m]ples, as no book (that I haue seene) hath hitherto.

His books were in the form of a dialogue between a Master and a Scholar, with the Scholar asking the leading questions in just the proper places. The books were well adapted to "such as shal lacke instructers."

The *Pathewaie to Knowledge* was dedicated to Edward VI with this explanation or apologia,

> Excuse me, Gentle Reader, if ought bee amisse straunge pathes are not trode[n] al truly at the first: the way muste needes be comberous, wher none

hathe gone before. . . . For neithe is my witte so finely filed, neither my learnyng so largely lettered neither yet my laisure so quiet and vncombered, that I maie performe iustely so learned a labour. . . . Yet may I thinke thus: This candle did I light: this light haue I kindeled: . . . I drew the platte [plan] rudelie wheron they maye builde, whom God hath indued with learnyng. . . . And this Gentle Reader I hartelie protest, where erroure hath happened I wishe it redreste.

As for the *Whetstone of witte*, Recorde is modest, but he felt that an algebra should be written and that it behooved him to do it. He says

> For better is it that a simple Coke doe prepare thy brekefast then that thou shouldest goe a hungered to bedde.

The *Grovnd of Artes* was a practical arithmetic. It found great use, as is witnessed by the fact that it had many editions over a period of a hundred and fifty years. It was quite in keeping with Recorde's purpose that the editors of later editions incorporated new things in them.

The *Pathewaie to Knowledge*, dedicated to Edward VI, was an introduction to geometry with definitions and conclusions in the first part, and proofs of the conclusions in the second. Third and fourth parts giving applications of these principles were promised but were never printed.

Recorde's astronomy, the *Castle of Knowledge*, dedicated to the Princess Mary, was both practical and theoretical. Dr. Taylor states that, "Like most physicians of his day, Recorde believed that the aspects of the heavens determined the correct times for taking medicines and other remedies, and he emphasized this equally with the needs of navigation for the study of astronomy." It has been claimed that Recorde's astronomy set forth the Copernican hypothesis, but Dr. Taylor says that Recorde "expressed himself cautiously . . . but held no brief for an unmoveable earth circled by the stars and planets." Recorde might well be excused for being cautious for it must be remembered that the date of Copernicus's great work was 1543, and that this had a preface explaining that the hypothesis that the earth moved round the sun was merely a convenience in computation. It was after Recorde's time that Kepler detected that this was an interpolation.

IV. Recorde's Algebra

The full title of Recorde's algebra is

> The whetstone
> of witte
> whiche is the seconde parte of
> Arithmetike: containyng the extrac
> tion of Rootes; the *Cossike* practise,
> with the rule of Equation: and
> the woorkes of *Surde*
> Nombers

Augustus de Morgan shows that this title is a play on words. The word *cos*, meaning a thing, was derived from the Latin *causa* by way of the Italian *cosa*. German and English writers used it to represent an unknown quantity. Consequently "Cossike practise" meant algebra. In Latin, the word *cos* means a grindstone. "Cossike practise" might be put into Latin as *cos ingenii* which in turn could be translated into English as the *Whetstone of witte*.

The preface is dated November 11, 1557. It should be remembered that at that time, the new year in England began March 25. There was time to print the volume before the end of 1557.

The book is dedicated to the "Venturers into Muscovia." In his introduction, Recorde says that he will

> ... shortly set forthe soche a booke of Nauigation as I dare saie shall partly satisfie and contente, not onely your expectation but also the desire of a greate nomber beside wherin I will not forgette specially to touche bothe the olde attempte for the Northlie Nauigators, and the later good aduenture.[4]

In writing this algebra, Recorde referred to the work of Johannes Scheubel (1494–1570), profes-

sor of mathematics in the University of Tübingen, whose *Algebrae compendiosa facilisque descriptio* was published in Paris in 1551. Recorde objects to Scheubel's classification of equations, substituting one which he considered much simpler. Having thus improved upon Scheubel, Recorde proceeded to use direct translations of Scheubel's verbal problems. In fact, not only was Recorde acquainted with Scheubel's work but he plagiarized a considerable part of it. This is true not only in the section on algebraic equations, but also in the work on surds.[5]

V. The Extraction of Rootes

The first part of the *Whetstone of witte* considers numbers: whole numbers and broken numbers, "abstracte" and "contracte," evenly even and evenly odd. These last were of the type 2^n and $2(2^n + 1)$. The ratios of numbers are classified with special names for each type. The subject of diametral numbers is treated in detail. A diametral number is the product of two integers which have the property that the sum of their squares is itself a perfect square. In other words, when a, b, and c are integers such that $a^2 + b^2 = c^2$, then ab is a diametral number. Recorde shows that diametral numbers must end in 0, 2, or 8. They must be divisible by 12 and they cannot themselves be square numbers.

Recorde devoted over fifty pages to the extraction of roots. Here he gives particular attention to what he calls "lower ma[t]ters in warre." For example,

> A citie should bee scaled, beyng double diched. And the inner diche .32. foote broade. And the walle .21. foote high. The captain commaundeth ladders to be made of that iuste lengthe, that maie reche from the utter brow of the inner diche, to the toppe of the wal.

> A bollette of yron of .7. inches diameter, doeth waie .27. pounds weighte: what shall be the diameter to that bollette that shall wai .125. pounde to the weighte?

VI. The Cossike Practice

The second part of the *Whetstone of witte* deals with algebraic numbers and with equations.

Following Scheubel's example, Recorde uses the symbol ♀ to denote a number, ⅇ for the unknown quantity or root, ⅄ for its square, and ⅇ for its cube. The fourth order of these numbers was indicated by repeating the symbol for a square, i.e., a square of a square. Recorde's predecessors had carried this system to great lengths, but Recorde outdid them by taking it to the eightieth order. It should be noted that in any equation every term, even the constant one, had its symbol. Like Scheubel, Recorde also made use of abbreviations for these quantities—N, Ra., Pri., Se., Ter., and so on (the number, a root, the first product, the second product, etc). As an aid to the reader, Recorde lists the symbols in order, numbering them 1, 2, 3, ... and says "By this table, maie you easily knowe the signe that shall serve for your newe somme in multiplication."

The Scholar had difficulty in grasping the fact that .12.⅄ multiplied by .6.ⅇ makes .72.ⅇ. He says

This passeth my cunnynge, for the findyng of the newe signe although the multiplication of the nombers be as easie as can be.

The Master answers

If you did well remeber what you haue learned before; the mater would not seme so harde.

In dealing with "Cossike numbers," Robert Recorde introduced two signs, familiar to German writers, and familiar also to readers of the *Grovnd of Artes*. He says

... touchyng these twoo signes + and − which bee the figures of more and lesse, you must giue regard whether thei bee like or unlike, in those numbers that must be added: For if thei be like in nombers of one denomination, then muste thei so remain as thei be. But if thei be vnlike, euermore abate the smaller nomber of theim that followe those unlike signes out of the greater, and sette doune the reste with the signe of the greater nomber.

The Scholar is then confronted with eight examples in which cossike numbers are to be added. These are the different arrangements of two basic problems.[6]

$$10x \pm 12 \qquad 10x \pm 8$$
$$\underline{4x \pm 8} \qquad \underline{4x \pm 12}$$

He says

Here haue I varied one example diuersely, to the intente you maie marke the vse of your rules in theim.

He explains the addition of $4x - 8$ to $10x + 12$ as follows:

this sum is not fully $4x$ but wanteth of it 8 and therefore if you put downe $4x$ fully, you must abate 8 out of the 12 in the larger summe.

Recorde had a liking for putting things that were to be memorized into rhyme. The results were interesting. Here is his verse for the rules of signs in multiplication and division.

who that will multiplie
or yet divide trulie:
shall like stille to haue more
and mislike lesse in store.

The Scholar summarized it in these words:

So meane you that like signes multiplied together, doe make more, or + and vnlike sines multiplied together doe yelde lesse or − ?

In Recorde's work with polynomials, no powers of the cossike number are omitted. For example, $8x^3 + 0x^2 + 0x + 64$ is to be divided by $2x + 4$.

Having practiced the fundamental operations with polynomials, the Master shows how to add fractions whose numerators and denominators are algebraic quantities. By his scheme, horizontal lines are drawn above and below the fractions that are to be added. The common denominator is written below the bottom line and the new numerators are written above the top line.

VII. "The Rule of Equation Commonly Called Algebers Rule"

> This Rule is called the Rule of *Algeber,* after the name of the inuentoure, as some men thinke . . . but of his vse it is rightly called the rule of *equation:* bicause that by the *equation* of nombers, it doeth dissolue doubtful questions: and vnfolde intricate ridles.

Recorde's statement about the use of equations in solving problems is involved but worth considering.

> When any question is propounded . . . you shall imagin a name for the nomber, that is to bee soughte, as you remember that you learned in the rule of false position. And with that nomber shall you procede, accordyng to the question, vntil you find a Cossike nomber equalle to that nomber that the question expresseth, whiche you shal reduce euer more to the leaste nomber.

The Scholar compares this with the rule of false, the method of solving problems by guessing the answer and then adjusting the guess until it fits the problem. The Master goes on

> . . . it mai bee thoughte to bee a rule of wonderful inuention that teacheth a manne at the firste worde to name a true nomber before he knoweth resolutely [surely] what he hath named. But bicause that name is common to many nombers [although not in one question] and therefore the name is obscure till the worke doe detect it, I thinke this rule might well bee called the rule of darke position, or of strange position, but not of false position.

> And for the more easie and apte worke in this arte wee doe commonly name that dark position .1. ꝛ and with it doe we worke as the question intendeth till we come to the equation.

At this point Recorde introduced the sign of equality, following it at once, as is shown in the accompanying facsimile, by a series of equations which he then considers in detail. In the case of $14x + 15 = 71$, he says

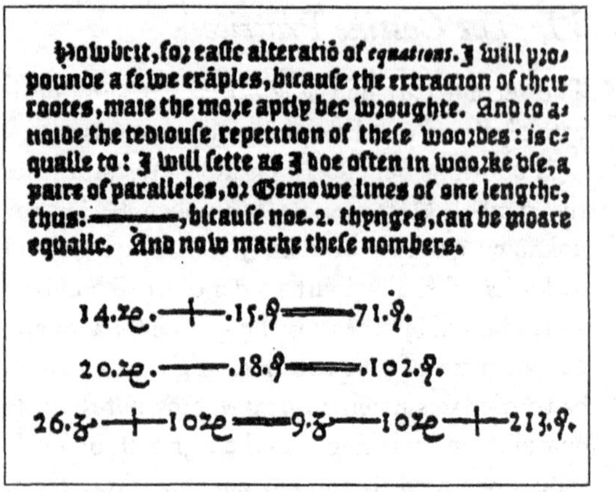
Howbeit, for easie alteratiō of *equations.* I will propounde a fewe exāples, bicause the extraction of their rootes, maie the more aptly bee wroughte. And to auoide the tediouse repetition of these woordes: is equalle to: I will sette as I doe often in woorke vse, a paire of paralleles, or Gemowe lines of one lengthe, thus: ═════, bicause noe .2. thynges, can be moare equalle. And now marke these nombers.

14.ꝛ.—+—.15.ϕ.═══71.ϕ.

20.ꝛ.———.18.ϕ.═══.102.ϕ.

26.ʒ.—+—10ꝛ═══9.ʒ.——10ꝛ—+—213.ϕ.

FIGURE I

> . . . you mai see one denomination on both sides of the *equation* which neuer ought to stand. Wherfore abating the lesser, that is 15, out of both nombers, there will remain
>
> $$14x = 56$$
>
> this is by reduction $x = 4$ according to the third common sentence in the *Pathewaie*—If you abate euen portions from thynges that bee equalle, the partes that remain shall be equall also.

Other "common sentences" are used in solving other equations.

When confronted with the equation $26x^2 + 10x = 9x^2 - 10x + 213$, the Scholar wonders whether to add $10x$ or to "abate them." The Master says

> In soche a case, you maie dooe either of bothe at your libertie and all will be to one ende. . . . And euermore when occasion serueth to translate nombers compounde, − on the one side is equalle to + on the other side.

Robert Recorde classifies equations into two groups:

1. Where one number is equal to another number,

2. Where one number is equal to two other numbers.

Here he has departed from Scheubel whose detailed classification can be represented as follows:

First type $bx = N$

Second type   (1) $ax^2 + bx = N$

              (2) $x + N = x^2$ or

                    $bx + N = x^2$

              (3) $bx + N = ax^2$

Third type $x^2 + N = bx$

Neither writer classified equations according to their degree. Recorde includes the equation $6x^3 = 24x$ under his first heading, one number equal to another number, but he shows that the answer is not 4 but the square root of 4. He did not recognize roots that were zero or negative.

No reasons or explanations are given. Quadratic equations are solved by completing the square.

When the Master has completed the solution of a number of equations, the Scholar says "I doe couette some apte questions, appertainyng to these equations." The Master supplies them. They are not easy.

In one of these problems two men have silk to sell, one man has 40 ells, the other 90. The first man's silk is of the poorer quality, so he has to sell a third of an ell more for an angel than does the second man. If their total receipts are 42 angels, how many ells of silk did each sell for an angel? The Scholar makes a false start. He lets his unknown quantity represent the first man's receipts. The Master does not approve and suggests that he divide each man's ells by the number he sold for an angel and the quotient will be the man's receipts. Working this way, the Scholar finally comes to the conclusion that one man sold three ells for an angel and the other three and one third, so their receipts were 12 angels and 30 angels. Not content with this, the Scholar reverses the problem. Instead of using x and $x + 1/3$, he uses $x - 1/3$ and x. This time he gets one root correctly, but he also finds the other one, namely 2/21, and says

> But how I maie frame that roote to agree to this question, I doe not see.

The Master is no help to him, except to say that

> . . . the forme of the question maie easily instruct you whiche of these .2. rootes you shall take for your purpose.

As an example of a question where both roots fit the problem, the Master gives the following:

> A gentilman, willyng to proue the cunnyng of a braggyng *Arithmetician*, saied thus: I haue in bothe my handes .8. crounes: But and if I accoumpte the somme of eche hande by it self seuerally and put therto the squares and the cubes of bothe, it will make in nomber 194. Now tell me (quod he) what is in eche hande: and I will giue you all for your laboure.

As would be expected, the answers 3 and 5 are the numbers in the man's hands.

In another case, the problem states that if 8 is added to a number and if 16 is taken from the square of the number, the product of the two results is 2560. This yields the equation $x^3 + 8x^2 - 16x = 2688$. The Scholar is baffled. It is above his cunning, for here two numbers are equal to two others, or one number is equal to three numbers. The Master does not solve the problem. One suspects that he concocted the question by starting with the result. He simply says that the required number is 12 and has the Scholar show that this answer fits the equation. At this point the Master says

> But to put you out of doubte, this equation is but a trifle to others that bee untouched.

There is no lack of variety in the problems. A man travels $1 1/2$ miles the first day and increases his day's journey by $1/6$ mile each day. How long does it take him to go 2955 miles? In another case, the day's mileage increases in a geometric progression.

A herald is offered a bribe if he will tell the number of his king's army. The problem goes on this way—

The Heraulte lothe to lease those giftes, and as lothe to bee vntrue to his Prince, diuiseth his answere, wiche was true, but yet not so plain, that the aduersarie could thereby vnderstand that whiche he desired. And that aunswere was this. Looke how many Dukes there are, and for eche of them, there are twise so many Erles. And vnder euery Erle, there are fower tymes so many souldiars, as there be Dukes in the field. And when the muster of the soldiers was taken, the .200. parte of them was .9. tymes so many as the nomber of the Dukes.

That is the true declaratio of eche number, quod the Heraulte: and I haue discharged my othe. Now guesse you how many of eche sorte there was.

A man dies leaving 72 crowns to his four children in this way: the second and third together were to have seven times as much as the first. The third and fourth were to have five times as much as the second. The first and fourth were to have twice as much as the third. It might be well to note that the children received the following sums: $4\frac{1}{2}$, $11\frac{1}{4}$, $20\frac{1}{4}$, 36.

A captain marshals his army in a square formation. When the square was of one size, he had 284 men too many, so he tried to arrange them in a square one man more on a side than before. This time he lacked 25 men. How many men did he have?

VIII. *The woorkes of Surde Nombers*

The third part of the *Whetstone of witte* is devoted to irrational numbers, which Recorde calls surds. He says

> Nombers *radicalle*, which commonly bee called nombers *irrationalle:* bicause many of them are soche, as can not bee expressed, by common nombers *abstracte*, nother by any certain ratiionalle nomber. Other men call them more aptly *surde* nombers.... A *surde* number is nothyng els, but soche a nomber set for a roote, as can not be expressed by any number absolute, as $\sqrt{10}$ or $\sqrt{18}$ or any nomber, that is not a square.

Surds are commensurable if they can be expressed as multiples of the same root; otherwise they are incommensurable.

Binomials made up of a rational number and an irrational one, are classified in two groups: "Nombers that be compounded with + be called Bimedialles and with − Residualles."

Accordingly, to divide by a binomial, dividend and divisor must be multiplied by the residualle of the divisor if the divisor is a bimedialle, or by the bimedialle of the divisor if the divisor be a residualle.

Recorde's square root sign was much like ours, but his signs for the cube root and the fourth root were perplexing. These are the symbols used by Scheubel, which he in turn had taken from the work of Rudolff (1525). When confronted with the symbol . $\mathcal{MW}$. for cube root and . $\mathcal{MV}$. for the fourth root, the Scholar speaks his mind.

> It were againste reason, to take treason for these signes, which be set voluntarily to signifie any thyng; although some tymes there bee a certain apte conformitie in soche thynges. And in these figures, the nomber of their minomes [i.e. upstrokes] seemeth disagreable to their order.

The Master replies:

> In that there is some reason to bee thewed [instructed]: for as .√. declareth the multiplication of a nomber, ones by itself; so .$\mathcal{MW}$. representeth that multiplication *Cubike*, in whiche the roote is represented thrise. And .$\mathcal{MV}$. standeth for .√.√. that is .2. figures of Square multiplication: and is not expressed with .4. minomes. For so should it seme to expresse moare then .2. *Square* multiplications. But voluntarie [i.e., arbitrary] signes, it is inough to knowe that thie doe signifie. And if any Manne can diuise other, moare easier or apter in use, that mai well be received.

IX. *Appraisal of the* Whetstone of witte

The chances are that the *Whetstone of witte* had few if any readers on the continent. Anyone interested in algebra would have found more satisfaction in Scheubel's concisely worded, elegantly printed treatise in Latin than in Recorde's popularization of it in English. So far as the equality sign was concerned, it is more than likely that

Recorde had no particular pride in his invention. It was a thing he had found convenient. On the other hand, his simplification of Scheubel's classification of equations, in which he took apparent satisfaction, is of little interest today.

The *Ground of Artes* was closely connected with practical affairs. The *Whetstone of witte* seems to contain no practical applicatons, the reason being that algebraic problems which we would solve by equations of the first degree were treated by the rule of false in Recorde's arithmetic. The *Whetstone of witte* certainly had a smaller public. Had Recorde been able to complete the books that were "partely to bee ended," the algebra might have had a wider circulation. As it was, the last of Recorde's books was the only one to have but a single edition.

Frequently the introduction of a book is best read after reading the book itself. So here, it seems appropriate to quote as a conclusion the verses on Recorde's title page:

> Though many stones doe beare great price,
> The whetstone is for exercise
> As neadfulle, and in woorke as straunge:
> Dulle thinges and harde it will so chaunge,
> And make them sharpe, to right good vse:
> All artesmen know, thei can not chuse,
> But vse his helpe: yet as men see,
> Noe sharpnesse semeth in it to bee.
> The *grounde of artes* did brede this stone:
> His Vse is great, and moare then one.
> Here if you list your wittes to whette,
> Moche sharpnesse therby shal you gette.
> Dull wittes hereby doe greatly mende,
> Sharp wittes are fined to their full ende
> Now proue, and praise, as you doe find,
> And to yourself be not vnkinde.

NOTES

1. Florian Cajori, *History of Mathematical Notation* (Chicago, 1928), Vol I.

2. See Chapter 6 of G. M. Trevelyan, *History of England* (New York: Doubleday Anchor Books, 1953), Vol. II.

3. E. G. R. Taylor, *Mathematical Practitioners of Tudor and Stuart England* (Cambridge, 1954).

4. The reference here is to expeditions sent out by the "mystery and Company of Merchant Venturers" with the co-operation of Sebastian Cabot, newly returned from Spain. The purpose was to find a route to India and to wider markets for English products. The 'olde attempte' was an expedition to find the North West Passage. The 'later good aduenture' was sent out in 1553 to find a North East passage to India. It reached Russia instead. By penetrating the White Sea to the place where the port of Archangel now stands, the leader went overland to Moscow, where he met the tsar and obtained trading concessions. He thus circumvented the Hanseatic League which held a monopoly of trade through the Baltic, and opened the way for later commercial agreements between Russia and England. The backers of this expedition received a charter for trade with Russia and became the Moscovy Company. Robert Recorde was one of their advisers, presumably in questions of navigation.

5. See Mary S. Day, *Scheubel as an Algebraist* (Contributions to Education No. 219 [Teachers College, Columbia University, 1926]).

6. For convenience, modern symbols are used in these illustrations and in the balance of this paper.

13

The Teaching of Arithmetic in England
from 1550 until 1800 as Influenced by Social Change

JAMES KING BIDWELL

*I*N 1542 THERE APPEARED IN LONDON a book by Robert Recorde entitled "The Grounde of artes: Teaching the worke and practice of Arithmetike, both in whole numbers and Fractions, after a more easyer and exacter form than any like hath hitherto been sette forthe." This publication was the first significant arithmetic book printed in English. Its printing marked the beginning of the dissemination in England of arithmetic techniques that we today would consider typical in our elementary schools. In the middle of the sixteenth century, however, these techniques were not widely known and were not taught in the vernacular English in the formal grammar schools.

The Sixteenth Century

The mathematics taught at that time was considered a part of the formal education of the classical "seven liberal arts" and was taught in grammar schools from Latin texts. Besides this, the so-called arithmetic was theoretical (what we would now call number theory) and was not computational. It amounted to classical mathematics essentially unchanged since the time of Greece. The teaching concerning numbers in English was quite limited. John Brinsley could say in the sixteenth century:

> In a word, to tell what any of these numbers stand for, or how to set down any of them; will performe

Reprinted from *Mathematics Teacher* 62 (Oct., 1969): 484–90; with permission of the National Council of Teachers of Mathematics.

> fully so much as is needfull for your ordinaire Grammar scholler. If you do require more for any, you must seeke Record's Arithmetique, or other like Author's and set them to the Cyphering school.[1]

And even this small amount of number work was done in the late afternoon of Saturdays or on half-holidays for a total of one hour a week.

> That arithmetic was not taught in the Latin Schools in order to make proficient reckoners is shown by the lack of practice problems in their textbooks; and, likewise, the lack of vital commercial problems of that day show that it was not taught in order to prepare for a business life.[2]

On the continent the development of Reckoning Schools coincided with the commercial development and the growth of the guilds. The demands for merchants trained in bookkeeping and reckoning, reading and writing became so great that the merchants themselves could not instruct their apprentices. So Reckoning Masters and eventually Reckoning Schools developed to fill this need. There was no corresponding development in England prior to the sixteenth century. No manuals for arithmetic are extant in English. It is clear, however, that Italian methods of accounts were known by the English merchant class. Charlton tells us:

> In 1476, for example, James Harrison was apprenticed to Christopher Ambrose, a Florentine by birth, who traded from Southampton and took English apprentices into his household where he undertook to introduce them to the mysteries of his trade.[3]

In the sixteenth century the growth of commerce, discovery of the larger geographic world, and increased "industrial" type production led to more and more technical training of the merchant class. Improved techniques in navigation, surveying, horology, cartography, gunnery, and fortification were needed, and all of these required good mathematics knowledge. Edmund Worsop, a land surveyor, wrote a book in 1582 with the fantastic title, worded as follows:

A Discoverie of sundrie errors and faults daily committed by Landmeaters, ignorant of Arithmeticke and Geometrie, to the damage, and prejudice of many of her Maiestris subjects, with manifest proofe that none ought to be admitted to that function, but the learned practitioners of those Sciences.[4]

Worsop demanded better training of surveyors and suggested that they be licensed to insure more effective land measuring practices. W. H. G. Armytage notes that "mining, navigation, river improvement and the building of Elizabethan country houses stimulated mathematics."[5] Others have mentioned the stimulus of the growth of coal mining. The managing of estates by the gentry also required much practical knowledge in order for the landowner to compete for his revenues and preserve his estate. None of these needs could be met in the existing formal schools or universities.

The increased power and restrictiveness of the guild supervision in the sixteenth century cut off the former educative function of the guilds. They instead turned to endow grammar schools, which did not provide at all for the apprenticeship program. So the work of providing arithmetic education fell to individual tutors, or as they were called in England, "mathematical practitioners." These tutors instructed and wrote manuals in English based on previous manuals or books in Latin or another foreign language, from which they had learned themselves. These men were not, of course, university trained. Among these men was Robert Recorde. E. G. R. Taylor tells us:

They were alamanack-makers, astrologers, retired seamen, surveyors, gunners, gaugers—in fact they were themselves mathematical practitioners who simply handed on their art. But, as might be expected, they all worked in close association with the instrument makers, and as the handling of instruments was the very badge of this new profession, it was quite usual for a teacher to design a novel one of his own. . . .[6]

To supply the needs of these technical occupations, more and more printed books dealing with arithmetic, accounts, and general mathematics were printed. Karpinski mentions that in the sixteenth century 45 English arithmetic editions were printed. He goes on to note that "popular interest in arithmetic and general instruction in the subject increased so rapidly after the sixteenth century that hundreds of books appeared (in Europe) to supply the new demand."[7] These books greatly varied in their effectiveness, but they all aimed at giving practical rules for solving problems that the man with commercial or technical interests might have to solve. They were, in fact, compendiums of different problem types. Since these books were generally adapted from earlier books, the standard evil of material, which was in these arithmetics simply because it had long been in arithmetics, was perpetrated. None of these books were, of course, designed for school use; rather, they were designed to be "do-it-yourself" books.

Since these newly printed books were for private use, the authors had to sell their content to the general public. The prefaces of some of these early arithmetics make interesting reading and point out the intended practicality of the arithmetic content:

Howe profitable and necessary this feat of Algorism is to all maner of persons, which have reckenyings or accountes, other to make, or else to receive, needyth no declaration. Neither is this arte only necessary to those, but also in maner to all manner of sciences and artificies.[8]

Arithmetique needes not the Logicians arguments, nor the Rhetoricians Eloquence to prove or parswade the vsefulnesse thereof to the world, every mans particular occasion, to vse it, is sufficient to satisfie any man in that point. . . .[9]

For if numbring be so common (as you grant it to be) that no man can do anything alone, and much lesse talk or bargain with others, but he shall still have to do with number; this proveth not number to be contemptible and vile, but rather right excellent and of high reputation, sith it is the ground of all mens affairs, . . . [10]

These books were intended to be complete in themselves, requiring no teacher; the authors tried various methods of exposition, each designed to make the work clear for the reader. They failed miserably if judged by today's teaching methods. The dialogue method was especially common and rules stated in verse were used frequently. Consider as one example this versified rule for the addition of fractions:

Addition of fractions and likewise subtraction
Requireth that first they all have like basses
Which by reduction is brought to perfection
And being once done as ought in like cases,
Then adde or subtract their tops and no more
Subscribing the basse made common before.[11]

The Seventeenth Century

In the seventeenth century, a movement developed in favor of further formal education along mathematical lines. However, more was said than done. Dr. John Pell wrote about reform of education in 1639; he envisioned a public mathematical library where anyone could study on his own. But suggestion of any change in grammar school teaching was lacking. Other plans to include arithmetic and mathematics as part of liberal education were put forth but came to nothing. But Taylor points out:

Even though all these plans failed of realization, they were symptomatic of a changing climate of thought and opinion, and the mathematical practitioners, even to the vulgar mind, were becoming distinguishable from the conjurors and the quack astrologers as useful members of society. Only a few university die-hards continued to maintain that mathematics was no study for a Christian man.[12]

But change came slowly even in urban centers, and the outlying areas were even further behind the needs of the time. Also, the clergy fought the advancement of mathematics and science in general; many of them still treated arithmetic, mathematics, chemistry, and the like as works of the devil.

In 1655 a book was published in London with the title: "An idea of Arithmetick at first Designed for the use of the Free-Schoole at Thurlow in Suffolk By R. B. Schoolmaster there." In 1668 John Newton published *The Scale of Interest: Or the Use of Decimal Fractions*. DeMorgan notes that this book "was expressly intended for a school-book, though it is a strange one for the time."[13] These editions mark the beginning of school texts in English. What schools were then beginning to teach the practical art of arithmetic?

The lack of mathematically trained personnel greatly affected navigation. The techniques of recent theoretical advances were beyond most shipmasters of the day. The pressures were such that King Charles ordered the development of a mathematical school at the Christ's Hospital School in 1672. This was to be a school for boys passing from ordinary grammar school at the age of $14^{1/2}$. The most immediate difficulty of the school was to find a master. The requirements were severe; the master was to have knowledge of Latin and mathematics and have experience at sea. The need for such schools for navigation as well as commercial interests led to more such schools. Even though these were schools for older boys who had completed a classical study, they led to further elementary study of arithmetic. In Scotland the demands also were met:

A feature of the last two decades of the seventeenth century was the establishment in Glasgow and Edinburgh of commercial schools and academies which were modelled on the 'Reckoning Schools' of Italy, Germany, France and the Netherlands. The first of these was started in Edinburgh in 1680. . . . In 1695 'a teacher of the art of navigation, bookkeeping, arithmetic and writing' was appointed by the Town Council of Glasgow.[14]

These private schools, offering a curriculum geared to the commercial needs of the times, were established quite naturally in industrial towns. The Manchester School was established in 1666, the Dartmouth School in 1679, and the Rochester School in 1701. Some of these schools lasted many years (some into the nineteenth century). Some existed only a few years and then died from lack of funds.

The Eighteenth Century

The development of specialized schools like Christ's Hospital Mathematical School led to the establishment of schools with a broader curriculum, still oriented to vocational training, but offering foreign languages and some classical studies. These private schools were in direct competition with the grammar schools, and the best of the new schools sent a certain percentage of graduates to the universities. This kind of private school was commonly called an "Academy." Nicholas Hans suggests that these schools were fashioned after the so-called courtly Academies that flourished in Germany, France, and England in the seventeenth century. They were designed to "prepare the noble youth for his profession as a courtier and soldier and introduced military subjects, mathematics, physical training and accomplishments." [15]

The curriculum of these private academies had four groupings of subjects: literary, mathematics-science, vocational-technical, and accomplishments–physical training.

> From textbooks on various subjects published by most of them, and from the description of some of the Academies, it is evident that their methods were not "bookish" but practical and whenever possible approximated to an actual situation of business or technical vocation. It was true not only of vocational and technical training; the same methods were applied in teaching mathematics and languages. [16]

Such academies were founded all during the eighteenth century, and although some were short-lived, approximately 200 such schools were operating in England throughout the latter part of the century.

Most of these schools drew their students from the lower middle class and craftsmen. Thus finally the needs of that group for educational opportunity were being met. In fact, the mathematical part of these schools was so vital a part of the needs of the times

> That when the Schism Act was drafted in 1714 to curb the dissenting academies, special exemptions were made in cases where 'any part of mathematics relating to navigation, or any mechanical art' was taught. [17]

Thus the need for individual tutors of mathematics and small schools for mathematics developed into a need for vocational-technical schools where classical studies formed only a small part of the curriculum. All of this occurred chiefly because the formal schools of the times were still too classically and religiously centered to meet the demands of social and industrial change.

The education in the grammar schools of the eighteenth century was the same as education in the grammar schools in the sixteenth century. These schools were bound by their foundation statutes and hence were unable to modify their classical structure. Some did add modern subjects such as elementary mathematics, French, and German. Some schools merely restricted their program to elementary education. Due to the competition from other private schools, their enrollments dropped and many went out of existence.

Even in the face of this demand for nonclassical education the classical teachers defended their cause, but from a new point of view:

> Joseph Cornish in *An attempt to Display the Importance of Classical learning* (1783), argued that most technical terms used in physics, botany, chemistry, astronomy, architecture, mechanics, mathematics, rhetoric and grammar 'and almost every art and science are of a Greek original and numerous others of a Roman'. That he should be justifying Greek on these grounds is significant. [18]

Those grammar schools that wanted to augment their curriculums created disputes that were taken to the law courts. Consider the Leeds Grammar School, which wished to include practical subjects like mathematics and foreign languages. The Governor of the school took the case to court in 1795. After ten years of indecision, in 1805 the court decided that:

> The intention of the founder was to establish a grammar school and . . . a grammar school was defined as an institution 'for teaching grammatically the learned languages.' The court could not sanction 'the conversion of that Institution by filling a school intended for that mode of Education with Schoolars learning the German and French languages, Mathematics, and anything save Greek and Latin.'[19]

In fact, it was only in 1840 that the Parliament passed the Grammar Schools Act, which allowed for modernization of the curriculum.

In the so-called public schools it was not any better. Classical studies dominated the education curriculum. This is illustrated by Curtis, discussing the writings of Dr. Thomas James, Headmaster of Rugby from 1778 to 1794:

> On holidays the boys attended school from ten to eleven in the morning and two to three in the afternoon. On half-holidays they attended from two to three o'clock. In these periods, the lower forms learnt writing and arithmetic, and the Vth form geography and algebra. It is probable that the latter subject was taught by Dr. James himself. At any rate he afterwards confessed that this weekly mathematical period 'wearied his body to excess and made it hot, or at any rate, perspire too much.'[20]

Curtis mentions that the same kind of arithmetic teaching was common in the grammar schools of the 1590s. Since the needs of the public centered on commercial subjects, the importance of these schools faded for all but the nobility and some of the gentry who desired to maintain the status quo.

Conclusion

We see therefore that the arithmetic taught in the eighteenth century was almost exclusively a vocational subject. Hence all the arithmetic the students ever learned was a set of rules that produced the answers to their problems. Texts had repeated editions with little changes in content and usually considerable additions of specialized commercial problem types.

No accomplished mathematicians wrote the arithmetics of the eighteenth century. DeMorgan, writing reasonably close to this time, greatly condemned the editions of Cocker's Arithmetic. Cocker's book had an enormous popularity, although it was one of the most offensive texts from a mathematical point of view. It was first published in 1677 and was still in print in the early nineteenth century. Hence DeMorgan was well acquainted with it in 1847, and after six pages of disparagement he notes: "I am of the opinion that a very great deterioration in elementary works on arithmetic is to be traced from the time at which the book called after Cocker began to prevail."[21]

Such books as Cocker's were common enough to cause widespread complaints about the mathematics they contained. DeMorgan also comments on the large number of criticizers who can offer nothing better. Cajori, writing in 1896, also comments on the same problem:

> To summarize, the causes which checked the growth of demonstrative arithmetic are as follows:
>
> (1) Arithmetic was not studied for its own sake, nor valued for the mental discipline which it affords, and was, consequently, learned only by the commercial classes, because of the material gain derived from a knowledge of arithmetical rules.
>
> (2) The best minds failed to influence and guide the average minds in arithmetical authorship.[22]

By the beginning of the nineteenth century, the dichotomy of teaching practices was at its worst. The grammar schools were slowly breaking from the classical narrowness on the one hand and the technical schools were teaching a rigid rule-dominated commercial arithmetic which had little mathematical content. Surely this was a low point in arithmetic teaching in England. In the nineteenth century, men like DeMorgan advocated and effectively initiated reforms that began the way towards the modern arithmetic teaching of today.

As the years from 1550 to 1800 are reviewed, even in this narrow field, the dynamics of interplay can clearly be seen between the demands of the rising middle class, the clergy, and the government's national needs. The emergence of arithmetic teaching can be seen as a slow and painful development. That is to say, the development of arithmetic practices followed the same patterns that all educational practices follow. When the change involves the integration of many different social points of view, it traditionally follows that the change is slow. Such is the way of all education.

NOTES

1. John Brinsley, *The Grammar Schoole*, ed. E. T. Campagnere (Liverpool: University of Liverpool Press, 1917), p. 26.

2. L. L. Jackson, *Educational Significance of Sixteenth Century Arithmetic* ("Contributions of Education," No. 8 [New York: Teacher's College, Columbia University, 1906]), p. 178.

3. Kenneth Charlton, *Education in Renaissance England* (London: Routledge and Kegan Paul, 1965), p. 258.

4. Edmund Worsop, *Discoverie of Sundrie Errours Committed of Landmeaters . . .* (London: 1582, University Microfilms, S.T.C. Case 404).

5. W. H. G. Armytage, *Four Hundred Years of English Education* (Cambridge: University Press, 1964), p. 9.

6. E. G. R. Taylor, *Mathematical Practitioners of Tudor and Stuart England* (Cambridge: University Press, 1954), pp. 9–10.

7. Louis Karpinski, *The History of Arithmetic* (Chicago: Rand McNally, 1925), p. 73.

8. Dorothy Yeldham, *The Teaching of Arithmetic Through Four Hundred Years* (London: Harrap, 1936), p. 11.

9. Edmund Wingate, *Arithmetique made easie* (London: 1630, Rare Book Room), preface.

10. Robert Recorde, *The Ground of Artes* (1646), preface.

11. T. Hylles, "The arte of vulgar arithmetique," 1600 as quoted in Augustus DeMorgan, *Arithmetical Books* (London: Tahlor and Walton, 1847).

12. E. G. R. Taylor, *op. cit.*, p. 82.

13. Augustus DeMorgan, *op. cit.*, p.46.

14. Duncan Wilson, *The History of Mathematical Teaching in Scotland* (London: University of London Press, 1935), pp. 26–27.

15. Nicholas Hans, *New Trends in Education in the Eighteenth Century* (London: Routledge and Kegan Paul, 1951), p. 64.

16. *Ibid.*, p. 68.

17. W. H. G. Armytage, *op. cit.*, p. 31.

18. *Ibid.*, p. 67.

19. S. J. Curtis, *History of Education in Great Britain* (London: University Tutorial Press, 1948), p. 60.

20. *Ibid.*, p. 62.

21. Augustus DeMorgan, *op. cit.*, p. 62.

22. Florian Cajori, *A History of Elementary Mathematics* (New York: Macmillan, 1896), p. 211.

14

Tangible Arithmetic
Napier's and Genaille's Rods

PHILLIP S. JONES

*H*AROLD LARSON COMMENTED that Napier's "bones" or "rods" are well known with several accessible articles on them when he published the drawing of Genaille's rods which is our Figure 4.[1] This is true, and this is our reason for showing without discussion (Fig. 1) the title page of the posthumous (1617) book in which Napier explained them, and a page from it (Fig. 2) which shows the four faces of a few of these bones as Napier designed them.

The word *Rabdologia* is probably compounded from Greek words meaning *a collection of rods.* William Leybourn, who published a translation, *The Art of Numbering by Speaking-Rods; Vulgarly Termed Napier's Bones,* in 1667, thought the latter part of *Rabdologia* was from *logos, speech,* rather than *logia, a collection.* Translations also appeared in Verona and Berlin in 1623 and a second Latin edition in 1626. Similar devices appeared in China later in the seventeenth century and again in the nineteenth.

The diagonal line separating the units and tens digits on the bones and the method of using the bones are counterparts of the popular *gelosia* or *jealousy* method of multiplication. This came into Europe from Arabic writers shortly after the introduction of the Hindu-Arabic numerals, and can be traced back to the Hindu Bhaskara (1150) or

Reprinted from *Mathematics Teacher* 47 (Nov., 1954): 482–87; with permission of the National Council of Teachers of Mathematics.

earlier. Wöpcke notes that the notion of a separate multiplication table for each of the nine digits can also be found in the fifteenth century writings of the Arab Alkalsadi, who "taught multiplication by separating the columns of the table of Pythagoras."[2] ("The table of Pythagoras" referred to the 9×9 or 10×10 square multiplication table which appeared in many books on the Hindu-Arabic arithmetic. Neither Pythagoras nor later Greeks used the Hindu-Arabic number system, and Pythagoras' greatest interest was in the "arithmetica" which today we call number theory. Nevertheless the term "table of Pythagoras" illustrates the importance of the Greek contributions to many areas of early mathematics.)

Although the bones and their use are well known today as enrichment-teaching aids especially related to multiplication, it is not so generally known that Napier designed special rods for square roots and cube roots. A later variation of these is shown at the top of our Figure 3. It is also interesting to note the connection of Napier's work with the spread of the idea of decimal fractions. Although Simon Stevin published his basic work in 1585, his notation (a small zero in a circle to mark the units place, a small one in a circle to mark the tenths place, etc.) was awkward. The first publication of a decimal point as we know it occurred in a 1616 translation into English of Napier's work on logarithms. On pages 21–22 of the *Rabdologia,* Napier discusses Stevin's work and uses a comma (as still used on the Continent) for our decimal

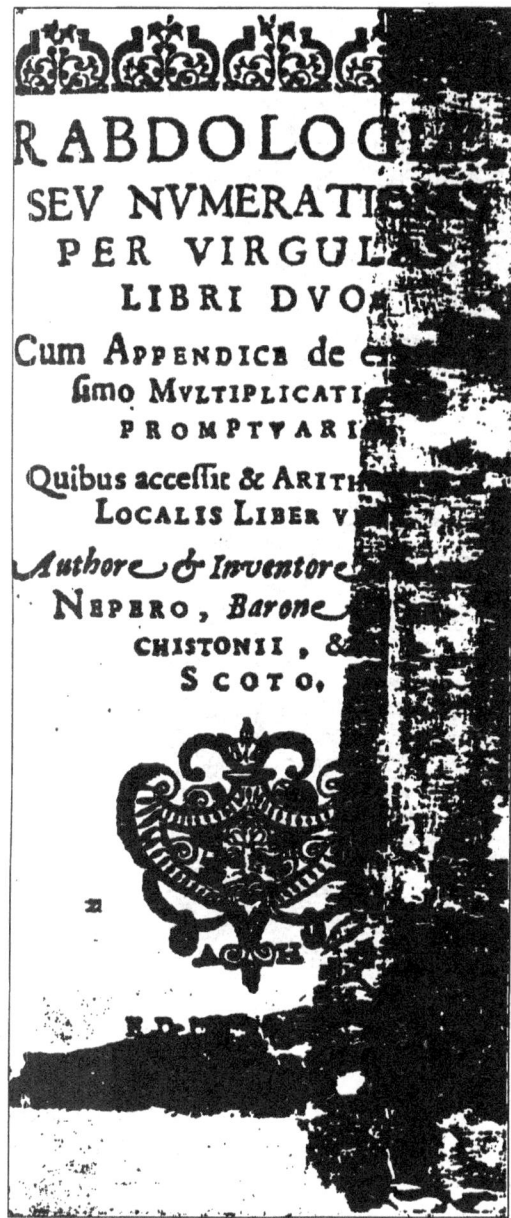

FIGURE 1

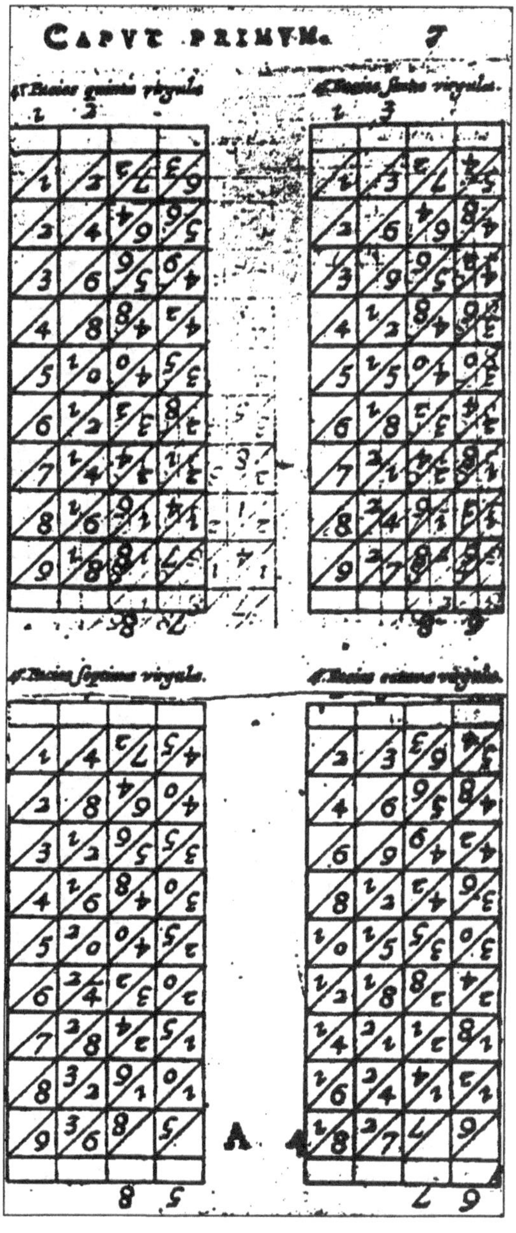

FIGURE 2

point. No doubt the popularity of Napier's works assisted in the spread of these improved notations as well as in the use of decimal fractions.

The *Rabdologia* is a small book (3 in. × 5.5 in.) of 154 pages. "Liber primus" tells how to construct the rods and to do multiplication, division, square, and cube root, and the rule of three with them. "Liber secundus" tells how to use them in computing the solutions to a number of different geometric problems relating to regular

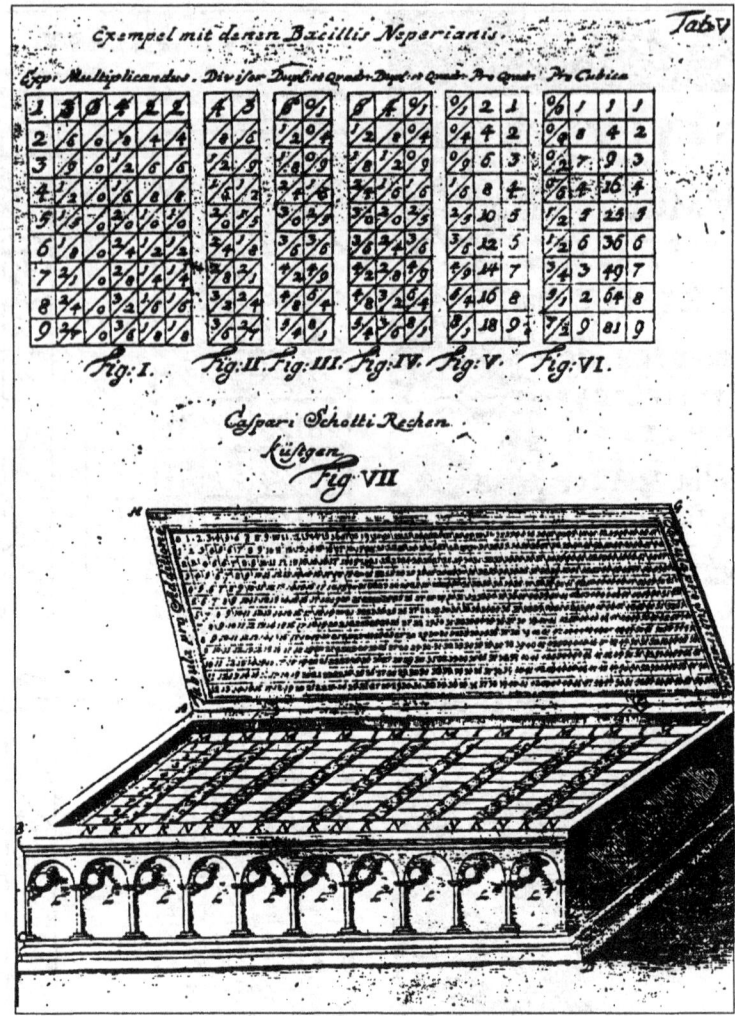

FIGURE 3

polygons and polyhedra and to weights and measures.

In an appendix beginning on page 91, Napier describes more complicated apparatus and procedures for further speeding up computations, especially trigonometric and astronomical computations.

Many variations on Napier's fundamental design grew up in the years after 1617. For example, about 1668 Gaspar Schott mounted cylinders, each of which carried Napier-type tables for 0–9, in a box as shown in our Figure 3. This figure is taken from Jacob Leupold's *Theatrum Arithmetico Geometricum* published in Leipzig in 1727.

A set of bones dating to about 1680 has the tables on flat sticks which had been described by Leybourn in 1667[3] and which were also advocated by Schott.[4]

Figure 4 is Professor Larsen's drawing of the "Réglettes Multiplicatrices" invented by Henri Genaille and "perfected" by Edouard Lucas in 1885.[5] Figure 5 is a photograph of a companion set of "Réglettes Multisectrices" in the University of Michigan library. To multiply 471,963 by 6 one uses the data between the horizontal lines bounding 6 on the left-hand or index scale in Figure 4. Going to the extreme right, one begins with the number on the "3" rod below the upper of horizontal lines bounding 6. This is 8.

One then proceeds to the number on the next rod which is pointed out by the left-hand vertex of the angle which includes 8. This next number is 7. From there, by going from right to left and always reading the number on the next rod at the vertex of the angle including the last number, we read: 7, 1, 3, 8, 2. Thus, finally, 471,963 × 6 = 2,831,778. As with Napier's bones, to multiply by 526 one would have to write down separately and then add the partial products of 471,963 by 6, 2, and 5 which may be read from the rods. On these rods the numbers in the narrow vertical columns opposite 6 are the units digit of the product by 6 of the number of the rod and this units digit plus 1, 2, 3,

4, 5. These latter are the numbers which might have been "carried" from an earlier multiplication on a previous rod. The vertex of the angle on the previous rod points out which of these is to be used in a particular case. (Professor Larsen writes that he believes one line is wrong in his drawing. Use this as a test of your understanding of the construction of the rods.)

The rods of Figure 5 are used similarly except that they are read from left to right and give the quotient. Thus 1,234,567,890 ÷ 6 = 0,205,761,315 with a remainder of 0 (the remainder is read from the right-hand index rod).

As you can see, these Genaille-Lucas rods actually eliminated one step in the use of Napier's rods, the adding in one's head of the amounts "carried" to the next rod at each step of the reading. The earlier variations we mentioned merely changed the shape, or size and arrangement of Napier's rods.

The next steps were to devise ways of multiplying by more than one digit at a time and then to simplify or eliminate the need to add separately the partial products. Several solutions to the former problem were proposed, some before Genaille's rods were invented. The latter problem was solved in part by M. Rous in 1869 by combining an abacus with a set of Napier's bones. Probably the last step in this process was L. Bollée's "Arithmografo" a device related to the rods but linking multiplication and addition semi-mechanically in such a way as to be considered, historically, a link between Napier's rods and true arithmetic machines. These latter have an interesting story of their own going back to Pascal and Leibnitz. Perhaps some of our readers have data on the latter or on other early machines, or on some of the variations and improvements on Napier's and Genaille's rods which we did not mention. In fact an exposition of Napier's procedures for division, square and cube root might be fun though perhaps they would only rarely be useable as teaching aids or enrichment and have no modern practical value.

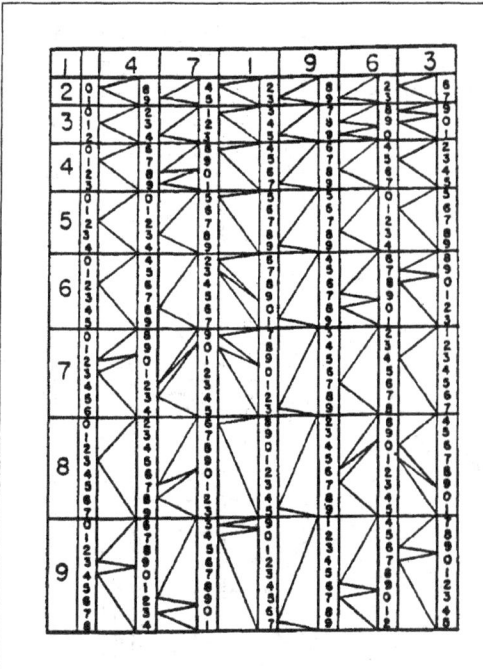

FIGURE 4

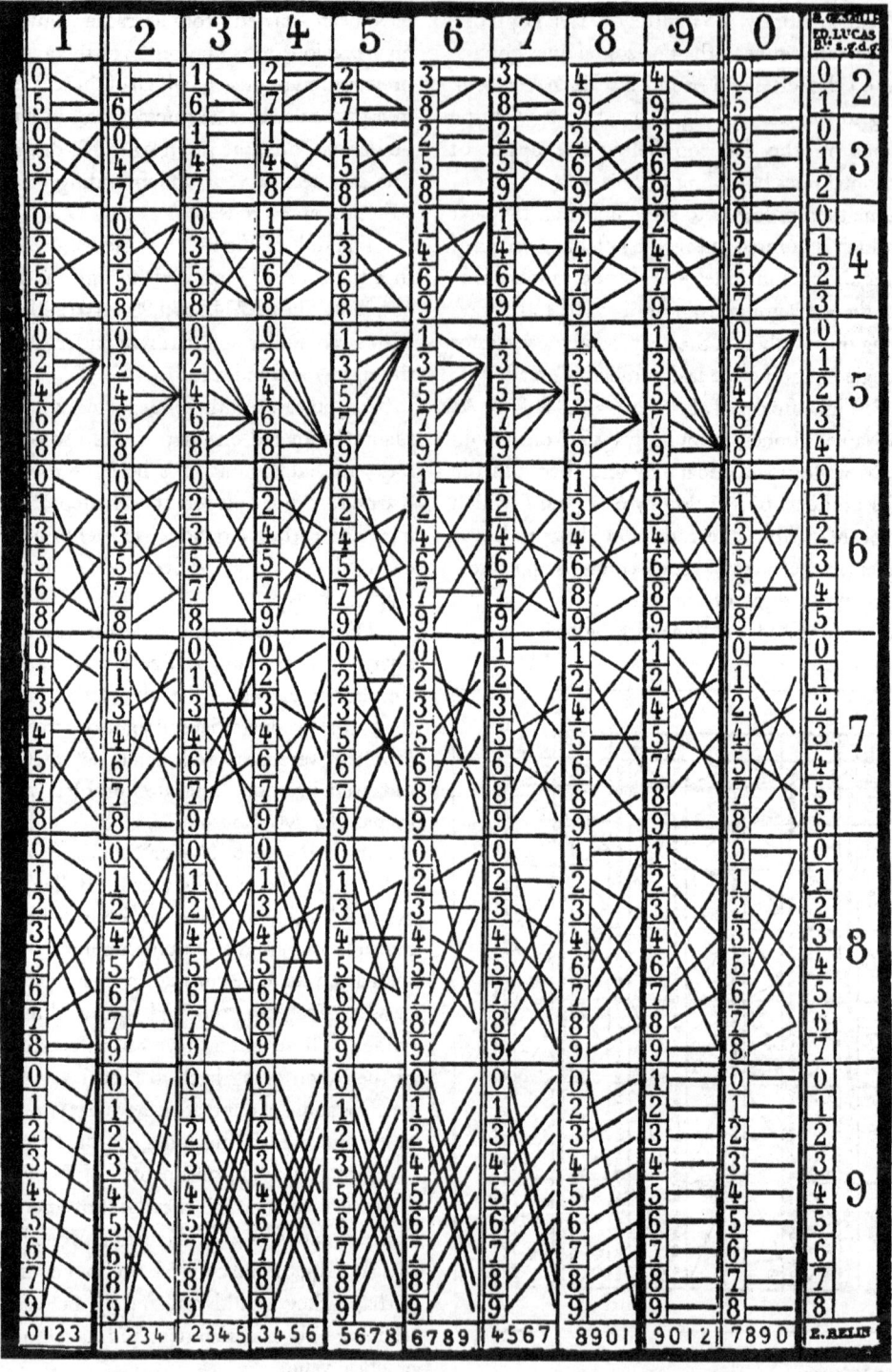

FIGURE 5

NOTES

1. H. D. Larsen, "Genaille's Rods," *American Mathematical Monthly,* Vol. 60 (Feb. 1953), pp. 140–141. He cited "The Pentagon," VIII (Spring 1949), pp. 98–100. We might add D. E. Smith, *History of Mathematics* (Boston: Ginn and Co., 1925), Vol II, pp. 202–203; D. E. Smith, *A Source Book in Mathematics* (New York: McGraw-Hill Book Co., Inc., 1929), pp. 182–185; Vera Sanford, *A Short History of Mathematics* (Boston: Houghton Mifflin Co., 1930), pp. 339–340.

2. F. Wöpcke in *Atti Accademia Pontificiana Nuovi Lincei,* 12 (1858–59), p. 245. This is cited in *Enciclopedia delle Matematiche Elementari e Complementi* (Milan, 1950), Vol. I, Parte 1, pp.415–416, which in turn makes much use of R. Mehmke, M. d'Ocagne, "Calculs Numériques," *Encyclopédie des Sciences Mathematiques* (Paris, 1908), Tome I, Vol. 4, Fascicule 2, pp. 230–234. The last two articles contain many references and were the source of much of the data compiled here.

3. E.M. Horsburgh, *Modern Instruments and Methods of Calculation. A Handbook of the Napier Tercentenary Exhibition* (London and Edinburgh, 1914), pp. 18–19.

4. Gaspar Schott, *Cursus Mathematicus* (Francofurti ad Moenum, 1674), p. 50.

5. H. D. Larsen, *loc. cit.*

15

Life and Times of Johann Kepler

BERNARD H. TUCK

$\mathcal{T}$HE YEAR 1571, when Johann Kepler was born, had little to mark it as exceptional in a time when events moved but little faster than in ages past. The shadows of the Dark and Middle Ages still spread the heavy, restraining cloak of authority over the lives of people. The advent of the printing press had made the expression of personal opinions more than personal, and sober thought must now be given to the expression of an idea. But it had also promulgated knowledge, and new discoveries were imminent. An interesting age into which to be born—and safe enough, if one used one's eyes much, and one's tongue but little.

The small town of Weil, Württemberg, in which Kepler was born, was enmeshed in the ideological difficulties of the Reformation, as was most of the north of Europe. Luther had died twenty-five years before, and Calvin but a scant seven years before.

In England, Elizabeth was not yet "mistress of the seas," and Shakespeare was an urchin of seven. In Italy a boy named Galileo was Shakespeare's match in age—and more than his match in curiosity.

Servetus had been burned at the stake some nineteen years before (we will have no shedding of blood, they said), and Giordano Bruno had found it expedient to leave Italy. Not for another twenty-nine years would Bruno suffer the same fate suffered by Servetus—in a different country—persecuted by advocates of a rival religious persuasion, but persecuted by authorities who knew how to build a fire just as hot as that of their ideological opponents.

Copernicus, Kepler's hero, was twenty-eight years in his grave. John Napier, gadfly of Popes, was twenty-one. His logarithms would appear too late to assist at the birth of Kepler's first two laws, but would do yeoman service in lightening the tasks of Kepler's later years.

This, then, in some small part, was the time of Kepler. The spirit of the Renaissance was struggling valiantly, but cautiously, with the institutions of privilege, tradition, and authority. And although the privileged and the supporters of institutions quarreled bitterly (and often violently) among themselves, in the face of innovations the political axiom, "Rumps together, Horns outward," was the rule of the times.

The proponents of the doctrine of *Reason* were themselves captives of habits and customs and beliefs of the centuries past (as, no doubt, are we). Their dedication to the seeking of a small grain of truth was accompanied by acceptance of age-old superstitions and judgments; their hopes were those of the eclectic, to modify, not revolutionize. The shoulders upon which they stood to see better the truth were the shoulders of the ancients—authority here, too, to be respected, but authority to be accepted with judicious restraint.

These people were, then, an intellectually active, vigorous, partly civilized people, ruled by custom, yet adventurous of spirit. Believing in law and discipline, they yet revolted occasionally against the rigidity of the law.

Reprinted from *Mathematics Teacher* 60 (Jan., 1967): 58–65; with permission of the National Council of Teachers of Mathematics.

Kepler's Early Life

The young Johann seems to have spent a joyless youth; yet when one notes the gentleness, courage, and freedom of manner and speech that were his as a young man and as an adult, one wonders if bare circumstance can merit such conclusions.

One of his forefathers had been knighted by the Emperor Sigismund in 1430 while on a campaign in Italy. Perhaps Kepler's father, Henry, a soldier of fortune, sought a like route to privilege and wealth, as did many in those days. He met with no success, but did succeed in drawing Kepler's mother from the young boy's side to keep him company in the Netherlands, leaving the youngster in the charge of his grandfather, burgomaster of the Free City of Weil of the Holy German Empire.

Kepler's mother, Catherine Guldenmann, daughter of the burgomaster, seems to have been undisciplined and uneducated as a girl—and later, a woman whose unbridled tongue won her much enmity and trouble. But, at the same time, one has indications of courage that might perhaps make one wish to reserve judgment about her character.

An infection with smallpox when Kepler was but four years of age left his eyesight much impaired, and added to his general weakness. He attended grammar school at Weil, and later (age six) moved with his parents to the small town of Leonberg. Here he attended the local Latin school and graduated at the age of thirteen.

Kepler spent the next two years at the seminary at Adelberg, where the rigorous discipline and renewed attacks of minor ailments left him even more unwell. However, his stay at the higher seminary at Maullbronn (about three years) brought him better health, mastery of Latin, and admittance to Tübingen University. The year was 1589; Kepler was eighteen.

Tübingen and Graz

Theology, philosophy, mathematics, and astronomy: the greatest of these was the first. Kepler prepared himself for the Protestant church. But the mind of a Kepler seems never to be the mind of the orthodox.

Mastlin, his revered professor of astronomy, taught Ptolemy's theory of the solar system but privately believed that the earth went around the sun. So did Kepler. Most unorthodox. Should a young man who seems unable to accept intellectually all the tenets of a religion be permitted the pulpit? Those who felt the heavy responsibility religion demanded in those times believed not. Much respected for his brilliance and for his character (flawed only by his liberal views), he was urgently recommended for a professorship of morals and mathematics at the Protestant Gymnasium (High School) in Graz, the capital of Styria in Austria.

Kepler was now twenty-three. His teaching duties in mathematics were trivial. He lectured on rhetoric and Virgil. He wrote the yearly almanac (*Calendar and Prognostications*) for five years, and gained an increase in salary thereby.

Astrology—"the foolish daughter of the respectable, reasonable mother of astronomy," he once called it, and again indicated that it was mostly foolishness that nonetheless permitted the astronomer to earn a livelihood—was a much respected art, and Kepler's good sense brought forth predictions that often came true. He gained much in respect, authority, and prestige. But, while playing with astrology, he was working, and working hard, on cosmology.

During the later part of 1595 his observations and work were being brought together, and he was eager to publish the results. He wrote Mastlin:

> I strive to publish them in God's honor who wishes to be recognized from the book of nature. But the more others continue in these endeavors, the more shall I rejoice; I am not envious. . . .

This eagerness for truth rather than place, for honesty rather than honors, was not at all the spirit of the time—nor is it entirely the spirit of today.

To the Baron von Herberstein
and
The Estates of Styria
Graz, May 15, 1596

What I promised seven months ago . . . .

His first book, *Mysterium cosmographicum*, was published.

The book was brilliant. The structures it brought forth had no lack of beauty. Mathematics and imagination and aesthetics were brought together in a manner that Aristotle himself would have approved. Perhaps this was its fatal flaw. It was, almost, more Aristotelian than Aristotle himself. It portrayed the power of an active mind, but not enough of nature.

The loveliest gem in Kepler's *Mysterium* was the hypothesis of the five regular polyhedra—or Platonic solids, as we often call them today. The distances between the planets were such that if the sphere containing the earth's orbit as a great circle should be inscribed within a dodecahedron, then the orbit of Mars would be a great circle on the sphere circumscribing the dodecahedron. And if the sphere of Mars' orbit should be circumscribed by a tetrahedron, the Jupiter's course would lie on the sphere circumscribing the tetrahedron. Should the sphere of Jupiter's orbit be circumscribed by a cube, the orbit of Saturn would lie on the sphere that encloses the cube. Uranus, Neptune, and Pluto were, of course, not yet known to exist, so Kepler then looked inward from the earth.

The orbit of the earth lies on a sphere that encloses an icosahedron, which itself envelops the sphere containing the orbit of Venus. And the sphere upon which lies the path of Venus circumscribes an octahedron which encloses the sphere upon which lies the road of Mercury as he goes around the sun.

There are but five regular convex polyhedra. (Kepler himself was to discover two of the four other regular polyhedra which exist.) So there could be but six planets. The orbits of the planets, being planar sections of spheres, must be circles, just as Aristotle said.

So *Mysterium cosmographicum* was published, the small town of Graz prepared itself for the winter, and life proceeded, for a little while, in its usual manner.

Kepler had reached the ripe old age of twenty-five. And bachelorhood, it seems, was a joyless estate.

Goodbye to Graz

The publishing of his first book may well mark the beginning of maturity in Kepler's life. Not, in truth, the maturity of solemn, heavy, cautious nature, but a maturity that finally accepts reverses with understanding and proceeds to make the best of a most imperfect world.

He fell in love with a twice-married noblewoman, Barbara Muller von Mühleck, some three years his junior. To prove his own noble descent he was forced to journey home and spend some seven months in order to procure the necessary papers. They were married February 9, 1597.

On August 4, 1597, Galileo wrote the "highly learned gentleman," thanking him for the copy of *Mysterium cosmographicum* that had been sent to him.

I would certainly dare to approach the public with my ways of thinking if there were more people of your mind. As this is not the case, I shall refrain from doing so. . . .

Yours in sincere friendship,
GALILACUS GALILAEUS
Mathematician at the Academy of Padua

To Galileo, October 13, 1597, Kepler sent a letter urging his fellow Copernican to proceed openly with his beliefs.

Be of good cheer, Galileo, and appear in public. If I am not mistaken, there are only a few among the distinguished mathematicians of Europe who would disassociate themselves from us. . . .

But Galileo chose a later day for his day of reckoning.

The next several years were years of trouble and worry for Kepler. His first son, Heinrich, died, and his wife was inconsolable. The young Archduke Ferdinand instituted such steps to force all Austria back to the Catholic church that Kepler himself had to flee to Hungary for a month. Not one to shift his beliefs for personal advantage, Kepler tried to prepare for the inevitable by writing his friend and teacher, Mastlin, entreating him for assistance in obtaining a position at his beloved Tübingen. Mastlin answered some of these pleas in an offhand manner, and some letters he answered not at all.

Another friend, however, did assist him in his difficulties. Johann Herwart von Hohnburg, a Bavarian diplomat, who had also been a student of Mastlin, suggested a possible collaboration with Tycho Brahe, and may even have suggested the same to Tycho. (Tycho had received and praised Kepler's work, while disagreeing with Kepler's conclusions.) At any rate, Tycho's invitation to visit him in Prague was accepted by Kepler. He stayed with the eminent astronomer from February to June, 1600, succeeding with great difficulty in gaining access to some mathematical data on the planets. It must have been a frustrated Kepler who returned to face the increasing difficulties at Graz.

The intolerance of the times made the end almost inevitable. On August 1, 1600, Kepler and many others were forced to leave Styria forever. His possessions had to be sold hastily, and heavy taxes took most of what he had left. His financial position was most precarious. His wife and her family were in little better position, since their fortune was in real estate, which brought little under conditions of forced sale. Kepler wrote:

> All this is rather hard. But I should not have believed that in the communion of brethren it is so sweet to suffer loss or insult for our faith and Christ's honor, and to abandon home, fields, and country ....

Later he was to write:

> I will not take part in the fury of the theologians. I will not stand as a judge over my brethren. ...

And later still, when refusing to abandon his Protestant beliefs for the privilege of continuing his work in peace and with "honor":

> I cling to the Catholic Church. Even if she rages and beats my heart I remain united with a heart full of love, so far as human weakness allows ....

An appointment at Tübingen was not forthcoming, so Kepler returned to Prague. With some reluctance, it seems, he became associated with Tycho Brahe. But even then he wrote affectionately to Mastlin asking for assistance for an appointment to the university.

The Tables of Tycho and the Years at Prague

The year with Tycho must have been one of continual frustration for Kepler. Tycho was most parsimonious with his astronomical observations. And Kepler was at his wit's end to accomplish all he wished to accomplish in the face of such obstruction. To Mastlin he wrote:

> His observations . . . are accessible to me, but first I had to promise to keep them secret. I have complied with this as far as it befits a philosopher . . . .

Tycho Brahe died in October 1601. Emperor Rudolph II appointed Kepler to the position of "Mathematician of His Holy Christian Majesty." Kepler's major work had begun.

The prelude to the Thirty Years War was being enacted. The troubled times would add to the burdens imposed by Kepler's personal difficulties, but his work would progress and would occupy him for the rest of his life.

Kepler spoke at Tycho's funeral, and later often expressed admiration for the work he had accomplished. He managed to gain custody of Tycho's *Rudolphine Tables*, and spent much of his time editing them and using them to work out the orbit of Mars. In any plane, at least two coordinates are needed to locate oneself: a street and an address, the intersection of two streets, an angle from a place along with a distance away, or (as with sailors) bearings from two objects. Kepler had only one in space, the sun. But Brahe's tables had such voluminous data on Mars that he hoped to use this planet for his second signpost. The difficulties were tremendous and took years to overcome.

At the beginning of the seventeenth century the earth-centered Ptolemaic theory was the theory of the day. But so inaccurate were the predictions that the theory permitted that many astronomers changed it about in minor ways to permit better, though still grossly inaccurate, predictions to be made. Thus Tycho accepted the idea that the sun circled the earth, but believed that the other planets circled the sun. This idea was an improvement on most, but was still inadequate to explain the positions of the planets as measured.

Kepler's almost mystic faith in the Copernican heresy (not yet so-called) urged him toward a hypothesis that would accept the main features of the Copernican scheme, but would improve upon it in such a way that the stars themselves would affirm the plan.

Copernicus believed that the stars were so far away that the distance from the earth to the sun was insignificant in comparison. Brahe did not, and since he could measure no parallax for any star, he disregarded Copernicus' theory, though it led to predictions which were more accurate than his own.

Kepler's first great stride was to abandon the accepted idea that the velocity of the planets is constant. He intuitively arrived at the idea that sources of velocity of planets were rays emanating from the sun. Therefore, as a planet moves farther from the sun, fewer rays reach it, and it slows down proportionally to the square of the distance. (Or, as would usually be stated, its velocity is inversely proportional to the square of its distance from the sun.) Almost eight years were spent trying various ovals and other geometrical figures which might match the observed positions of Mars. At last Kepler tried one which worked, the ellipse.

Kepler published his findings, along with many other observations, in *The New Astronomy*. The place, Prague; the year, 1609; the reaction, far from earth-shaking, the publication causing scarcely a tremor through the intellectual life of the day.

When Copernicus published his *Books on the Revolutions* in 1543, only a few days before his death, his sun-centered ideas were not taken too seriously by the intelligentsia of the day. Its tables and drawings were copied assiduously, his methods of measuring distances evoked the greatest admiration, his standards revolutionized ideas as to what astronomers could do; but since his theory did not explain the erratic movement of, say, Mars, much better than other theories, nor much more accurately, his sun-centered hypothesis was accepted by only a few during the following decades.

Kepler's book caused much less stir among the learned. His position as the court mathematician seemed to earn him more respect than did his work. Yet, as he later wrote as part of as poetic an outburst as may be found in the history of science *(Harmony of the Worlds)*, this was in no way a matter of concern:

> I am writing a book for my contemporaries or—it does not matter—for posterity. It may be that my book will wait for its readers for a hundred years. Has not God himself waited for 6000 years for an observer?

Kepler's first two laws of planetary motion are

1. *The planets move about the sun in elliptical orbits with the sun at one focus.*
2. *The radius vector joining a planet to the sun sweeps over equal areas in equal intervals of time.*

These laws seem to have been little noticed by the great Galileo. Full of the importance of his discovery of the "Medician Stars" (the four largest satellites of Jupiter), he wrote the following to Kepler in 1610:

> What do you say to the main philosophers of our school, who, with the stubbornness of vipers, never

wanted to see the planets, the moon, or the telescope, although I offered them a thousand times to show them the planets and the moon.... With logical reasons, as if they were magic formulas, he wanted to tear the planets from the heavens and dispute them away. . . .

Kepler's work at the time encompassed much more than astronomy and astrology. His work on optics, ephemerides, gravity, and tides, and his mathematics took much of his time. And his private life was none too smooth.

Kepler was forty-one. The year was 1612. One of his sons died and his wife passed away. Prague itself was a battleground, and subject to the roughness of the soldiers in the town. Kepler's salary was always in arrears, and his patron, the emperor, pushed aside by his brother, died later in the year. Matthias, Rudolph's successor, confirmed him in his office as "Imperial Mathematician," and granted him permission to seek employment outside the court.

Kepler left Prague for Linz.

The Years at Linz

Leaving his remaining two children with relatives, Kepler took a teaching position in a small college in Linz. This small Austrian city, on the banks of the Danube, remained his home for over fourteen years. Here he remarried (and fathered seven more children, five of whom survived), was denied communion because of his liberal views, and labored industriously at his work.

His work on cubical contents of wine casks, initiated by an unusually heavy vintage year, contained some of the early ideas of the integral calculus. He continued his work on Tycho's tables and published, in 1619, *Harmony of the Worlds*. This publication contained, along with other observations and conjectures, his third, and most beloved, law of planetary motion:

3. The square of the time of one complete revolution of a planet about its orbit is proportional to the cube of the orbit's semi-major axis.

The stage, of course, had been set for Newton in the earlier book containing the first two laws of planetary motion (as well as by Galileo's two laws of motion), but as an embellishment, Kepler's third law was a pretty thing.

During all this time, and more, Kepler was much troubled by charges of witchcraft leveled against his mother. During January 1616 he wrote to the authorities of Leonberg, hoping to settle the affair once and for all. But the proceedings dragged on. Despite all his efforts, and despite the ridiculous nature of the charges, his mother was imprisoned in August 1620, and remained imprisoned for fourteen months.

Kepler, admitting to the Duke of Württemberg that his mother had for years been quite difficult, talkative, malicious, and a maker of trouble, pressed with every means at his command for her release and, later, her acquittal. Since one of her associates had left a thumb in the rack, there can be no doubt that he was very much concerned about her fate.

Kepler spent almost a year in Guglingen trying to save his mother from torture and execution. Her famous son could do nothing to save her from being threatened in the torture chamber, but here she resolutely refused to confess and was released. She died six months later, leaving the learned mathematician with more rumors to combat than those which would have him half popish, half Lutheran, half Calvinist, and a complete eclectic in religious beliefs. Yet the fact that Kepler continued to take delight in his work can scarcely be refuted. His letters well reflect his pride and happiness in his many occupations.

Kepler's *De Cometis* of 1619 was dedicated to James I of England. Although invited to come to England, he did not accept. Perhaps he refused, as he had refused a post in Bologna, because of a fear that freedom of expression would be denied him. Yet the city of Linz itself was becoming more and more untenable.

Protestant rule of Linz itself was overturned in 1620, and a growing oppression of Protestants began. Kepler, however, was permitted to go on with his work. He did much work on Napier's logarithms, and continued his work in astronomy.

However, the climate of intolerance at Linz eventually made work impossible for Kepler. In 1626 his private library was sealed up and it became evident that it would be impossible to print the *Rudolphine Tables* in Linz, as had been commanded by the Emperor Ferdinand II. He therefore saw his wife and children safely away and left for Ulm, lived through a siege for fourteen days, had type set for the tables at his own expense, and had them published. He wrote the following to a friend:

> I long for a place where I can teach them. If possible, in Germany; if not there, in Italy, France, Belgium, or England; but only if an adequate salary is available for the stranger. . . .

The next several years were spent mostly upon work connected with the *Rudolphine Tables,* and with his personal affairs—his children, his wife, his never-collected salaries. In 1628 the emperor found him worthy of favor, and Count Albert of Friedland and Sagan granted him a yearly allowance and promised him a printing press and a quiet place in Sagan. Kepler's life seemed almost secure.

The good fortune, however, was short-lived. The 11,817 gulden owed Kepler by the court were not forthcoming. The work on a supplement to the tables (movements of the planets for 1631) was completed, and other work of like nature was brought to a conclusion.

Kepler left Sagan in the fall for Leipzig, and thence to Regensberg, hoping he might receive his salary from the German Reichstag. The bitterness of the weather during the journey, which he made on horseback, brought him to Regensberg with little life left to live.

Kepler died here, after a short illness, on November 15, 1630. "Your, mine, our Sun," wrote a friend. "The Sun of all astronomers has set."

BIBLIOGRAPHY

1. Baumgardt, Carola. *Johannes Kepler: Life and Letters.* New York: Philosophical Library, 1952.

2. Eves, Howard. *An Introduction to the History of Mathematics* (rev. ed.). New York: Holt, Rinehart & Winston, 1964.

3. Gade, John Allyne. *The Life and Times of Tycho Brahe.* Princeton, N.J.: Princeton University Press, 1947.

4. Gebler, Karl von. *Galileo Galilei.* London: C. Kegan Paul & Co., 1897.

5. Hogben, L. T. *Science for the Citizen.* New York: W. W. Norton & Co., 1938.

6. "Kepler, Johannes," *Encyclopaedia Britannica* (1964).

7. Kuhn, Thomas S. *The Copernican Revolution.* New York: Modern Library, 1957.

8. Physical Science Study Committee. *Physics.* Boston: D. C. Heath & Co.

Editor's Note: For further information on Kepler's work, see:
Tyco & Kepler, The Unlikely Partnership that Changed our Understanding of the Heavens. Kathy Ferguson (2002).

Multiplication Algorithms of the Fifteenth and Sixteenth Centuries

The introduction of algorithmic computing schemes using "Hindu-Arabic" numerals in fifteenth century Europe was a complex and often confusing process. Algorithms for the multiplication of two multi-digit numbers had to build upon the basic multiplication facts and insure the proper preservation of place-value in the computation. Luca Pacioli in his *Summa de Arithmetica Geometria Proportioni et Proportionalita* (1492) described eight different algorithms for multiplication. Four of these methods are illustrated below for finding the product of 1234 and 56789.

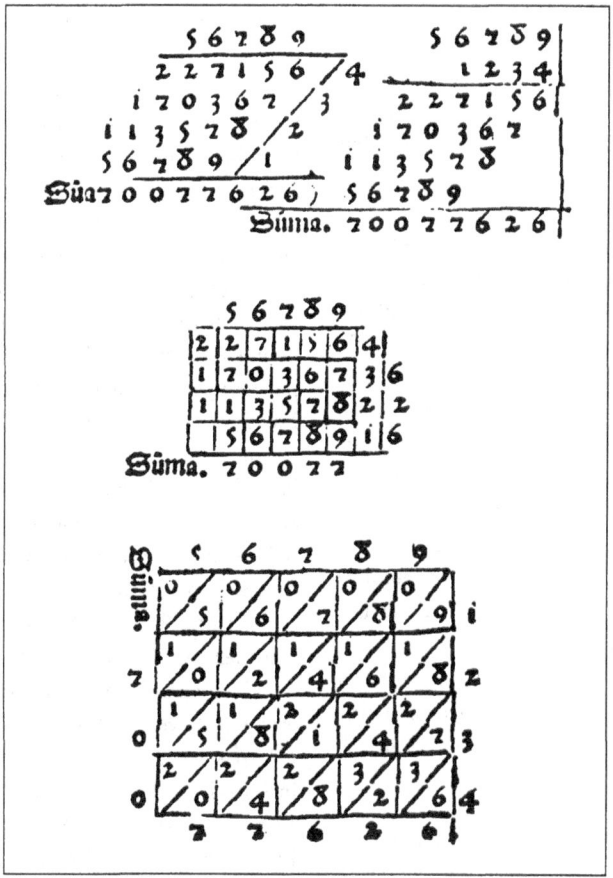

16

Simon Stevin and the Decimal Fractions

D. J. STRUIK

I

SIMON STEVIN, the Flemish-Dutch engineer and mathematician, is best known in the history of mathematics as the inventor of the decimal fractions. This is substantially true, even if we acknowledge the fact that decimal fractions were used before Stevin. It was primarily through Stevin's little book, really no more than a pamphlet, entitled *De Thiende (The Tenth)*, published at Leiden in 1585, that decimal fractions became a regular part of the curriculum in arithmetics.

This pamphlet was republished, in Dutch, French, English, and Latin, several times during Stevin's lifetime and shortly thereafter. Its text has now been made available again in a photostatic reproduction, together with a contemporary English translation by Richard Norton, as part of the second volume of the new edition of Stevin's principal works, prepared under the auspices of the Physics Section of the Royal Netherlands Academy of Sciences.[1] This edition contains an introduction devoted to the history of decimal fractions which gives us an opportunity to evaluate Stevin's achievement in this area.

Simon Stevin was a native of Bruges; it is fairly certain that 1548 was the year of his birth. He started as a bookkeeper, but left Flanders in the period of unrest and persecution which opened the war with Spain. In 1581 we find him at Leiden, and from that time until his death he lived in Holland. Here, like thousands of other immigrants from the Southern Netherlands, he had to find a living under new circumstances. With his particular talents as engineer, accountant, and mathematician, his welcome in the new and striving republic was assured. For many years he was an adviser to Prince Maurice of Orange, one of the most accomplished military commanders of his age, whom he served not only as military engineer, but also as a tutor in different fields of mathematics. Much of Stevin's later work bears the imprint of these tutoring sessions, which covered a wide field of practical and applied geometry, trigonometry, perspective, and bookkeeping. After his death at La Haye in 1620, his admirer Albert Girard, also an immigrant mathematician and engineer, edited his collected mathematical works in French.[2] This edition was published as a big folio in 1634, and has been the main source of our knowedge of Stevin as a mathematician until 1958.[3] The new edition of *The Principal Works* is planned in five volumes, of which Volume I, dealing with Mechanics, and Volume II, dealing with Mathematics, have already appeared. When the five volumes are published they will offer an excellent insight into the status of the exact sciences in Europe in the period just before Fermat and Descartes opened the new era of calculus and coordinate geometry.

Reprinted from *Mathematics Teacher* 52 (Oct., 1959): 474–78; with permission of the National Council of Teachers of Mathematics.

A decimal fraction has a positive integral power of 10 in its denominator, and can be written in many ways, e.g.: 71/100, 0.71, 0,71, or .71, to use notations in use at the present time. It is not easy to say where decimal fractions were first used in any systematic way, but the priority seems to go to the Chinese.[4] With them we find the decimal place-value notation as early as the fourteenth century B.C. and the use of decimal fractions in metrology as early as the third century A.D. Liu Hui, who lived in this period, expresssed a length of 1.355 feet as 1 chhih, 3 tshun, 5 fen, 5 li. The same Liu Hui also performed root extractions in decimals; in this process, which we find in the Middle Ages in use among Arabic, Jewish, and Latin authors, one writes, for instance, $\sqrt{17} = \sqrt{170,000}/100 = 412/100$. When we come to the Sung Dynasty, we find computation with decimal fractions already well developed; Yang Hui, in 1261, multiplies 24.68 by 36.56 and finds 902.3008. From the Chinese and the Indians the decimal notation came to the Islamic writers; our example of $\sqrt{17}$ is found in the writings of Al-Nasawi (Persian, ca. 1030), who translated 412/100 again into sexagesimal fractions as 4°7'12", meaning 4 + 7/60 + 12/3600. The familiarity of Yang Hui with decimal fractions was shared, one and a half centuries later, by the Persian astronomer Jamshid Al-Kashi,[5] who multiplied 25.07 by 14.3 to obtain 358.501. Whether the first notion of decimal fractions in Europe came through contact with the Orient or whether it arose spontaneously is difficult to say, but there is little doubt that wherever computations in the decimal system were used on a wide scale, decimal fractions were sooner or later bound to appear through the logic of the computations themselves. As it was, it is rather astonishing that with the decimal system in use since the Stone Age in so many parts of the world, the regular use of this system for fractions appeared at so late a period.

Readers may ask: if decimal fractions appeared outside of China at such a relatively late date, how did people get along before their introduction?

The answer is that they used either fractions like 3/7, 7/13, etc., with any kind of denominator, or the sexagesimal fractions based on the number 60. An example of a quite complicated reckoning of the first kind is Archimedes's approximation of π as a ratio between 3 10/71 and 3 1/7; examples of sexagesimal reckoning can be found on cuneiform clay tablets dating back to Iraq (Mesopotamia) of the fourth millennium B.C. Ptolemy, in his astronomical handbook known as the *Almagest* (ca. A.D. 150), also used sexagesimal fractions. And so do we when we express the magnitude of an angle as 29°51'32"; here minutes and seconds are expressed in the sexagesimal system, while the numbers 29, 51, 32 are written in the decimal system. Variations of these two methods of the fractional calculus are known to have existed, such as the ancient Egyptian method of expressing fractions as the sum of unit fractions (fractions such as $1/3$, $1/7$, with numerator 1). Strictly speaking, there was, and still is, another way of coping with the problem, popular wherever fractional calculus is considered difficult, and that is the avoidance of all fractions by the choice of an appropriate scale of measurement, e.g., 60, so that $1/4$ is expressed by 15, $2/5$ by 24, etc. We do the same when we say 475 m instead of .475 km, or 3 quarts instead of $3/4$ gallon. We shall see how all these concepts played a role in Stevin's work.

2

During the fifteenth and sixteenth centuries, European computers began to use the Hindu-Arabic system of decimal notation—that is, our present positional system—with ever-increasing efficiency, stimulated by the spread of a mercantile civilization. Gradually the influence of the system began also to be felt in computation with fractions.

We can clearly see this process going on in the trigonometric tables of the Nuremberg astronomer and mathematician Regiomontanus, who died in 1476, and whose books and tables were standard even in the days of Stevin. In those days, and also much later until the eighteenth century, the sines, as well as the tangents and the other trigonometric entities, were conceived as lines and not as ratios, so that they were expressed in terms of a circle radius R of given length. In the sine table of Regiomontanus we find $R = 60,000$, so that the sine of 30° is 30,000; later he used $R = 6,000,000$. Both data show the influence of the sexagesimal system. (Ptolemy's tables are based on $R = 60$.) The different sines are thus expressed as integers. But Regiomontanus also had a tangent table in which the tangent of 45° is 100,000, so that here $R = 10^5$, and he had another table with $R = 10^7$. The decimal system thus became established as the base for the computation of trigonometric tables. The great tables of Rhaeticus (1551), which contain to seven decimal places the values of all six trigonometric functions for angles ascending by 10" intervals, are based on $R = 10^7$.[6]

These tables contain no decimal fractions in the strict sense of the word. Actual fractions based on powers of 10 as denominator occur in some of the many books on arithmetic which appeared during the sixteenth century. For instance, we find, in a book of 1530 written by the widely read German teacher of arithmetic, Christopher Rudolff, a table for compound interest, in which the values of $375 \times (1 + 5/100)^n$ for $n = 1, 2, \cdots,$ 10 are written in a form which differs from our present notation only by the use of a vertical dash instead of a point as decimal separatrix, e.g., 413|4375 for $n = 2$.[7] There are several similar cases, but none of these authors used decimal fractions consistently, and where they used them they varied their notation.

Stevin was the first in the Occident to divest the decimal fraction of its casual character. Appealing to the learned as well as the practical man, to the teacher as well as the merchant and the wine gauger, he advertised the advantages of his notation as "teaching how to perform with an ease, unheard of, all computations necessary between men by integers without fractions." In doing this he used a notation which reminds us strongly of the sexagesimal one; where we write 47.58, he wrote 47⓪, 5①, 8②, where the unit ⓪ is called "commencement," the tenth ① is called "prime," the hundredth ② is called "second," etc. When we write 27.847 + 37.675 + 875.782 = 941.304, Stevin wrote

	⓪	①	②	③
	2	7	8	4	7
	3	7	6	7	5
8	7	5	7	8	2
9	4	1	3	0	4

In a similar way he deals with subtraction, multiplication, and division—all performed "without fractions." He ends his pamphlet with a plea to introduce the decimal system also in the measurement of lengths, areas, volumes, etc.

Stevin had the correct idea, but his notation seems clumsy to us and less elegant than that which Rudolff used half a century earlier. The circle notation was taken from the Italian mathematician Bombelli, who had used a similar notation in his *Algebra* of 1572 for the powers of the variable (anticipating our exponents in x^1, x^2, x^3, etc.). He might have used the sexagesimal notation in the form 47°5'8'', for our 47.58, as some of his followers did; he himself exchanged his notation occasionally and wrote 732② for our 7.32. His notation may have had some advantages for inexperienced pupils, since the circle notation allows intermediate steps: 7⓪5①8② plus 4⓪7①5② is equal to 11⓪12①13②, which reduces to 11⓪13①3②, and this again to 12⓪3①3②. Stevin could also do away with zeros: 2③7⑤ means .00207. But the notation remained unwieldy, and Stevin's work would not have had its lasting influence if it had not been for Napier and his logarithms.

3

Logarithms, invented by John Napier, the Scottish nobleman, were first presented in his Latin *Descriptio* of 1614. These first logarithms are not the logarithms that we use, but are certain numbers defined with the aid of sines, which in Napier's exposition were based on $R = 10^7$. This edition of 1614 had no decimal fractions. These appear in a 1616 English translation of the *Descriptio*, with a point as decimal separatrix. This notation was adopted by Napier in his Latin *Rabdologia* of 1617, the book in which he showed how to perform computations with his "rods," the so-called "rods of Napier." Here Napier quotes Stevin's *Arithmetica Decimalis* and proposes the notation of 1993,273 (with point or comma) for 1993 273/1000, although he also uses 821, 2′5″ for 821 25/100. Then, in the posthumous *Constructio* of 1619, the notation has become consistent: "whatever is written after the period is a fraction." Thus 25.803 means 25 803/1000.

The great tables of logarithms, based on 10, which now appear, take the decimal notation of fractions for granted, with dot or period. With tables of such logarithms, in which the decimal part of such numbers as 43, 430, 4300 is the same, the decimal notation of fractions is only natural. Henry Briggs, in his table of 1624, and Adrian Vlacq, in his tables of 1627, use this notation consistently, and from then on the decimal fractions with comma or dot as separatrix were generally accepted, at any rate in computations with logarithms.

Stevin's, Napier's and Briggs's contributions were combined in two Dutch books by the surveyor Ezechiel De Decker, entitled *The New Arithmetic, First Part* and *Second Part*[8] (1626, 1627). Here we find together Stevin's *Thiende*, Vlacq's translation of the *Rabdologia*, and the Briggsian logarithms of all integers from 1 to 100,000. These two books by De Decker are a kind of glorification of the triumph of the decimal system. They stress three essential aspects of this victory: the Hindu-Arabic notation with the modern digits, the decimal fractions, and the logarithms to base 10. One change was still due, although it was already implied in the whole framework of the system, namely the rewriting of the trigonometric tables to a unit

$R = 1$, a thing that even Stevin did not do. The systematic introduction of this unit $R = 1$ had to wait until Leonard Euler's *Introductio in Analysin Infinitorum* of 1748, and from that time on trigonometric entities were no longer considered as line segments but as dimensionless ratios.

4

The triumph of the decimal fractions through the work of Stevin and Napier did not mean that notation and use became immediately standardized. There were loyal Stevin followers who preferred his notation, or a slight modification of it. As late as 1739 we find the Abbé Deidier teaching that decimal fractions should be written as $89 \cdot 5^{\mathrm{I}} \, 2^{\mathrm{II}} \, 7^{\mathrm{III}} \, 6^{\mathrm{IV}}$ or 895276^{IV}; he used, however, the ordinary point notation for logarithms. Many such inconsistencies remained, again until Euler, in his *Introductio* of 1748, standardized our present notations.

The use of the decimal notation for weights and measures, also proposed by Stevin, had to wait in the West until the French revolution introduced meters, ares, and liters, and also standardized the monetary system.[9] We know that this was only partially accepted in England and in the United States, though the United States introduced a decimal monetary system as a result of the efforts of Robert Morris, Thomas Jefferson, and Alexander Hamilton.[10] As to angular measurement, the struggle is still on, and when we use our ordinary notation of degrees, minutes, and seconds, we pay our respects to a system which can boast of an age of five thousand years, certainly one of the longest ages in our whole scientific heritage.

NOTES

1. *The Principal Works of Simon Stevin.* Amsterdam: Swets and Zeitlinger. Vol. I, *Mechanics,* edited by E. J. Dyksterhuis, v + 617 pp. (1955); Vol. II, *Mathematics,* edited by D. J. Struik, in two parts, v + 976 pp. (1958).

2. Albert Girard (1595–1632) is best known as the author of the *Invention nouvelle en algebre* (1629, repub-

lished 1884), in which we find the theorem that an algebraic equation of degree n has n roots (Girard's formulation is different).

3. There have been two facsimile editions of *The Tenth* before 1958, one by H. Bosmans of the Dutch edition of 1585 (La Haye: Anvers, 1924), and one by G. Sarton of the French edition of 1585 (*Isis* 33 (1935)).

4. J. Needham, *Science and Civilisation in China*, vol. III. Cambridge: Cambridge University Press, 1959. The first section of this standard work deals with mathematics.

5. *D.G. Al-Kaši Ključ Arifmetiki, Traktat ob Okružnosti*, translated and edited by B. A. Rozenfel'd (Moscow, 1956), especially p. 62; Y. Mikami, *The Development of Mathematics in China and Japan* (Leipzig, 1913), especially p. 26. This is the way in which Al-Kashi multiplies 25.07 by 14.3 to get 358.501:

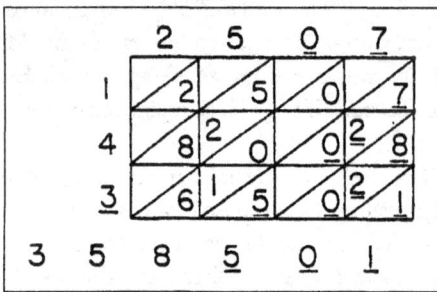

(The integers are in black and the fractional parts in red. The digits are Arabic ones, which differ from our present digits, the latter being in use only since the time of the European Renaissance.)

6. These tables were expanded by Valentin Otho into the *Opus Palatinum* of 1596, a classic among the tables which has many values in ten decimals and the sines up to 15 decimals.

7. *Exempel Buechlin Rechnung belangend darbey* (Augsbury, 1530). The place where decimal fractions are introduced has been more than once reproduced, e.g., in D. E. Smith, *History of Mathematics*, vol. II, p. 241.

8. *Eerste Deel van de Nieuwe Telkonst* (Gouda, 1626); *Tweede Deel van de Nieuwe Telkonst* (Gouda, 1627). The second part was always known, but the first part disappeared until it was rediscovered by M. van Haaften in 1920; see *Nieuw Archief voor Wiskunde*, 15 (1928), pp. 49–54; 31 (1942), 59–64. This discovery has shown that it was not Vlacq in 1628, but De Decker in 1627, who first published a complete table of logarithms.

9. The Chinese, as mentioned in the text, had a decimal system at least as early as the third century A.D.

10. The dollar with its decimal division was introduced by the Coinage Act of 1792, sponsored by Hamilton. This act was preceded by a resolution of Congress of 1785, the result of a report to the President of Congress by Robert Morris (1782), endorsed by Jefferson. See A. Nussbaum, *The History of the Dollar* (New York, 1957, viii + 308 pp.), Chapter II; C. D. Hellman, "Jefferson's efforts toward the decimalization of the U.S. weights and measures," *Isis*, 16 (1931), pp. 266–314.

HISTORICAL EXHIBIT 7

Mathematical Considerations on the Trajectory of a Cannon Ball

Early cannons were quite primitive and their range limited but as the technology of warfare improved by the fifteenth century, they were hurling their projectiles beyond the view of their gunners. The questions of the range of a cannon shot became important. Early mathematical models for the trajectory of a cannon ball were mostly speculative. The first such model to be used was a right triangle with the path of the shot tracing out the hypotenuse until a maximum height is reached at which time the shot would fall vertically downwards completing a leg of the right triangle. In this instance, the length of the horizontal leg supplied the range of the cannon shot. By the early sixteenth century, this model was modified to include a circular arc connecting the ascending and descending paths of a cannon ball.

Niccola Tartaglia investigated the paths of cannon shot and published his finding in *Nova Scientia* [The New Science], 1537. While Tartaglia related the properties of a shot to the angle of barrel elevation, he still employed a line-arc-line model for trajectory. Galileo Galilei took up the challenge of determining the geometrical path of a cannon ball trajectory. Through experimentation, he determined that if a body is projected horizontally, its descent in successive time intervals t_1, t_2, $t_3 \ldots t_k$ would be in the ratios given by the number sequence 1, 4, 9, 16, $\ldots d_k$. Galileo concluded that the path would be a parabola (i.e., $d = kt^2$). Evangelista Torricelli (1608–1647), a pupil of Galileo, refined these theories further to place the science of artillery on a firm mathematical basis.

17

Viète's Use of Decimal Fractions

CARL B. BOYER

$\mathcal{T}$HERE IS NO SUCH PERSON as the inventor of decimal fractions. It is well known that the earliest appearances of decimal fractions were incidental and inadvertent and that their use developed slowly.[1] The positional system had entered mathematics in conjunction with the sexagesimal system of numeration—in Babylonia of some four thousand years ago—and sexagesimal fractions had preempted the field of accurate computation until some four hundred years ago. Even when the principle of local value came to be associated with a decimal system of integers, more than a thousand years ago, the decimal fraction did not form part of the new system (generally referred to as Hindu-Arabic). The fractional domain continued to be made up of two principal parts. The layman still generally made use of the common or vulgar fractions (occasionally also of Egyptian unit fractions); the applied mathematician adhered to the established use of sexagesimals so single-mindedly that these latter came to be known as *fractiones astronomiae or fractiones physicae*. Occasional use of the equivalent of decimal fractions can nevertheless be found during the early centuries of our era. In the third century in China[2] the use of decimal metrological units can be regarded as a forerunner of decimal fractions, and square roots were found by a process tantamount to the rule

$$\sqrt{a} = \frac{\sqrt{10^{2n}a}}{10^n} .$$

Reprinted from *Mathematics Teacher* 55 (Feb., 1962): 123–27; with permission of the National Council of Teachers of Mathematics.

Rules such as this were known to the Hindus and were used in medieval Europe, keeping alive the spore from which the systematic use of decimal fractions developed during the sixteenth century.

Trigonometry has been a fertile source for innovation in computation, and here, too, one finds germs of the decimal idea prior to the definitive development. In order to avoid difficulties inherent in the "astronomer's fractions," it became customary in early modern trigonometry to replace Ptolemy's radius of 60 by a still larger integer, such as 10^8. In this case the trigonometric lines could be expressed to a high degree of approximation without resorting to fractions of any kind, and their modern equivalents are easily read off today by a simple shift of the decimal point eight places to the left. That is, $\sin 45°$ would in this case appear as 70,710,678 instead of 0.70710678. Nevertheless, the concept of decimal fraction was in these cases not specifically invoked before the sixteenth century. Occasionally a closer approach to the decimalization of fractions is found in problems involving division by multiples of ten. In 1492 Pellos, for example, carried out the division of 5836943 by 30 through a process akin to our shift of a decimal point.[3] He first separated the final digit from the others by a dot, divided the left-hand number by three to obtain 194546, and then expressed the remainder as $^{23}/_{30}$. The form of the answer betrays clearly that he was not thinking of a decimal fractional algorithm.

It is common knowledge that the first book written with the purpose of popularizing the decimal fraction was published by Stevin in 1585 as *La*

Disme or *De Thiende*. Despite the clumsy notation Stevin employed, he has been referred to repeatedly as the inventor of the decimal fraction, and his little work has been so thoroughly described, and so frequently reproduced,[4] that there is no need to duplicate this effort. It will be well to remark, however, that Stevin's point of view is not ours. Instead of believing that he was introducing a new type of fractional calculation, Stevin claims in *La Disme* that he has done away with the need for fractions of any sort. The first part of Stevin's work opens with, "The First Definition: Disme is a kind of arithmetic, invented by the tenth progression, consisting in characters of ciphers, whereby a certain number is described and by which also all accounts which happen in human affairs are dispatched by whole numbers without fractions or broken numbers."[5] That is, Stevin regards his tenths, hundredths, etc., as units of a new order of magnitude, similar to the subdivisions in the metric system. His successors, however, soon shifted from this view to the modern concept of decimal fraction.

Stevin is by no means the only claimant for the invention. David Eugene Smith wrote, "If one man were to be named as the best entitled to be called the inventor of decimal fractions, Rudolff might properly be the man."[6] Christoff Rudolff operated with decimal fractions fifty-five years before Stevin, using a vertical bar as a separatrix, but he made no special plea for the decimal fraction. Pages of his *Exempel Buechlin* (1530) and *Kunstliche Rechnung* (1626) have been reproduced by Smith and others. Smith went further and reproduced facsimile pages from other original sources in order to illustrate "some of the steps in the invention of the decimal fractions that have not been recognized as fully as they deserve." Among these is, of course, a page from Adam Riese, *Rechnung auff der Linien und Federn* (1522), for the name Adam Riese has been a household term in Germany indicating correctness of computation. There is also a page from Pellos' arithmetic of 1492. Overlooked entirely, in Smith's account, is a celebrated

mathematician whose contribution to the rise of decimal fractions has too often been neglected by historians, and it is the purpose of this note, through a brief excerpt from Viète's work and the facsimile reproduction of a page, to call attention to the fact that, Stevin excepted, no one made a more significant contribution to the rise of decimal fractions than did Viète, the leading mathematician of the sixteenth century.

In 1579 Viète published a work entitled *Canon mathematicus* in which, following the practice of his day, he adopted a radius so large that for ordinary trigonometric purposes fractions are avoided. With a radius of 100,000,000 parts, he gives the sine of 45°, for example, as 70,710,678. However, in a supplementary work, *Universalium inspectionum ad canonem mathematicum, liber singularis*, bound with the *Canon* and of the same date, one finds not only the systematic use of decimal fractions, but also a strong plea for their adoption throughout mathematics. Sarton years ago remarked that "Viète renounced sexagesimal fractions in favor of decimal ones," but Sarton's statement was submerged in a glowing eulogy of Stevin's *La Disme*.[7] Incidental allusions to Viète's contribution have appeared since but they have not attracted attention.[8] In the thought that seeing might lead to believing, we therefore reproduce[9] page 51 of Viète's *Universalium inspectionum*. Here more than two dozen decimal fractions, in an eminently clear and appropriate notation, appear on a single page. Among these is the approximate value

$$314{,}159{,} \tfrac{265.36}{}$$

for the circumference of a circle whose diameter is 200,000. Slight modifications of this form appear elsewhere in the book, and on page 69 it is published simply as

$$314{,}159{,}265{,}36.$$

Sometimes a vertical bar replaces the comma as a decimal separatrix, as on page 65 when the apothem of a regular polygon of 96 sides in a circle of radius of 100,000 is given as

$$99,946 \mid 458,75.$$

Forms of the decimal fractions in this work, both in the supplement and in the *Canon* itself, may differ slightly, but their use is far from incidental. Viète, on page 17 of *Universalium inspectionum*, wrote: "Finally, sexagesimals and sixties are to be used sparingly or never in mathematics, and thousandths and thousands, hundredths and hundreds, tenths and tens, and similar progressions, ascending and descending, are to be used frequently or exclusively."

A biographer of Stevin wrote some years ago that "if one understands by inventor the one who crystallizes, in an attractive and harmonious form, the timid and scattered attempts of his predecessors, if one understands by inventor the one who proposes the systematic use rather than one who had employed them incidentally, Stevin is incontestably the inventor of decimal fractions."[10] And another expositor has categorically attributed to Stevin the first systematic use of decimal fractions to the exclusion of ordinary fractions.[11] It is clear that the name of Viète could very appropriately be substituted for that of Stevin in either case without falsifying the proposition, and Viète's work antedated that of Stevin by half a dozen years.

Viète was a deep and original research mathematician who did not institute a campaign to popularize decimal fractions, nor did his works enjoy a circulation commensurate with his reputation. Thus it was that the world of science adopted decimal fractions chiefly after Stevin, in 1585, had explained their use in language readily comprehended by the multitude. By this time decimal fractions were in the air, and their use was not in all cases traceable to Stevin. Kepler, for example, believed that the use of decimal fractions was to be attributed to Jobst Bürgi's trigonometry. The spread of decimalization in fractions was spurred not only through trigonometry, but also by the seventeenth-century invention of logarithms, in which Bürgi shared. Notations similar to Viète's were among the varied forms which were preferred to the clumsy indices of Stevin, but even to this day uniformity has not been achieved. The use of a comma as a decimal separatrix can be traced back to G. A. Magini's *De planis triangulis* of 1592, and a year later the decimal dot appeared in Clavius' *Astrolabium*.[12] Each of these notations is with us today, the former chiefly on the European continent, and the latter among Anglo-Saxons. No single notation can be said to have won the field, and no one mathematician can claim the decimal fraction as his own. But any list of names of men who shared in the invention must provide an honorable place for François Viète. Decimal fractions were by no means the chief contribution of Viète to mathematics; their use was but a small part of the work of one whom the late E. T. Bell appropriately described[13] as "the first mathematician of his age to think occasionally as mathematicians habitually think today."

NOTES

1. An excellent account by D. J. Struik of the early history of decimal fractions, together with further references, is to be found in *The Principal Works of Simon Stevin*, edited by D. J. Struik, II (Amsterdam: 1958), 373–85.

2. Joseph Needham, *Science and Civilisation in China*, III (Cambridge: 1959), 46, 82, Cf Wang Ling, "The development of decimal fractions in China," *Acts of the International Congress of the History of Science*, VIII (1956), Vol. I, 13–17.

3. David Eugene Smith, "The invention of the decimal fraction," *Teachers College Bulletin*, No. 5 (1910), 11–21.

4. See especially George Sarton's two articles, "Simon Stevin of Bruges," *Isis*, XXI (1934), 241–303, and "The first explanation of decimal fractions and measures," *Isis*, XXIII (1935), 153–244. For further references see the work of Struik cited in Note 1, or his article "Simon Stevin and the decimal fractions," *The Mathematics Teacher*, LII (1959), 474–78.

5. *The Principal Works of Simon Stevin*, II 403.

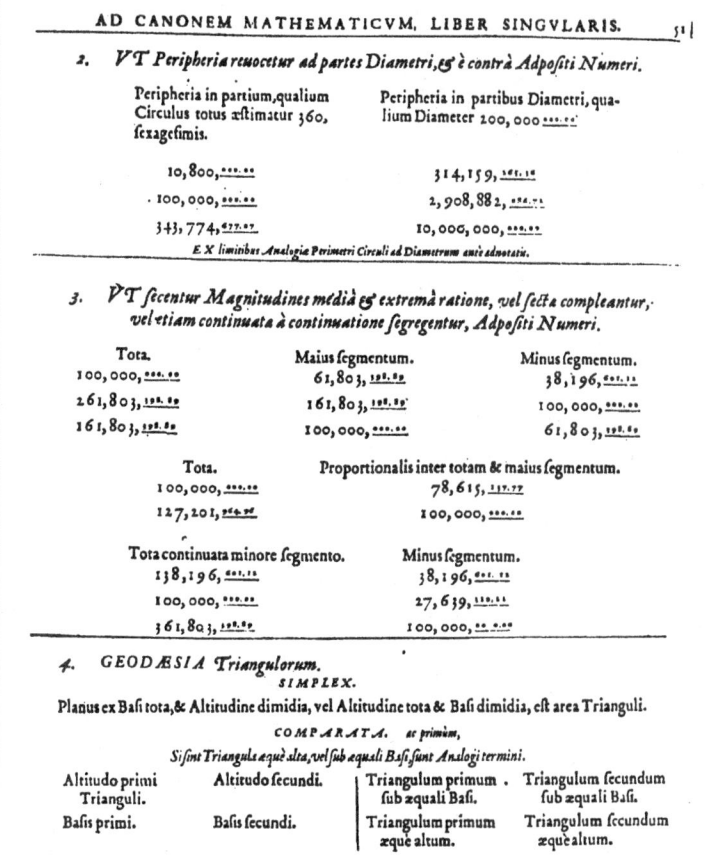

Reproduction of page 51 of Viète's Universalium inspectionum, *through the courtesy of the Burndy Library.*

6. Smith, *op. cit.* This view is echoed in Carl N. Shuster, *A Study of the Problems in Teaching the Slide Rule* (New York: 1940), p. 9.

7. *Isis,* XXIII (1935), 173–74.

8. Casual notice of Viète's use of decimals is found in Florian Cajori, *History of Mathematical Notations* (Chicago: 1928–29), II, 316, and in *Encyclopédie des Sciences Mathématiques,* I (1), p. 53, note 180, and in Struik's article cited in Note 1. In the *Encyclopaedia Brittanica* article in "Mathematics, History of," Otto Neugebauer has Viète share with Stevin credit for the development of a consistent decimal place-value notation. Forceful statements of Viète's role are found in K. Hunrath, "Zur Geschichte der Decimalbrüche," *Zeitschrift für Mathematik und Physik,* XXXVIII (1893), Hist. lit. Abt., pp. 25–27, and Pierre Dedron and Jean Itard, *Mathématiques et mathématiciens* (Paris: *c.* 1959), p. 290.

However, I have nowhere seen a reproduction of a page from Viète's work showing his use of decimal fractions.

9. We have used, through the kindness of Colonel Bern Dibner, the copy of Viète's *Canon* at the Burndy Library, and the accompanying reproduction has been provided through the courtesy of the Burndy Library.

10. Robert Depau, *Simon Stevin* (Bruxelles: 1942), p. 62.

11. Henri Bosmans, *La Thiende de Simon Stevin* (Andvers and La Haye: Édition de la Société des Bibliophiles Anversois, No. 38, 1924), 1–5.

12. Jekuthiel Ginsburg, "On the early history of the decimal point," *Scripta Mathematica,* I (1932), 84–85, 168–69.

13. E. T. Bell, *The Development of Mathematics* (New York: 1940), p. 99.

18

John Napier
and His Logarithms

C. B. READ

*I*T IS DOUBTFUL that any teacher would today try to introduce the subject of logarithms other than by means of some use of exponents. It is, then, indeed hard to realize that Napier constructed his logarithms before the concept of exponents as we know them was developed.

John Napier was born in 1550, probably when his father was about sixteen years of age. He is known to have matriculated as a student at St. Andrews University, but records fail to show that he graduated. His first published work was *A Plaine Discovery of the Whole Revelation of St. John*. However, it appears that from an early age he was interested in mathematics as well as theology. He apparently did some early work in algebra and in arithmetic. Toward the end of the sixteenth century, Napier became perturbed at the fact that scientific progress was hindered by the great labor involved in numerical calculation. He deliberately attacked the problem of developing some means of lessening the work involved. He devised several mechanical aids, among these being sets of small rods often called "Napier's bones." In his approach it seems evident that Napier was (at least at first) interested in trigonometric computation, for he deals almost exclusively with sines.

The great contribution of Napier was the invention of what we now know as logarithms, al-

though at first these were called by Napier "artificial numbers" or simply "artificials." The number corresponding to the logarithm was called a "natural number."

The exact date of the discovery, if indeed one can say there was a definite date of discovery or invention, is uncertain. Kepler tells us that in 1594 Tycho Brahe had information from a Scotch friend that there was a possibility of publication of the material. Actually, the computation of the table, or "canon" as the author called it, probably took several years. There are two books which are explanatory of the system, generally abbreviated as the *Constructio* and the *Descriptio*. Of the two, the *Mirifici Logarithmorum Canonis Constructio* is the most important, in fact the most important of all Napier's works.

The *Contructio* presents clearly his original conception of logarithms. The "canon," with instructions for its use, was published in 1614. The method of construction, although probably written several years earlier, was not published until 1619. It was published in Edinburgh, some two years after Napier's death (which may have been hastened by the strain involved in the development and computation of the "canon"). In the body of this book, the term "artificial number" is frequently but not exclusively used instead of "logarithm." The term "logarithm" is used in the title page, the headings, and in the appendix, which discusses the advantages of logarithms to the base ten.

Reprinted from *Mathematics Teacher* 53 (May, 1960): 381–85; with permission of the National Council of Teachers of Mathematics.

This original work is relatively rare, few writers having had the privilege of examining a copy. In addition, the fact that the work is in Latin makes it less easy for a present-day student or teacher to follow, even if a copy were available. An English translation was published by W. R. Macdonald in 1889.

The book consists of sixty numbered paragraphs, occupying, in the English translation, less than fifty pages. (This does not include the appendix, which is entitled "On the construction of another and better kind of Logarithms, namely one which the Logarithms of unity is 0," nor a supplement dealing with methods of solving spherical triangles with the use of logarithms.) The entire book needs to be studied to understand Napier's approach; a few necessarily brief comments may be helpful.

The first numbered paragraph defines a logarithmic table as " . . . a small table by the use of which we can obtain a knowledge of all geometrical dimensions and motions in space, by a very easy calculation."

As the treatment proceeds, the author finds it necessary to define arithmetical and geometrical progressions. Since accuracy is needed, he suggests taking large numbers for a basis, "but large numbers are most easily made from small by adding ciphers." He then says "these large numbers may again be made still larger by placing a period after the number and adding ciphers." It is found necessary to explain that 10000000.04 is the same as 10000000 $^4/_{100}$. It is perhaps not generally known that Napier was the first to use our present notation for decimal fractions. It is not impossible that this was a by-product of his working out the invention of logarithms, although in another work Napier makes reference to the contributions of Simon Stevin. However, in spite of the simplicity of the notation proposed by Napier, it did not come into general use until long after his death.

Napier next presents a discussion of the accuracy of working with his artificial numbers—a discussion which has much resemblance to our present-day work on the accuracy of computation with approximate numbers.

To follow the text, it is necessary to understand that Napier's canon was not a table of logarithms of numbers as we know it (that is, of equally spaced numbers) but of "sines of arcs" for every minute from 0 to 90 degrees. It may be easier to understand Napier's approach if it is known that at this period the "sine of an arc" (which we call the sine of an angle) was the length of a line equal to the half-chord of a double central angle in a circle whose radius was unity. Hence Napier's explanations frequently represent both sines and logarithms by lines. Instead of calling the radius unity, he adds seven ciphers. From this radius he suggests subtracting its 10000000th part, obtaining 9999999; then from this he subtracts the 10000000th part, obtaining 9999998.0000001; again subtracting the 10000000th part he obtains 9999997.0000003. Continuing this process he creates, as he calls them, "a hundred proportionals." The "hundredth proportional" is 9999900.0004950. The resulting table is called the "First Table."

For his "Second Table" Napier points out the difficulty of forming fifty proportional numbers between the first and last numbers of his First Table (10000000.0000000 and 9999900.0004950). However, as he puts it, "A near and at the same time an easy proportion is 100000 to 99999," which will yield "sufficient exactness." So for the Second Table he starts by "adding six ciphers to radius and continually subtracting from each number its own 100000th part," obtaining fifty other proportional numbers, the last being 9995001.222927 (an unfortunate error in computation—the correct value being 9995001.224804).

The Third Table "consists of 69 columns, and in each column are placed 21 numbers, proceeding in the proportion which is easiest, and as near as possible to that subsisting between the first and last numbers of the Second Table." For ease in calculation, the proportion is considered to be 10000 to 9995.

Essentially these three tables constitute a table of sines, or natural numbers, "progressing in geometrical proportion" with no two numbers differing by more than unity (after rounding). Napier now proceeds to show how to place, beside the sines, or "natural numbers" (which decrease geometrically), their logarithms, or "artificial numbers" (which increase arithmetically). To do this he makes use of the concept of a "geometrically moving point" which, when approaching a fixed point, has its "velocities proportionate to its distances from the fixed point." With this concept we have essentially Napier's definition of a logarithm: "that number which has increased arithmetically with the same velocity throughout as that with which the radius began to decrease geometrically."

Having shown that "nothing is the logarithm of the radius," he discusses "the limits of the logarithm," which covers roughly what we would call the maximum error in a computed logarithm. He then develops certain laws relating to proportions, expressed in terms of logarithms. In modern notation, for example, if $a : b = b : c$ and we know the logarithms of any two of the three quantities, we may obtain the logarithm of the third; similarly for $a : b = c : d$. The explanation proceeds to show how to find logarithms of sines outside the limits of the table. There are numerous examples.

The above explanation seems quite involved. It may be of some help to note that in the First Table, the logarithm of 10000000 is zero; the logarithm of the next number 9999998.0000001 is 1; that of 9999997.0000003 is 3, and of 9999900.0004950 is 100. Rounding for ease in computation, gives 100 as the logarithm of 9999900. Then, using a new ratio for the Second Table, more values can be inserted.

Napier then used existing tables of sines (disregarding decimal points which would appear in modern tables). He chose the number in his tables nearest the desired sine, and by use of "limits of its logarithm" (essentially interpolation) found the desired logarithm. Paragraph 43 demonstrates the method of finding the logarithm when the given

sine is 9995000.000000. The nearest sine in the Second Table is 9995001.222927, from which he obtains, in modern notation: log 9995000.000000 = 5001.2485387.

Paragraph 58 shows that if the logarithm of "all arcs not less than 45 degrees are given, the logarithms of all less arcs are very easily obtained," thus paralleling our present-day tables which tabulate values to 45 degrees, then use the complementary angle. Paragraph 60 gives details for forming a logarithmic table—employing 45 pages, each capable of holding sixty lines of figures, and with seven columns of figures on each page. Once the pages are prepared, reference is made to paragraphs 49 and 50, which explain in detail how the calculations are performed.

Although it is very unlikely that anyone at the present time would wish to make up a "canon" of logarithms using Napier's method, there are several points of definite historic interest. First, with Napier's logarithms, as the natural numbers *decrease*, their logarithms *increase*. The logarithm of 10,000,000 is zero, the logarithm of 9,995,000 is 5,001.2, the logarithm of 9,900,000 is 100,503.3, the logarithm of 4,998,609 is 6,934,250, etc. Second, in the sense that we know it, Napier's logarithms actually had no base. Third, it is interesting to note that although decimal fractions were used in the construction of the canon, the logarithms given are large whole numbers—for example, paragraph 57 computes the logarithm of an arc of 34° 40' as 5,642,242. (For brevity Napier speaks of the logarithm of an arc rather than, as in modern usage, of the logarithm of the sine of an angle.) Finally, we note with interest that Napier's construction or demonstration was based on the concept of moving points, using what might be termed arithmetical and geometrical motion. Arithmetical motion implies constant velocity; Napier's geometrical motion involved variable velocity, continually decreasing.

By no means is it to be implied that credit should be withheld from Napier. On the other hand, when we read that Napier invented logarithms, it should be recognized that his canon, or table, differed in several important essentials from what we now recognize as a logarithmic table, and that his method of construction differs markedly from the explanation now given (as, for example, in a course in calculus) relative to the computation of logarithms. Finally, what we now often call natural or Naperian logarithms are not identical with those first developed by John Napier of Merchiston.

NOTE

* See, however, D. E. Smith, *History of Mathematics*, II (Boston: Ginn and Co., 1925), pp. 238 and 244.

HISTORICAL EXHIBIT 8

The Evolution of Algebraic Symbolism

Since the time of the ancient Egyptians and Babylonians, mathematical problems, even situational problems, were completely written out in words as were their solution procedures. This phase of algebraic thinking is usually called the "rhetorical stage." Due to printing and the repeated use of certain terms and words, mathematicians began to use abbreviations to form mathematical relationships. At first, each mathematician or local group of mathematicians had their own system of symbolization but gradually the symbols as well as the procedures became standardized. Below are various examples of how different mathematicians employed symbols to express the modern equation $4x^2 + 3x = 10$.

Nicolas Chuquet	(1484)	$4^2 \, p3^1$ égault 10^0
Vander Hoecke	(1514)	$4 \, Se + 3$ Pri dit is ghelijc 10
F. Ghaligai	(1521)	$4 \, \square \, e \, 3c° - 10$ numeri
Rudolff	(1525)	Sit $4 \, \mathcal{Z} + 3 \mathcal{e}$ aequatus 10
Jean Buteo	(1559)	$4 \, \Diamond \, p3 \, [\, 10$
R. Bombelli	(1572)	$4 \, p \, 3$ equals á 10
Simon Stevin	(1585)	$4 \, ② + \, 3 \, ①$ egales 10
Ramus and Schoner	(1586)	$4 \, q \longmapsto 3 \, \mathcal{R}$ aequatus sit 10
François Viète	(c1590)	$4Q + 3N$ aequatur 10
Thomas Harriot	(1631)	$4aa + 3a ≡ 10$
René Descartes	(1637)	$4ZZ + 3Z \, ∞ \, 10$
John Wallis	(1693)	$4XX + 3X = 10$

19

Projective Geometry

MORRIS KLINE

*I*N THE HOUSE OF MATHEMATICS there are many mansions and of these the most elegant is projective geometry. The beauty of its concepts, the logical perfection of its structure and its fundamental role in geometry recommend the subject to every student of mathematics.

Projective geometry had its origins in the work of the Renaissance artists. Medieval painters had been content to express themselves in symbolic terms. They portrayed people and objects in a highly stylized manner, usually on a gold background, as if to emphasize that the subject of the painting, generally religious, had no connection with the real world. An excellent example, regarded by critics as the flower of medieval painting, is Simone Martini's "The Annunciation." With the Renaissance came not only a desire to paint realistically but also a revival of the Greek doctrine that the essence of nature is mathematical law. Renaissance painters struggled for over a hundred years to find a mathematical scheme which would enable them to depict the three-dimensional real world on a two-dimensional canvas. Since many of the Renaissance painters were architects and engineers as well as artists, they eventually succeeded in their objective. To see how well they succeeded one need only compare Leonardo da Vinci's "Last Supper" with Martini's "Annunciation."

The key to three-dimensional representation was found in what is known as the principle of projection and section. The Renaissance painter imagined that a ray of light proceeded from each point in the scene he was painting to one eye. This collection of converging lines he called a projection. He then imagined that his canvas was a glass screen interposed between the scene and the eye. The collection of points where the lines of the projection intersected the glass screen was a "section." To achieve realism the painter had to reproduce on canvas the section that appeared on the glass screen.

Two woodcuts by the German painter Albrecht Dürer illustrate this principle of projection and section. In "The Designer of the Sitting Man" the artist is about to mark on a glass screen a point where one of the light rays from the scene to the artist's eye intersects the screen. The second woodcut, "The Designer of the Lute," shows the section marked out on the glass screen.

WOODCUT I

WOODCUT 2

Of course the section depends not only upon where the artist stands but also where the glass screen is placed between the eye and the scene. But this just means that there can be many different portrayals of the same scene. What matters is that, when he has chosen his scene, his position, and the position of the glass screen, the painter's task is to put on canvas precisely what the section contains. Since the artist's canvas is not transparent and since the scenes he paints sometimes exist only in his imagination, the Renaissance artists had to derive theorems which would specify exactly how a scene would appear on the imaginary glass screen (the location, sizes, and shapes of objects) so that it could be put on canvas.

The theorems they deduced raised questions which proved to be momentous for mathematics. Professional mathematicians took over the investigation of these questions and developed a geometry of great generality and power. Let us trace its development.

Suppose that a square is viewed from a point somewhat to the side [Figure 1]. On a glass screen interposed between the eye and the square, a section of its projection is not a square but some other quadrilateral. Thus square floor tiles, for instance, are not drawn square in a painting. A change in the position of the screen changes the shape of the section, but so long as the position of the viewer is kept fixed, the impression created by the section on the eye is the same. Likewise various sections of the projection of a circle viewed from a fixed position differ considerably—they may be more or less flattened ellipses—but the impression created by all these sections on the eye will still be that created by the original circle at that fixed position.

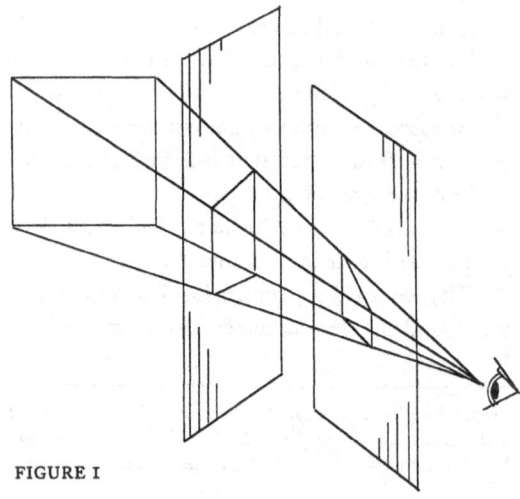

FIGURE I

"The Annunciation" by Simone Martini is an outstanding example of the flat, stylized painting of the medieval artists. The figures were symbolic and framed in a gold background.

"The Last Supper" by Leonardo da Vinci utilized projective geometry to create the illusion of three dimensions. Lines have been drawn on this reproduction to a point at infinity.

Drawing by da Vinci, made as a study for his painting "The Adoration of the Magi," shows how he painstakingly projected the geometry of the entire scene before he actually painted it.

To the intellectually curious mathematicians this phenomenon raised a question: Should not the various sections presenting the same impression to the eye have some geometrical properties in common? For that matter, should not sections of an object viewed from different positions also have some properties in common, since they all derive from the same object? In other words, the mathematicians were stimulated to seek geometrical properties common to all sections of the same projection and to sections of two different projections of a given scene. This problem is essentially the one that has been the chief concern of projective geometers in their development of the subject.

It is evident that, just as the shape of a square or a circle varies in different sections of the same projection or in different projections of the figure, so also will the length of a lone segment, the size of an angle, or the size of an area. More than that, lines which are parallel in a physical scene are not parallel in a painting of it but meet in one point; see, for example, the lines of the ceiling beams in da Vinci's "Last Supper." In other words, the study of properties common to the various sections of projections of an object does not seem to lie within the province of ordinary Euclidean geometry.

Yet some rather simple properties that do carry over from section to section can at once be discerned. For example, a straight line will remain a line (that is, it will not become a curve) in all sections of all projections of it; a triangle will remain a triangle; a quadrilateral will remain a quadrilateral. This is not only intuitively evident but easily proved by Euclidean geometry. However, the discovery of these few fixed properties hardly elates the finder or adds appreciably to the structure and power of mathematics. Much deeper insight was required to obtain significant properties common to different sections.

The first man to supply such insight was Gérard Desargues, the self-educated architect and engineer who worked during the first half of the seventeenth century. Desargues's motivation was to help the artists; his interest in art even extended to

writing a book on how to teach children to sing well. He sought to combine the many theorems on perspective in a compact form, and he invented a special terminology which he thought would be more comprehensible than the usual language of mathematics.

His chief result, still known as Desargues's theorem and still fundamental in the subject of projective geometry, states a significant property common to two sections of the same projection of a triangle. Desargues considered the situation represented here by two different sections of the projection of a triangle from the point O [Figure 2]. The relationship of the two triangles is described by saying that they are perspective from the point O. Desargues then asserted that each pair of corresponding sides of these two triangles will meet in a point, and, most important, these three points will lie on one straight line. With reference to the figure, the assertion is that AB and $A'B'$ meet in the point R; AC and $A'C'$ meet in S; BC and $B'C'$ meet in T; and that R, S, and T lie on one straight line. While in the case stated here the two triangle sections are in different planes, Desargues's assertion holds even if triangles ABC and $A'B'C'$ are in the same plane, e.g., the plane of this paper, though the proof of the theorem is different in the latter case.

The reader may be troubled about the assertion in Desargues's theorem that each pair of corresponding sides of the two triangles must meet in a point. He may ask: What about a case in which the sides happen to be parallel? Desargues disposed of such cases by invoking the mathematical convention that any set of parallel lines is to be regarded as having a point in common, which the student is often advised to think of as being at infinity—a bit of advice which essentially amounts to answering a question by not answering it. However, whether or not one can visualize this point at infinity is immaterial. It is logically possible to agree that parallel lines are to be regarded as having a point in common, which point is to be distinct from the usual, finitely located points of

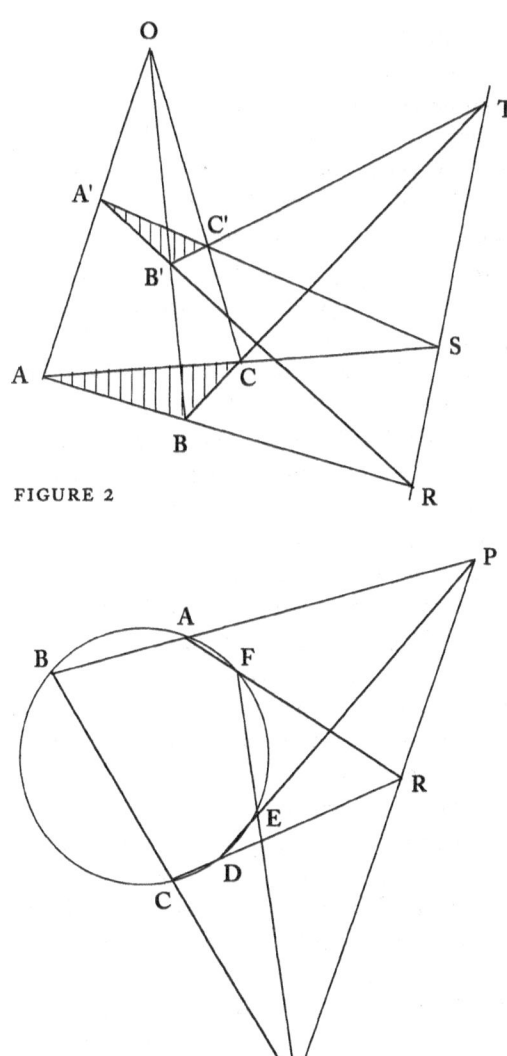

FIGURE 2

FIGURE 3

These conventions or agreements not only are logically justifiable but also are recommended by the argument that projective geometry is concerned with problems which arise from the phenomenon of vision, and we never actually see parallel lines, as the familiar example of the apparently converging railroad tracks reminds us. Indeed, the property of parallelism plays no role in projective geometry.

At the age of 16 the precocious French mathematician and philosopher Blaise Pascal, a contemporary of Desargues, formulated another major theorem in projective geometry. Pascal asserted that if the opposite sides of any hexagon inscribed in a circle are prolonged, the three points at which the extended pairs of lines meet will lie on a straight line [Figure 3].

As stated, Pascal's theorem seems to have no bearing on the subject of projection and section. However, let us visualize a projection of the figure involved in Pascal's theorem and then visualize a section of this projection [Figure 4]. The projec-

FIGURE 4

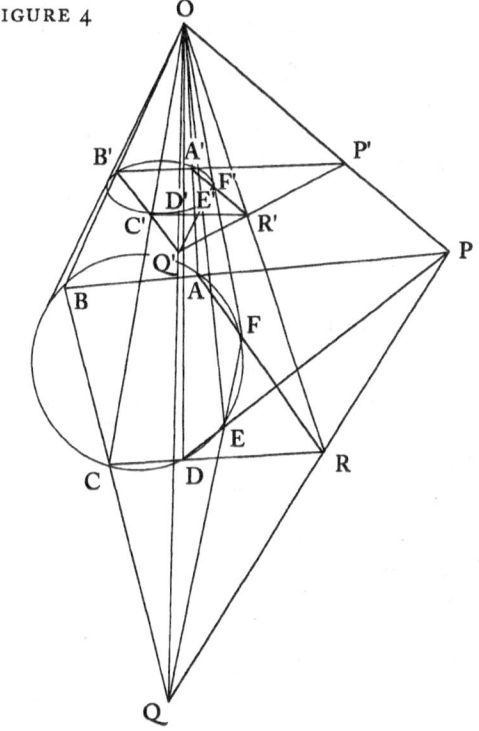

the lines considered in Euclidean geometry. In addition, it is agreed in projective geometry that all the intersection points of the different sets of parallel lines in a given plane lie on one line, sometimes called the line at infinity. Hence even if each of the three pairs of corresponding sides of the triangles involved in Desargues's theorem should consist of parallel lines, it would follow from our agreements that the three points of intersection lie on one line, the line at infinity.

tion of the circle is a cone, and in general a section of this cone will not be a circle but an ellipse, a hyperbola, or a parabola—that is, one of the curves usually called a conic section. In any conic section the hexagon in the original circle will give rise to a corresponding hexagon. Now Pascal's theorem asserts that the pairs of opposite sides of the new hexagon will meet on one straight line which corresponds to the line derived from the original figure. Thus the theorem states a property of a circle which continues to hold in any section of any projection of that circle. It is indeed a theorem of projective geometry.

It would be pleasant to relate that the theorems of Desargues and Pascal were immediately appreciated by their fellow mathematicians and that the potentialities in their methods and ideas were eagerly seized upon and further developed. Actually this pleasure is denied us. Perhaps Desargues's novel terminology baffled mathematicians of his day, just as many people today are baffled and repelled by the language of mathematics. At any rate, all of Desargues's colleagues except René Descartes exhibited the usual reaction to radical ideas: they called Desargues crazy and dismissed projective geometry. Desargues himself became discouraged and returned to the practice of architecture and engineering. Every printed copy of Desargues's book, originally published in 1639, was lost. Pascal's work on conics and his other work on projective geometry, published in 1640, also were forgotten. Fortunately a pupil of Desargues, Philippe de la Hire, made a manuscript copy of Desargues's book. In the nineteenth century this copy was picked up by accident in a bookshop by the geometer Michel Chasles, and thereby the world learned the full extent of Desargues's major work. In the meantime most of Desargues's and Pascal's discoveries had had to be remade independently by nineteenth-century geometers.

Projective geometry was revived through a series of accidents and events almost as striking as those that had originally given rise to the subject.

Gaspard Monge, the inventor of descriptive geometry, which uses projection and section, gathered about him at the Ecole Polytechnique a host of bright pupils, among them Sadi Carnot and Jean Poncelet. These men were greatly impressed by Monge's geometry. Pure geometry had been eclipsed for almost 200 years by the algebraic or analytic geometry of Descartes. They set out to show that purely geometric methods could accomplish more than Descartes's.

It was Poncelet who revived projective geometry. As an officer in Napoleon's army during the invasion of Russia, he was captured and spent the year 1813–14 in a Russian prison. There Poncelet reconstructed, without the aid of any books, all that he had learned from Monge and Carnot, and he then proceeded to create new results in projective geometry. He was perhaps the first mathematician to appreciate fully that this subject was indeed a totally new branch of mathematics. After he had re-opened the subject, a whole group of French and, later, German mathematicians went on to develop it intensively.

One of the foundations on which they built was a concept whose importance had not previously been appreciated. Consider a section of the projection of a line divided by four points [Figure 5]. Obviously the segments of the line in the section are not equal in length to those of the original line. One might venture that perhaps the ratio of two segments, say $A'C'/B'C'$, would equal the corresponding ratio AC/BC. This conjecture is incorrect. But the surprising fact is that the ratio of the ratios, namely $(A'C'/C'B') / (A'D'/D'B')$, will equal $(AC/CB) / (AD/DB)$. Thus this ratio of ratios, or cross ratio as it is called, is a projective invariant. It is necessary to note only that the lengths involved must be directed lengths; that is, if the direction from A to D is positive, then the length AD is positive but the length DB must be taken as negative.

The fact that any line intersecting the four lines OA, OB, OC, and OD contains segments possessing the same cross ratio as the original segments suggests that we assign to the four projection lines meeting in the point O a particular cross ratio, namely the cross ratio of the segments on any section. Moreover, the cross ratio of the four lines is a projective invariant, that is, if a projection of these four lines is formed and a section made of this projection, the section will contain four concurrent lines whose cross ratio is the same as that of the original four [Figure 6]. Here in the section $O'A'B'C'D'$, formed in the projection of the figure $OABCD$ from the point O'', the four lines $O'A'$, $O'B'$, $O'C'$, and $O'D'$ have the same cross ratio as OA, OB, OC, and OD.

The projective invariance of cross ratio was put to extensive use by the nineteenth-century geometers. We noted earlier in connection with Pascal's theorem that under projection and section a circle may become an ellipse, a hyperbola, or a parabola, that is, any one of the conic sections. The geometers sought some common property which would account for the fact that a conic section always gave rise to a conic section, and they found the answer in terms of cross ratio. Given the points O, A, B, C, D, and a sixth point P on a conic section containing the others [Figure 7], then a remarkable theorem of projective geometry states that the lines PA, PB, PC, and PD have the same cross ratio as OA, OB, OC, and OD. Conversely, if P is any point such that PA, PB, PC, and PD have the same cross ratio as OA, OB, OC, and OD, then P must lie on the conic through O, A, B, C, and D. The essential point of this theorem and its converse is that a conic section is determined by the property of cross ratio. This new characterization of a conic was most welcome, not only because it utilized a projective property but also because it opened up a whole new line of investigation on the theory of conics.

The satisfying accomplishments of projective geometry were capped by the discovery of one of the most beautiful principles of all mathematics— the principle of duality. It is true in projective geometry, as in Euclidean geometry, that any two points determine one line, or as we prefer to put it,

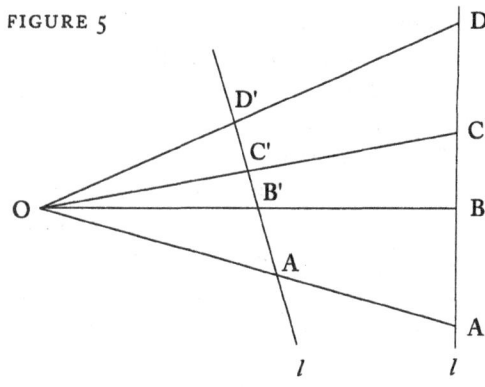

FIGURE 5

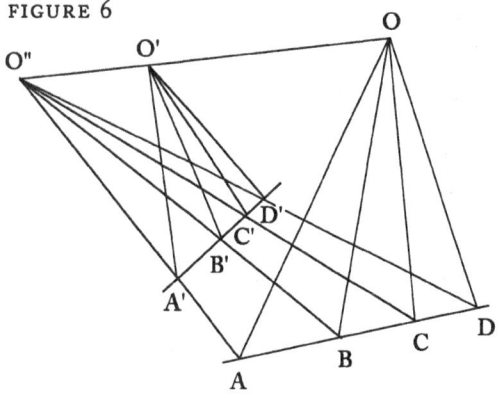

FIGURE 6

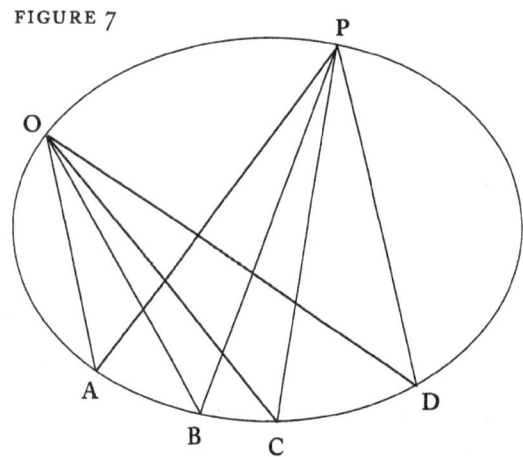

FIGURE 7

any two points lie on one line. But it is also true in projective geometry that any two lines determine, or lie on, one point. (The reader who has refused to accept the convention that parallel lines in Euclid's sense are also to be regarded as having a point in common will have to forego the next few paragraphs and pay for his stubbornness.) It will be noted that the second statement can be obtained from the first merely by interchanging the words point and line. We say in projective geometry that we have dualized the original statement. Thus we can speak not only of a set of points on a line but also of a set of lines on a point [Figure 8]. Likewise the dual of the figure consisting of four points no three of which lie on the same line is a figure of four lines no three of which lie on the same point [Figure 9].

Let us attempt this rephrasing for a slightly more complicated figure. A triangle consists of three points not all on the same line and the lines joining these points. The dual statement would read: three lines not all on the same point and the points joining them (that is, the points in which the lines intersect). The figure we get by rephrasing the definition of a triangle is again a triangle, and so the triangle is called self-dual.

Now let us rephrase Desargues's theorem in dual terms, using the fact that the dual of a triangle is a triangle and assuming in this case that the two triangles and the point O lie in one plane. The theorem says:

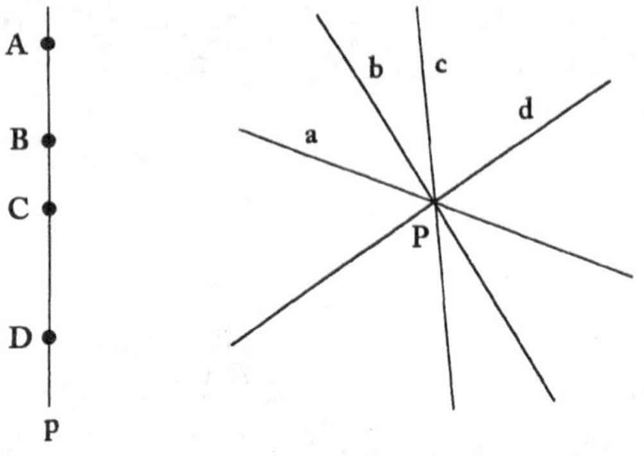

"If we have two triangles such that lines joining corresponding vertices pass through one point O, then the pairs of corresponding sides of the two triangles join in three points lying on one straight line."

Its dual reads:

"If we have two triangles such that points which are the joins of corresponding sides lie on one line O, then the pairs of corresponding vertices of the two triangles are joined by three lines lying on one point."

We see that the dual statement is really the converse of Desargues's theorem, that is, it is the result of interchanging his hypothesis with his conclusion. Hence by interchanging point and line we have discovered the statement of a new theorem. It would be too much to ask that the proof of the new theorem should be obtainable from the proof of the old one by interchanging point and line. But if it is too much to ask, the gods have been generous beyond our merits, for the new proof can be obtained in precisely this way.

Projective geometry also deals with curves. How should one dualize a statement involving curves? The clue lies in the fact that a curve is after all but a collection of points; we may think of a figure dual to a given curve as a collection of lines. And indeed a collection of lines which satisfies the condition dual to that satisfied by a conic section turns out to be the set of tangents to that curve [Figure 10]. If the conic section is a circle, the dual figure is the collection of tangents to the circle [Figure 11]. This collection of tangents suggests the circle as well as does the usual collection of

FIGURE 8

FIGURE 9

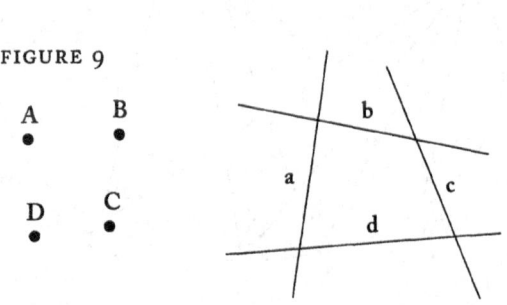

FIGURE 10

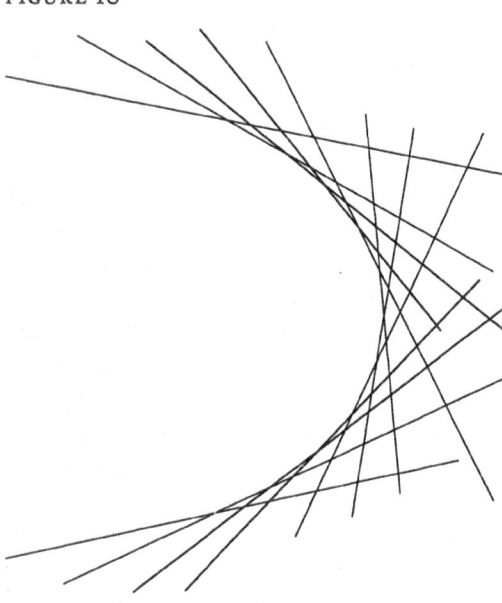

FIGURE 10

FIGURE 11

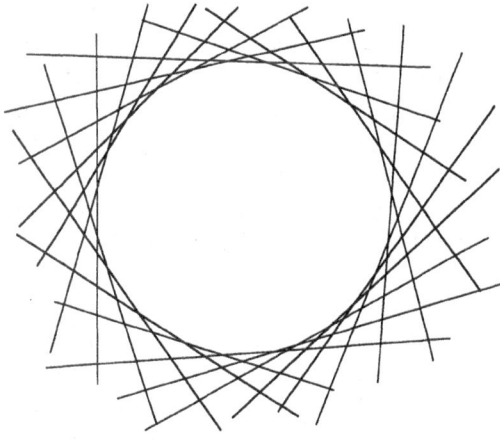

which join *C* and *D* and *F* and *A* join in a point *R*. The three points *P*, *Q*, and *R* lie on one line *l*."

Its dual reads:

"If we take six lines, *a*, *b*, *c*, *d*, *e*, and *f*, on the line circle, then the points which join *a* and *b* and *d* and *e* are joined by the line *p*; the points which join *b* and *c* and *e* and *f* are joined by the line *q*; the points which join *c* and *d* and *f* and *a* are joined by the line *r*. The three lines *p*, *q*, and *r* lie on one point *L*."

The geometric meaning of the dual statement amounts to this: Since the line circle is the collection of tangents to the point circle, the six lines on the line circle are any six tangents to the point circle, and these six tangents form a hexagon circumscribed about the point circle. Hence the dual statement tells us that if we circumscribe a hexagon about a point circle, the lines joining opposite vertices of the hexagon, lines *p*, *q*, and *r* in the dual statement, meet in one point [Figure 12]. This dual statement is indeed a theorem of projective geometry. It is called Brianchon's theorem, after Monge's student Charles Brianchon, who discovered it by applying the principle of duality to Pascal's theorem pretty much as we have done.

It is possible to show by a single proof that every rephrasing of a theorem of projective geometry in accordance with the principle of duality must lead to a new theorem. This principle is a remarkable possession of projective geometry. It reveals the symmetry in the roles that point and

FIGURE 12

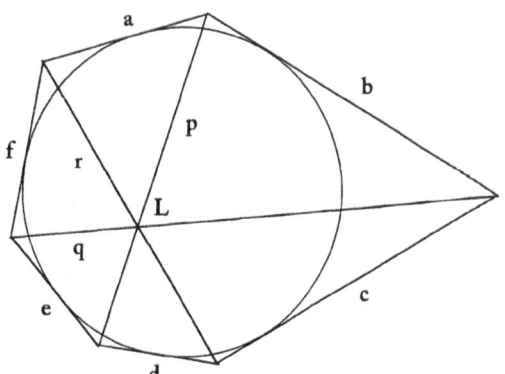

points, and we shall call the collection of tangents the line circle.

Let us now dualize Pascal's theorem on the hexagon in a circle. His theorem goes:

"If we take six points, *A*, *B*, *C*, *D*, *E*, and *F*, on the point circle, then the lines which join *A* and *B* and *D* and *E* join in a point *P*; the lines which join *B* and *C* and *E* and *F* join in a point *Q*; the lines

line play in the structure of that geometry. The principle of duality also gives us insight into the process of creating mathematics. Whereas the discovery of this principle, as well as of theorems such as Desargues's and Pascal's, calls for imagination and genius, the discovery of new theorems by means of the principle is an almost mechanical procedure.

As one might suspect, projective geometry turns out to be more fundamental than Euclidean geometry. The clue to the relationship between the two geometries may be obtained by again considering projection and section. Consider the projection of a rectangle and a section in a plane parallel to the rectangle [Figure 13]. The section is a rectangle similar to the original one. If now the point O moves off indefinitely far to the left, the lines of the projection come closer and closer to parallelism with each other. When these lines become parallel and the center of the projection is the "point at infinity," the rectangles become not merely similar but congruent [Figure 14]. In other words, from the standpoint of projective geometry the relationships of congruence and similarity, which are so intensively studied in Euclidean geometry, can be studied through projection and section for special projections.

FIGURE 13

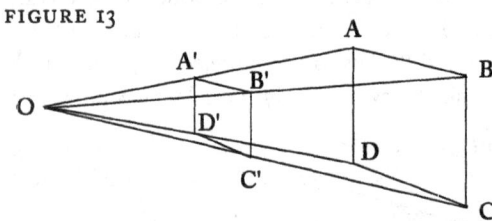

FIGURE 14

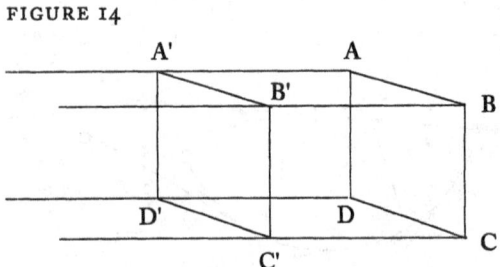

If projective geometry is indeed logically fundamental to Euclidean geometry, then all the concepts of the latter geometry should be defined in terms of projective concepts. However, in projective geometry as described so far there is a logical blemish: our definition of cross ratio, and hence concepts based on cross ratio, rely on the notion of length, which should play no role in projective geometry proper because length is not an invariant under arbitrary projection and section. The nineteenth-century geometer Felix Klein removed this blemish. He showed how to define length as well as the size of angles entirely in terms of projective concepts. Hence it became possible to affirm that projective geometry was indeed logically prior to Euclidean geometry and that the latter could be built up as a special case. Both Klein and Arthur Cayley even showed that the basic non-Euclidean geometries could be derived as special cases of projective geometry. No wonder that Cayley exclaimed: "Projective geometry is all geometry!"

It remained only to deduce the theorems of Euclidean and non-Euclidean geometry from axioms of projective geometry, and this geometers succeeded in doing in the late nineteenth and early twentieth centuries. What Euclid did to organize the work of three hundred years preceding his time, the projective geometers did recently for the investigations which Desargues and Pascal initiated.

Research in projective geometry is now less active. Geometers are seeking to find simpler axioms and more elegant proofs. Some research is concerned with projective geometry in n-dimensional space. A vast new allied field is projective differential geometry, concerned with local or infinitesimal properties of curves and surfaces.

Projective geometry has had an important bearing on current mathematical research in several other fields. Projection and section amount to what is called in mathematics a transformation, and it seeks invariants under this transformation. Mathematicians asked: Are there other transformations more general than projection and section whose invariants might be studied? In recent times one new geometry has been developed by pursuing this line of thought, namely, topology. It would take us too far afield to consider topological transformations. It must suffice here to state that topology considers transformations more general than projection and section and that it is now clear that topology is logically prior to projective geometry. Cayley was too hasty in affirming that projective geometry is all geometry.

The work of the projective geometers has had an important influence on modern physical science. They prepared the way for the workers in the theory of relativity, who sought laws of the universe that were invariant under transformation from the coordinate system of one observer to that of another. It was the projective geometers and other mathematicians who invented the calculus of tensors, which proved to be the most convenient means for expressing invariant scientific laws.

It is of course true that the algebra of differential equations and some other branches of mathematics have contributed more to the advancement of science than has projective geometry. But no branch of mathematics competes with projective geometry in originality of ideas, coordination of intuition in discovery and rigor in proof, purity of thought, logical finish, elegance of proofs, and comprehensiveness of concepts. The science born of art proved to be an art.

20

Pisa, Galileo, Rome

EDMOND R. KIELY

*I*N PROFESSOR W. R. CARNAHAN's article, "Mathematics an Essential of Culture," which appeared in the November, 1950 issue of this magazine, there occur certain oft-repeated statements which, in the light of recent research, appear in need of revision. These statements concern Galileo's experiments with falling bodies from atop the Tower of Pisa and Galileo's clash with ecclesiastical authorities. As a matter of fact the reiterated statements concerning Galileo and Aristotle, and Galileo and the Catholic Church tend to obscure Galileo's real merits as a scientist. Even in Professor Carnahan's article we do not have as detailed a discussion of Galileo's contributions to the science of mechanics, on which his right to a place in the history of science properly rests, as we should have expected.

Concerning "the famous gravity demonstration at the Tower of Pisa" the writer would like to recommend an admirable little book by Professor Cooper.[1] There one learns that the story of these experiments was first written by Viviani, Galileo's earliest biographer, about twelve years after Galileo's death and more than sixty years after the assumed date (ca. 1590) of the experiments. Since Galileo meticulously committed to writing and published[2] the results of his physical experiments and deductions it is strange that he should have omitted any mention of these "famous" experiments. It is true that in Galileo's writings we find

the material on which, in all probability, Viviani built his story. In criticizing Aristotle's ideas on motion Galileo asks the following question in *De Motu:*

> . . . if two stones were flung at the same moment from a high tower, one stone twice the size of the other, who would believe that when the smaller was half-way down the larger had already reached the ground?[3]

Again in Galileo's last and most important work, *Discorsi e Dimostrazioni Mathematiche intorno a Due Nuove Scienze*, we are told in the words of Salviati, one of the characters carrying on the dialogue:

> Aristotle says "An iron ball of one hundred pounds, falling from a height of one hundred cubits, reaches the ground before a one-pound ball has fallen a single cubit." I say that they arrive at the same time. You find on making the experiment, that the larger precedes the smaller by two finger-breadths; that is, when the larger one has struck the ground, the other is short of it by two fingers. Now you would not conceal behind these two fingers the ninety-nine cubits of Aristotle."[4]

Nowhere does Galileo state that he himself carried out such experiments. But we know definitely that others did.[5]

However, more important than the actual performance of such experiments is the defamation which the story throws on Aristotle as a physicist. Nowhere in Aristotle's works can be found the statement attributed to him by Galileo through the mouth of Salviati and which has been continuously associated with the various ramifications of Viviani's story. Arguments against Aristotle are usually based on quotations from his *Physica* and *De Caelo* where he proposes proofs for the non-

Reprinted from *Mathematics Teacher* 45 (Feb., 1952): 173–82; with permission of the National Council of Teachers of Mathematics.

existence of a void or vacuum. But as Hardcastle pointed out:

> What he [Aristotle] taught was that the terminal velocity of a heavy body was greater than the terminal velocity of a light body. . . . Aristotle throughout treats only of the motion of projectiles, and of that only in a resisting medium, and then only of that part of the vertical motion when the projectile has attained that constant speed known to ballisticians as "terminal velocity" which can be as readily observed in rising smoke as in falling rain.[6]

One from many possible quotations from Aristotle should suffice to substantiate this interpretation:

> We see the same weight or body moving faster than another for two reasons, either because there is a difference in what it moves through, as between water, air and earth, or because, other things being equal, the moving body differs from the other owing to excess of weight or of lightness. Now the medium causes a difference because it impedes the moving thing, most of all if it is moving in the opposite direction, but in a secondary degree even if it is at rest, and especially a medium that is not easily divided, i.e., a medium that is somewhat dense.[7]

It is true there are passages in Aristotle's physics which seem to offer indirect evidence in proof of Galileo's viewpoint. But it must be remembered that, first and foremost, Aristotle was a realist, and not a dreamer, as he has been sometimes portrayed. From ordinary experience it should have been obvious to him that similar bodies of different weight would fall through a given distance in practically the same time. It is difficult to imagine that the man whose physical theories have recently been so highly estimated by Professor Boyer[8] could have decided otherwise and with such a discrepancy from reality. It may be pointed out here that Aristotle's meaning in many passages has been obscured. Through errors in the copying of early manuscripts and the difficulty of determining the actual meaning of many of the technical terms employed a variety of translations is possible, as may be seen by even a comparison of the translations given here with those in The Loeb Classical Library editions.

In any examination of Aristotle's physical theories it is important to bear in mind, as Professor Boyer points out, that to the Greeks "*physis* meaning the essence or nature of things, was more concerned with the explanation of essential properties and relations than with quantitative description." Professor Boyer's estimate of Aristotle's scientific achievements is of importance in the light of later developments in the physical sciences:

> The physical science of Aristotle is a coherent and systematic treatment which, while less accurate than that of Archimedes, is of far wider scope. In method there is a striking similarity between these two great scientists of antiquity. Both men began with careful observation unaided by the use of instruments, framed inductive generalizations and with these as premises built a deductive science. But while Archimedes, a mathematician, limited himself to a few observations comparable in exactitude to the axioms of Euclid's *Elements* of geometry, the philosopher Aristotle surveyed the whole of nature so that he might disclose the rational order in cause and purpose. For this reason Aristotle placed the inner consistency of his system above accuracy in detail, and in his eager search for certainty he failed to exercise the needed suspension of judgment.[9]

Criticisms of Aristotle's statements on falling bodies and other sections of his physics did not begin with Galileo, they date back at least as far as the period of Philoponus in the sixth century. The modern concept of the content of physics as a science which seeks proximate causes in contrast to Aristotle's concept which embraced ultimate as well as immediate causes came prominently to the fore during the thirteenth century. Two important innovations of this century were the increased emphasis on the mathematical or quantitative aspects of the subject matter of physics and the systematic development of the experimental method for the verification of scientific hypotheses. The earliest important depositions on these fundamental changes are to be found in the writings of Bishop Grosseteste of Lincoln,[10] for many years prominently connected with Oxford University. But it was in connection with the quantitative study of motion that the most fundamental rupture occurred between the advocates of the new physics and the adherents of Aristotelian physics. Through the critical attacks on Aristotelian ideas on motion, carried on particularly at the Universities of Oxford and Paris by such men as Occam, Buridan, Duns Scotus, Suiseth, Bradwardine, Roger Bacon, Albert of Saxony, and Bishop Nicholaus of Oresme, there arose the mathematical theory of accelerated motion which reached its final form with the development of the calculus by Newton and Leibniz, and also the theory of impetus which Descartes was later to develop into the theory of inertia.[11] In the meantime, during the fifteenth and sixteenth centuries, the study of natural phenomena by the so-called adherents of Aristotelian-Scholasticism had sunk to a low ebb and as a result the whole science of Aristotle was repudiated. As Whittaker, the well-known British mathematical-physicist, has expressed it:

> From the fourteenth century onwards, Scholasticism was decadent and by the end of the sixteenth it had become thoroughly debased. The love of nature that had been so vital in Aristotle had almost perished; the practice of observation and experiment, on which he and St. Thomas had so strongly insisted, was neglected save by a few solitary workers; and the degenerate Schoolmen occupied themselves with futile subtleties that bore no relation to life and reality.[12]

Or again quoting Professor Boyer:

> By the time of Galileo and Newton the science of Aristotle had been thoroughly discredited. It became the fashion to ridicule Aristotelian science for its errors, more putative than real.[13]

The fashion still seems to persist, but in view of the high esteem in which the work of Aristotle is held by many modern historians of the physical sciences and mathematics, future critics of Aristotelian physics should be deterred from endeavoring to bury "The Philosopher" beneath hundred-pound balls dropped from atop the Tower of Pisa. All the more so when it is kept in mind that the greatest derider of Aristotle in the seventeenth

century was himself a rather haphazard physicist even by medieval standards. It is rather difficult to overlook Galileo's disregard of experimental verification as exemplified in the following:

> Galileo worked out his physics by thought, by correct reasoning and mathematics, not by induction from experiments. During his days at Pisa, before he went to Padua, he wrote: "But, as ever, we employ reason more than examples (for we seek the causes of effects, and these are not revealed by experiment)." Galileo liked to use what he called "Thought Experiments," imagining the consequences rather than observing them directly. Indeed, when he described the motion of a ball dropped from the mast of a moving ship, in his *Dialogue on the Two Great Systems of the World*, he then had the Aristotelian, Simplicio, ask whether he had made an experiment, to which Galileo replied: "No, and I do not need it, as without any experience I can affirm that it is so, because it cannot be otherwise."[14]

Before considering Galileo's astronomical work it is necessary to clarify the meaning of scientific truth as developed by Grosseteste and the Medieval Schoolmen who broke away from Aristotelian concepts of science in the thirteenth century. To them physics or more generally the physical sciences, as such, were no longer concerned with ultimate reality. All questions on primal substances, essences, and ultimate causes were relegated to the science of metaphysics. The objective of the physical sciences was henceforth to answer questions on the immediate efficient causes of phenomena. Briefly, the usual order of the process by which such answers were to be determined consists, in the first place, in intelligent observation. The second step is the setting up of hypotheses as a result of inductive thought on these observations. As part of such hypotheses or auxiliary to them, there may be set up a mathematical or mechanical model to simplify and clarify the hypotheses. The third step consists in the execution of experiments (controlled if possible) which test the hypotheses directly or deductions from them. The more experimental evidence acquired in favor of the hypotheses the stronger the hypotheses become. One piece of confirmed experimental evidence which contradicts a hypothesis is sufficient to cause either the complete rejection or the modification of the hypothesis. Finally, a hypothesis which has been proved to be in agreement with experimental evidence or, as more generally happens, a combination of such hypotheses, helps to formulate by logical deduction into an intelligible system or set of laws a series of physical facts which in themselves are seemingly isolated. These laws should be suggestive of further experimentation and expansion in accordance with the deductions which can logically be derived from them; such deductions are usually expressible in mathematical form. Our scientific truths are contained in such laws which, because of the method of derivation, can never be final or absolute but must always remain subject to revision. It is also possible to set up hypotheses on purely intuitional ideas, but the evaluation of all hypotheses depends primarily on the amount of agreement between them and the phenomena which they are supposed to represent as disclosed by experimentation.

The history of scientific progress or man's endeavor to improve his knowledge of nature is marked by the tombstones of abandoned hypotheses. What was thought to be true at one period is later shown, usually as a result of better experimentation, to be either totally at variance with the natural phenomena—in which case the former knowledge must be considered as false—or only partially in agreement with phenomena—in which case there is proportionally partial truth. A scientific theory is true only in so far as it remains uncontradicted by experimentation. Two or more hypotheses on the same phenomena could, in this sense, be considered as simultaneously true. Since experimentation depends on measurement and measurement of its nature is subject to error, scientific knowledge, as now conceived, cannot aspire to produce the complete truth about any natural phenomena.[15] If two or more hypotheses appear to agree equally well with observable facts or, as the Greeks said, "save the phenomena," a scien-

tist is likely to choose the simpler hypothesis, not because it is more significant of reality but because it more easily leads to new experiment. A simpler hypothesis is not necessarily in better agreement with the actual facts than a more complex one; neither is the simpler mathematical relationship of a physical law necessarily closer to actuality than a more complex one, even though this idea, Pythagorean in origin, has often supplied the necessary impetus to formulate new hypotheses.[16]

In 1609 when Galileo's astronomical work really began he had a choice of three hypothetical systems recently proposed as replacements for the Ptolemaic system. Up to the fifteenth century Europeans, in general, found that the geocentric system, as explained in such popular treatises as the *Sphere* of Sacrobosco, was sufficiently accurate for the needs of the times. But by the sixteenth century matters had considerably changed. The tremendous impetus given to navigation in the previous century had made acute the need for more accurate astronomical data. Developments in surveying and navigation instruments had reciprocally aided in the production of better instruments for astronomical observation. The work of the Portuguese school of navigation, established in 1416, and the astronomical studies of the mid-Europeans under the influence of Cardinal Nicholaus de Cusa, Purbach, and Regiomontanus had prepared the way for new developments. Another impetus was given by the need felt by the Catholic Church authorities for a reform of the calendar. It was becoming more and more apparent that the complicated Ptolemaic system was unsatisfactory.

Such were the conditions under which Copernicus undertook to promulgate once more the heliocentric hypothesis of Aristarchus. He was actuated primarily by the desire to provide a simpler mathematical model than that of the eighty-odd spheres necessary for explanation and calculation in the Ptolemaic system. For Copernicus, the heliocentric system represented nothing more than a simplified mathematical hypothesis of thirty-four spherical orbits which would also serve to save the phenomena. For him it was not the result of induction from observational data. He spent many years of his life pondering over this mathematical hypothesis but did little practical observation. Whereas in terms of Aristotelian physics there had been a plausible explanation of the Ptolemaic system, Copernicus was unable to offer any acceptable scientific explanation of his system either in terms of the anti-Aristotelian physics of motion or in terms of the older concepts of motion. For this reason and also on religious grounds Tycho Brahe, the greatest astronomical observer of the ensuing period, rejected the Copernican hypothesis, but since he also saw defects in the Ptolemaic system, he established a compromise system of his own. Kepler, the great computer and enthusiastic follower of Pythagorean number mysticism, who fell heir to the observational data of Brahe was an ardent supporter of the heliocentric system, but found it necessary to change Copernicus' circular orbits into elliptical ones so that the mathematical model would be in better agreement with the calculations he had laboriously worked out from the data of Brahe. All three systems still needed explanation and confirmation in terms of anti-Aristotelian physics. Let us now examine what Galileo actually did with these hypotheses.

Galileo claimed and most of his biographers to date have also claimed that, as a result of his astronomical observations and experiments on motion, he had proved the Aristarchian-Copernican hypothesis to be scientifically true. If this were so he should have been able to produce experimental data either in proof of the two main physical facts implied by the hypothesis, namely, that the earth revolved daily on its own axis and that it revolved annually round the sun, or at least in proof of deductions from these facts. Such proofs, as given by Galileo, are to be found in his *Dialogo sopra i due Massimi Sistemi del Mondo, Tolemaico e Copernicano*.[17] In 1597 Galileo stated in a letter to Kepler that he was a firm believer in the Copernican system; therefore, *The Two Systems* may justly

be considered the result of thirty-three years of research with the objective of establishing the scientific truth of the hypothesis. These proofs as given by Galileo break down into three groups. The first group is in terms of the physics of motion. They consist, in the main, of arguments against Aristotelian ideas on motion which had been bandied across Europe for centuries since the rise of anti-Aristotelian physics. Despite his work on dynamics Galileo was unable to furnish a convincing argument in favor of either motion of the earth because he had failed to bring to completion the theory of impetus. With regard to Galileo's proofs from motion Butterfield points out:

> In his mechanics he was a little less original than most people imagine, since, apart from the older teachers of the impetus theory, he had had more immediate precursors, who had begun to develop the more modern views concerning the flight of projectiles, the law of inertia and the behavior of falling bodies. He was not original when he showed that clouds and air and everything on the earth—including falling bodies—moved around with the rotating earth, as part of the same mechanical system. . . . His system of mechanics did not quite come out clear and clean, did not even quite explicitly reach the modern law of inertia, since even here he had not quite disentangled himself from obsessions concerning circular motion.[18]

Galileo's second group of proofs are in terms of his telescopic discoveries. From observations of sun-spots he concluded that the sun was revolving on its axis. Parenthetically it may be mentioned that the revolution of the sun on its own axis was not part of the Copernican hypothesis. Galileo discovered that the moons of Jupiter revolved round that planet and explained the phases of Venus in terms of its revolution round the sun. Granted that his explanations of these phenomena were correct, they are nothing but proofs from analogy so far as the motions of the earth are concerned and could not be accepted as proofs from experiment in any system of physical sciences. Galileo knew only too well that these first two groups of proofs would not and could not be acceptable as scientific proofs for the Copernican hypothesis, and therefore, as he says himself, he reserved to the end of his book his clinching argument for the rotation of the earth. This, his main proof, turns out to be nothing more than the erroneous doctrine that the main factor in the production of the tides was the diurnal rotation of the earth.

> I have been induced upon no slight reasons to omit these two conclusions (having made withal the necessary presupposals) that in case the terrestrial Globe be immoveable, the flux and reflux of the Sea cannot be natural; and that, in case these motions be conferred upon the said Globe, which have been long since assigned to it, it is necessary that the Sea be subject to ebbing and flowing, according to all that which we observe to happen in the same.[19]

It should be borne in mind that Galileo arrived at this conclusion despite the fact that Kepler and others had, years before, suggested lunar influences as the main cause of the tides. As an argument against one of these lunar advocates Galileo writes:

> To that Prelate I would say that the Moon moveth every day along the whole Mediterrane, and yet its waters do not rise thereupon, save only in the very extream bounds of it Eastward and here to us at Venice.[20]

Is it then any wonder that the following estimate of Galileo's astronomical work is to be found in the *Encyclopaedia Britannica:*

> The direct services of permanent value which Galileo rendered to astronomy are virtually summed up in his telescopic discoveries. To the theoretical perfection of the science he contributed little or nothing.[21]

Butterfield's recent conclusion on the value of Galileo's proofs is unequivocal:

> At the end of everything Galileo failed to clinch his argument—he did not exactly prove the rotation of the earth—and in the resulting situation a reader could adopt his whole way of looking at things or could reject it *in toto*—it was a question of entering into the whole realm of thought into which he had transposed the question.[22]

As an astronomer Galileo ignored the work of Kepler whose first and second mathematical hypotheses of planetary motion were made known in 1609, to be followed by a third ten years later. He also ignored the life-time accumulation of observational data of Brahe, and as Dreyer points out:

> In the whole book [*The Two Systems*] there is no allusion whatever to the Tychonic system although it is scarcely too much to say that about the year 1600 nobody, whose opinion was worth caring about, preferred the Ptolemaic to the Tychonic system.[23]

As a matter of historic fact the heliocentric hypothesis was not even completely plausible until Newton had explained it in terms of his synthesis of mathematics, mechanics, gravitation through a hypothetical medium, and Kepler's elliptical orbits. Newton's explanation of the solar system appeared in 1687 in his famous *Principia*. The first experimental proof that the earth was revolving on its own axis occurred, more or less accidentally, when the French astronomer Jean Richer found that his pendulum clock, regulated for Paris, lost about two and a half minutes per day when set up at Cayenne, South America, in 1671. The change in the time of the pendulum oscillation could be explained in terms of a revolving, oblate, spheroidal earth flattened at the poles, an idea which both Newton and Huygens had derived by mathematical deduction. The conclusion from Richer's experience was contradicted in the ensuing years by the Cassini surveys, but was later reaffirmed by the more elaborate surveys carried on by the French Academy of Science in Peru, Ecuador, and Lapland during the years 1735 to 1743. Experimental proof of the earth's rotation round the sun was a more difficult problem. The first tangible evidence for it came when the astronomer Bradley had observed and calculated the phenomena of aberration of light in 1729.

In view of all this evidence it is not surprising that theologians of the seventeenth century should have been reluctant to give unequivocal consent to Galileo's arguments in *The Two Systems* as proof of the scientific truth of the Copernican hypothesis.

This is not the place to reiterate at length the facts leading to Galileo's condemnation by an ecclesiastical commission. If any reader wishes a clear, concise account of the whole affair he will find it in an article by Father Conway.[24] If he wishes for more details he has but to delve into the works of Von Gebler[25] and Favaro.[26] Here it will suffice to state that Galileo was twice tried by Church tribunals, the first time at his own request. As a result of the 1616 trial he was admonished privately by Cardinal Bellarmine, at the request of Pope Paul V, to abandon his opinions on the heliocentric system. To this admonition Galileo aquiesced. In 1633 he was publicly condemned after the publication of *The Two Systems*. The primary reason for the first trial was Galileo's provocative attitude on the question of interpretation of the Scriptures. Professor Carnahan's statement that "Theologians discovered that their doctrines and Galileo's science were not in agreement" is a strange twist to put on the controversy, but again, oft-repeated like the Tower of Pisa story. Far nearer the truth is the statement by Strong in his study of the development of science in the sixteenth and seventeenth centuries:

> The furore raised against Galileo in connection with the Copernican hypothesis cannot be taken to be the attitude of the Church against the mathematical-physical sciences in general. The pursuit of physical science involved Galileo in controversies with men opposed to his physical views, but this was not a clash over the religious consequences involved in astronomical questions.[27]

The results of both trials show that the examiners were satisfied that Galileo's proofs were insufficient to raise the Copernican hypothesis to the level of a scientific truth. With these decisions the calmer judgments of modern scientists must corroborate.

If Galileo had presented his scientific theories without being so arrogant as regards the positiveness of his proofs, and had avoided being drawn into the trap of Scriptural controversy, set up by some of his more wily opponents after the publication in 1613 of his *Istoria e Dimostrazioni intorno alle Macchie Solari*[28] in which he first publicly stated his decided Copernican views, the trial of 1616 would probably never have occurred. And if there had been no directive from the Congregation of the Index, as a result of the 1616 trial, that Copernicus' book was to be banned until its hypothetical nature was made clearer—which correction was carried out within a short while—the inquiry of 1633 before the Inquisition would have been unnecessary. At the second trial Galileo was accused of breaking his promise of 1616, since any unintelligent person could see that *The Two Systems* in no way favored the hypothetical nature of the Copernican system, despite the ruse which Galileo used in trying to cover up his real opinions on the subject. That there are reasonable grounds for this assumption may be seen from the attitude of Cardinal Bellarmine who was regarded as anti-Copernican. In a letter written to the Carmelite, Father Foscarini, an advocate of Copernicanism, shortly before the 1616 trial he states:

> I say that when there shall be a real demonstration that the sun stands in the center of the universe and the earth revolves around it, it will then be necessary to proceed with great consideration in explaining those passages of Scripture which seem to be contrary to it, and rather to say we do not understand them, than to say that a thing which is demonstrated is false. But I will not believe that there is such proof until it is shown to me; nor is it the same thing to show that the phenomena are saved by assuming that the sun is in the center and the earth revolves round it, and to show that in reality the sun is in the center and the earth revolves.[29]

These are the words of an outstanding theologian of the period but surely, no one can read into them that he is trying to direct the course of scientific investigation so as to make it conform to a religious belief; rather is it the statement of a careful scientist. Similar opinions can be found in the letters of many prominent Catholic ecclesiastics of the period.

Besides, since the publication of Copernicus' *De Revolutionibus Orbium Coelestium* in 1543 there had been no opposition to the heliocentric hypothesis on the part of the Catholic Church; whatever opposition existed came from individuals of various religious persuasions. In fact many of the higher Catholic clergy were known to favor it openly. Was it not through the instrumentality of the Catholic clergy that the Aristarchian heliocentric system had been revived, first by Cardinal Nicholaus de Cusa in his *De Docta Ignorantia* (ca. 1450) and again, nearly a century later, by the canon Copernicus? When Galileo visited Rome in 1611 to demonstrate his telescopes he was enthusiastically received by the Church authorities and especially by the future Pope Urban VIII. It is also important to notice that, despite the 1616 trial, Galileo's ecclesiastical supporters did not abandon him; he even became the recipient of a life pension from his very personal friend Urban VIII in 1624, a reward for his scientific endeavors, setting the precedent, in modern times, for scientific stipends! There was no action on the part of the Catholic Church during this period to show that it was in opposition to scientific progress. Any unbiased history of the period shows the opposite to be true. Nevertheless Galileo knew that, unless he could produce the experimental evidence, anything beyond a hypothetical statement of the Copernican system would run counter to the then accepted interpretation of the Scriptures. He also knew that there was nothing in Catholic theology or tradition against the Church's sanctioning a change in interpretation if necessity arose. What really led to Galileo's troubles was the method he adopted to bring about this change. Like many another highly gifted man who lacked tact he found himself before long antagonizing people who were formerly his best friends. If he had approached the matter cautiously and with less demand that his reasoning be accepted as demonstrative, which it was not, Cardinal Bellarmine and others would undoubtedly have used their influence to allow Galileo to pursue his astronomical work unmolested by any clerical interference, and the Church authorities would have been his best supporters if he had ever

reached the stage of producing what the Cardinal called "a real demonstration," that is an actual proof by experimentation of some deduction from the hypothesis.

It is often assumed that the philosophical system of Aristotle and St. Thomas Aquinas was irreconcilable with the researches undertaken by Galileo. But just as Aristotle was admittedly more realistic in certain respects than Galileo, so also St. Thomas might have given him timely—and modern—advice on the necessity of distinguishing between a scientific hypothesis and a scientific demonstration, as the following excerpt shows:

> The assumptions made by the astronomers are not necessarily true. Although these hypotheses seem to be in agreement with the observed phenomena we must not claim that they are true. Perhaps one could explain the observed motion of the celestial bodies in a different way which has not been discovered up to this time.[30]

A recent statement of Whittaker on the influence of the work of Brahe and Kepler is also apposite here:

> At this point it may be observed that, while the Scholastic cosmology was thereby completely disproved and overthrown, there was nothing in the new methods and discoveries that was inherently irreconcilable with the Scholastic metaphysics; the whole of Tycho's and Kepler's work might conceivably have been absorbed into the philosophy of the Schoolmen by a peaceful and conservative revolution. If this had happened, we in the twentieth century should have been spared the necessity of readjusting our position by a movement back towards Aristotelianism.[31]

But this did not happen because of the lack of true Aristotelianism at the time. Statements found in many of our popular histories of science to the effect that the experimental work of Galileo was completely at variance with the methods of the Aristotelian philosophers of the Italian universities of the seventeenth century are a calumny on Aristotle. A professor of natural philosophy who would refuse to look through a telescope, as some of Galileo's scientific opponents did, should not for that reason be classified as a follower of Aristotle. Not only was Aristotle one of the greatest exemplars of the use of deductive logic but, and this is too often forgotten, he was also well aware of the value and pitfalls of induction, as a reading of the first section of his *Metaphysics* will show. His insistence on the necessity of observation and experiment may be seen from what he has to say of those who

> . . . had certain predetermined views and were resolved to bring everything into line with them. . . . As though some principles did not require to be judged from their results, and particularly from their final issue! And that issue, which in the case of productive knowledge (i.e. in the case of art) is the product, in the knowledge of nature is the unimpeachable evidence of the senses as to each fact.[32]

Why then do we continue to associate the name of Aristotle with a system of natural philosophy which had become atrophied by lack of experimentation? The conflict between Galileo and his opponents in the realm of natural philosophy was caused by an absence of true Aristotelian principles and this was in no way confined to the members of one religious group. Even Galileo admitted that Aristotle would undoubtedly change his ideas about certain physical phenomena if he were then living and could have participated in the experimental work being carried on.

Milder judgments on the decisions of the Inquisition in 1633 than that implied by Professor Carnahan have long since been passed by men of high standing in the scientific world who cannot be accused of bias toward the Catholic Church. Thomas Huxley, in a letter to Professor Mivart in 1885 writes:

> I have looked into the matter [the Galileo controversy] when I was in Italy, and I have arrived at the conclusion that the Pope and the College of Cardinals had rather the best of it.[33]

The following from the latest edition of Dampier shows the change in attitude from the unwarranted statements of condemnation of the Catholic Church to be found in so many of the nineteenth century histories of science, and unfortunately still appearing from time to time.

> In spite of Whewell's clear and fair account of the incident some more recent writers have made too much of the persecution of Galileo for his Copernican views. As Whitehead [A. N.] says "In a generation which saw the Thirty years War and remembered Alva in the Netherlands, the worst that happened to men of science was that Galileo suffered an honorable detention and a mild reproof, before dying peacefully in his bed."[34]

Butterfield concluding his essay on the science of this period says:

> Aristotelian physics were clearly breaking down, and the Ptolemaic system was split from top to bottom. But not until the time of Newton did the satisfactory alternative system appear. . . . The long existence of this dubious, intermediate situation brings the importance of Sir Isaac Newton into still stronger relief. We can better understand also, if we cannot condone, the treatment which Galileo had to suffer from the Church for a presumption which in his dialogues on *The Two Principal World-Systems* he had certainly displayed in more ways than one.[35]

The man who, in Professor Carnahan's words, had to live "the last ten years of his life under the sternest requirement to say nothing, write nothing, think nothing of which entrenched authority does not approve" produced during that time the treatise on which his proper claim to fame rests, that is, the *Two New Sciences*, a compendium of his life's work on mechanics and allied topics. He also had time to work in conjunction with Torricelli and Viviani, and to write numerous letters on various scientific topics.

NOTES

1. Lane Cooper, *Aristotle, Galileo and The Tower of Pisa* (Cornell University Press, 1935).

2. Among the few exceptions is *De Motu*, written about 1590 but not published until 1883.

3. Galileo Galilei, *Le Opere di Galileo Galilei* (ed. A. Favaro), (Florence: 1890–1907). Reprint 1929; Vol. 1, p. 263. Trans. by Cooper, *op. cit.*, p. 83.

4. *Ibid.*, Vol. 8, p. 109. Trans. from *Dialogues Concerning Two New Sciences* by H. Crew and A. De Salvio (New York: The Macmillan Co., 1914), pp. 64–65. Hereafter referred to as *Two New Sciences*.

5. See Cooper, *op. cit.*, for details and also interesting examples of variations of the story. See also, H. Butterfield, *The Origins of Modern Science* (New York: The Maximillan Co., 1951), p. 59 ff.; H. T. Pledge, *Science Since 1500* (London: Ministry of Education, 3rd reprint, 1947), p. 61.

6. J. H. Hardcastle, Correspondence Section, *Nature* (June 25, 1914), p. 429. See also (January 22, 1914), pp. 584–85.

7. Aristotle *Physica*, 4, c. 8, 215ª 25–31. Trans. by Hardie and Gaye, *The Works of Aristotle* (ed. W. D. Ross), Oxford University Press, 1930.

8. Carl B. Boyer, "Aristotle's Physics," *The Scientific American* (May, 1950), pp. 48–51. See also by the same author "Quantitative Science without Measurement: The Physics of Aristotle and Archimedes," *The Scientific Monthly* (May, 1945), pp. 358–64.

9. *Ibid.*, p. 50.

10. G. Sarton, *Introduction to the History of Science* (Baltimore: The Williams and Wilkins Co.), Vol. 2, p. 583 ff.

11. P. Duhem, *Études sur Léonardo de Vinci*, Vol. 3, "Les Précurseurs de Galiléi" (Paris, 1913). A. Koyré, *Études Galiléennes* (Paris, 1939). Carl B. Boyer, *The Concepts of the Calculus* (New York: Hafner Co., 1949), p. 71 ff. H. Butterfield, op. cit., Ch. 1.

12. E. Whittaker, "Aristotle, Newton, Einstein," *Science* (Sept. 17, 1943), p. 250.

13. Carl B. Boyer, *op. cit.* (8), p. 50.

14. I. B. Cohen, "Galileo," *Scientific American* (August, 1949), p. 45. For further elaboration on Galileo's "Thought Experiments" see, E. A. Burtt, *The Metaphysical Foundations of Modern Physical Science*, London, 1949, p. 65 ff. and H. Butterfield, *op. cit.*, p. 61. It is strange to find A. Koyré in agreement with Galileo's attitude in the case mentioned above. Mathematical deduction of itself is not sufficient to establish the truth of natural phenomena. See his article "Galileo and the Scientific Revolution of the Seventeenth Century," *The Philosophical Review*, July, 1943, p. 347.

15. " . . . no contingent, hypothetical laws, however wide, can offer an ultimate explanation of concrete facts.

Laws are but the expression of the *modus agendi*, the manner of acting, of causes or combinations of causes. And we shall not have fully explained any concrete fact in the universe until we know why the agencies of nature act according to those widest laws—why, for example, matter gravitates, or why life comes only from life, or why natural causes act uniformly." P. Coffey, *The Science of Logic* (New York: Longmans, Green and Co., 1918), Vol. 2, p. 240.

16. For further analysis of scientific truth, see E. F. Caldin, *The Power and Limits of Science* (London: Chapman and Hall, 1949).

17. G. Galilei, *op. cit.*, Vol. 7, English translation by Thomas Salusbury, *Mathematical Collections and Translations*, Vol. 1 (London: 1661). Hereafter referred to as *The Two Systems*.

18. H. Butterfield, *op. cit.*, p. 54.

19. G. Galilei, *op. cit.* (17), Salusbury translation, p. 380.

20. *Ibid.*, p. 383.

21. Agnes M. Clerke, "Galileo Galilei," *Encyclopaedia Britannica* (14th ed., 1946), vol. 9, 980.

22. H. Butterfield, *op. cit.*, p. 54.

23. J. L. E. Dreyer, *History of the Planetary Systems from Thales to Kepler* (Cambridge University Press, 1906), p. 416.

24. Pierre Conway, O.P. "Aristotle, Copernicus, Galileo," *The New Scholasticism*, Vol. 23, Nos. 1 and 2, 1949.

25. Carl Von Gebler, *Galileo Galilei und die römische Curie* (Stuttgart: 1876). English translation by Mrs. G. Sturge, London, 1879. This was the first historical research to make use of all the Vatican records relative to Galileo's trials.

26. See note 3. This work, in twenty volumes, contains all Galileo's scientific treatises and available letters together with numerous other letters and documents relative to Galileo's work.

27. E. W. Strong, *Procedures and Metaphysics* (California University Press, 1936), p. 137.

28. G. Galilei, *op. cit.*, Vol. 4.

29. D. Berti, *Copernico e le vicende del sistema Copernicano in Italia* (Rome, 1876), p. 123.

30. St. Thomas Aquinas, *Commentary on Aristotle's De Caelo II*, lectio 17, no. 2 (Rome, 1886).

31. E. Whittaker, *op. cit.*, p. 251.

32. Aristotle, *op. cit.*, De Caelo III, 306ª.

33. Thomas H. Huxley, *Life and Letters* (ed. L. Huxley), London, The Macmillan Co., 1900, Vol. 2, p. 122. It is interesting to compare this with Huxley's statement on the same subject in his essay "Descartes' Discourse on Method" written in 1870.

"At that time, physical science suddenly strode into the arena of public and familiar thought and openly challenged not only Philosophy and the Church, but that common ignorance which often passes by the name of Common Sense. The assertion of the motion of the earth was a defiance of all three, and Physical Science threw down her glove by the hand of Galileo. . . . Charity children would be ashamed not to know that the earth moves; while the Schoolmen are forgotten; and the Cardinals—well, the Cardinals are at the Oecumenical Council, still at their old business of trying to stop the movement of the World." *Collected Essays*, New York, D. Appleton and Co., 1894, Vol. 1, pp. 179, 180.

34. Dampier, W. C., *A History of Science*, Cambridge University Press, 1949, p. 113.

35. Butterfield, H., *op. cit.*, p. 55.

Editor's Note: In 1992 Pope John Paul II speaking on behalf of the Catholic Church expressed regrets as to how the "Galileo Affair" was handled and acknowledged scientific errors on the part of the tribunal that originally tried Galileo. Since the appearance of this article, an extensive collection of research materials have been compiled on Galileo's life and work. See for example:

Galileo: Pioneer Scientist. Stillman Drake (1990).
The Church and Galileo. Ernan McMullin (ed.) (2005).
Behind the Scenes at Galileo's Trial. Richard Blackwell (2006).

HISTORICAL EXHIBIT 9

Torricelli's Wine Glass

The seventeenth and eighteenth centuries were times of great mathematical and scientific exploration and experimentation. At times, the experimenters were confronted by surprising and confusing results. One such instance of this happening was incurred by the Italian physicist/mathematician, Evangelista Torricelli (1608-1647) in his investigation of the solid of revolution formed revolving the positive branch of the rectangular hyperbola y = 1/x around its horizontal asymptote, the x-axis. A trumpet-like figure evolves a hyperboloid of revolution. Due to its shape, it was popularly called "Torricelli's trumpet" and "Torricelli's Wine Glass" or, attesting to its otherworldly properties, "Gabriel's Horn." To his amazement, Torricelli, in 1641, found that this solid possessed a finite volume but had an infinite surface area! Using the methods of modern calculus, we can easily affirm Torricelli's findings. Consider the solution shown below:

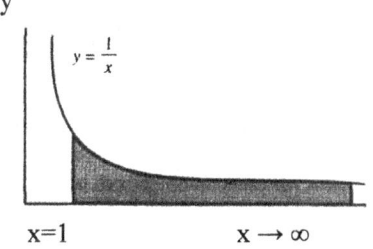

"Torricelli's Wine Glass"

The volume of the solid of revolution V:

$$V = \lim_{a \to \infty} \int_1^a \pi/x^2 \ dx \quad = \lim_{a \to \infty} \ \pi \left(1 - 1/a\right) = \pi$$

and the surface area, S:

$$S = \lim_{a \to \infty} \int_1^a 2\pi \frac{\sqrt{1 + (1/x)^4}}{x} \ dx \ > \ \lim_{a \to \infty} \int_1^a 2\pi \ dx/x = 2\pi \ \ln a =$$

$$\lim_{a \to \infty} \left(2\pi \ \ln a\right) = \infty$$

Torricelli communicated his discovery to Marin Mersenne (1588-1648), a mathematical correspondent who in turn dispersed it to the wider European scientific community. Mathematicians such as Christiaan Huygens (1629-1695) and René de Sluze (1622-1685) began examining other solids of revolution for such unusual traits. In particular, they sought a "bottomless wine glass," a solid with an infinite volume but a finite surface area. Of course, to their disappointment, they never found it. Torricelli's discovery is described today as "the painter's paradox."

21

Analytic Geometry: The Discovery of Fermat and Descartes

CARL B. BOYER

UNTIL a century ago the discovery of analytic geometry was ascribed categorically to the father of modern philosophy—René Descartes. *Proles sine matre creata*, Chasles called it, but this famous characterization is now quoted only to be refuted. The offspring is now recognized not only as having arrived as twins—for Pierre de Fermat had developed substantially the same method at about the same time—but also as having descended from a long line of legitimate antecedents. Recently attempts have been made to shift responsibility for the birth from the shoulders of Descartes and Fermat to those of one or more of the ancestors—Harriot or Cataldi or Ghetaldi or Viète or Benedetti or Oresme or Apollonius or the two thousand year old grandfather Menaechmus. In the case of any one of these individuals arguments may be advanced in his support. These can be based upon a definition of analytic geometry which is so framed as to include the work of the candidate in question and to exclude that of his rivals with respect to priority. The plausibility of the arguments is easily enhanced through an ingenious interpretation of the man's statements in terms of later symbolisms and concepts, thus implying a specious modernity of viewpoint. Few branches of mathematics lean as heavily upon a felicitous choice of notations as do those two powerful tools of science which we call

analytic geometry and the calculus. It is in fact precisely the algorithmic nature of these subjects which represents the secret of their effectiveness. In view of this, great caution must be exercised in superimposing modern notations upon ancient treatises. Conventional symbolisms have been devised ad hoc to express contemporary concepts which have developed slowly and laboriously, and through repeated use such notations have come to be associated implicitly with the ideas which they are now intended to represent. Projection of such notations into antiquity leads easily to a corresponding anachronism with respect to concepts, as the history of analytic geometry clearly demonstrates.

The query, "What constitutes analytic geometry?" has been variously answered. It is frequently characterized roughly as the combination of algebra and geometry. The utter inadequacy of this statement becomes apparent when one recalls that the ancient Pythagorean solution of quadratic equations through the application of areas would fall within this description and yet bears no significant resemblance to our analytic methods. Nor is one justified in holding that analytic geometry consists simply of the introduction of analytic methods of reasoning into the subject matter of geometry. This would result in applying the name to work which is far removed from Cartesian geometry. Plato is commonly regarded as having suggested the use of the analytic type of argument—of proceeding from the conclusion to the premise—although in a broad sense it can be said to have been

Reprinted from *Mathematics Teacher* 37 (Mar., 1944): 99–105; with permission of the National Council of Teachers of Mathematics.

employed much earlier. Plato, however, had no vision of the part this was to play later in Cartesian geometry.

The use of coordinate systems often has been regarded as the feature distinguishing analytic methods from other approaches in geometry. That this is a necessary aspect of the subject can scarcely be denied, but in no sense is it to be regarded as a sufficient criterion. The ancient Egyptians apparently made use of schemes of rectangular coordinates in their cadastral surveys; and Hipparchus applied such methods more systematically, both to the geography of the Mediterranean region and in the construction of star charts. In the field of geometry coordinate methods were certainly used by Apollonius in his study of the properties of conic sections, if indeed these may not have been employed in this connection a century earlier by Menaechmus. Yet the point of view and the method of attack of Apollonius and Menaechmus were essentially different from those indicated by Fermat and Descartes. A closer examination of their work will reveal that the coordinate frame was in every case simply an auxiliary construction superimposed a posteriori on a given curve in order to *study* its properties. It was not used as a means a priori of *defining* the locus or of determining the generation of the curve. That is, Menaechmus, in attempting to solve the problem of the duplication of the cube by means of the continued proportion of Hippocrates (expressed in modern symbolism

as $\dfrac{a}{x} = \dfrac{x}{y} = \dfrac{y}{b}$), had no conception of the

relationships $ay = x^2$, $y^2 = xb$, and $ab = xy$ as in themselves *defining* certain plane loci, but sought rather to *discover* curves otherwise defined, the points of which should possess these properties. This search led him, as is well known, to the sections of a cone. The very fact that conics were known in Greek geometry as "solid loci" indicates that they were defined stereometrically rather than analytically. More than 600 years later Pappus spoke likewise of "linear loci"—i.e., curves other

than straight lines, circles, and conics—as "being generated from more irregular surfaces and intricate movements." They were not given by equations. Once a curve had been thus defined, it was of course not difficult to discover innumerable relationships in terms of lines associated with the curve. Such properties when symbolically expressed may in a sense be regarded as equations corresponding to the curves, and for this reason the origin of analytic geometry recently has been claimed for Greek antiquity. But according to this point of view the Babylonians possessed analytic geometry by 2000 B.C.! They were familiar with the relationship $4r^2 = c^2 + 4(r - s)^2$, where s is the sagitta of any chord of length c in a circle of radius r. This relationship *may* be regarded—although there is no evidence that the Babylonians did so—as the equation defining the circle in terms of the rectangular coordinates c and s. Thus interpreted, the Babylonians would be regarded as having used analytic geometry. Similarly the Euclidean propositions on the straight line and the circle can be looked upon as equations of these curves, and the *Elements* would then be an example of analytic geometry. One will protest that Euclid did not make use of the modern algebraic symbolism which is used in equations. This is true, but then neither did Menaechmus before him nor Apollonius somewhat later. Moreover, they are alike in that invariably *the equations are derived from the curve*, and not conversely. It is true that Apollonius and other Greek geometers used the stereometric origin of the conics only so far as was necessary to deduce a single fundamental plane property (equation) for each curve, and that this latter result was then made basic in establishing further properties, such as those of the asymptotes. Nevertheless, it is significant that they felt it necessary to give the curves a geometrical definition distinct from that of points the coordinates of which satisfy a given equation. In fact, by "finding" a conic for which a given relationship held (i.e., satisfying a given equation) they understood not plotting it by points but by determining it as a section of a cone. While one may safely assert that in a broad sense Menaechmus and Apollonius made use of analytic methods, it is nevertheless not correct to say that they understood the fundamental principle of analytic geometry—that, in general, an equation in two variables defines a plane curve in a given coordinate system such that all pairs of values satisfying the equation are coordinates of points on a curve; and, conversely, all points on this curve have coordinates which satisfy the given equation. In part the failure of Menaechmus and Apollonius to appreciate this principle may have resulted from the weakness of contemporary algebra. It may also have been the consequence of a peculiar deficiency in Greek thought—the lack in mathematics of the notion of a variable. The paradoxes of Zeno had left a profound impression which caused the generic ideas of change and variability to be related to metaphysics. Geometric magnitudes were static and continuous; algebraic quantities were discrete constants. The symbols in the *Arithmetic* of Diophantus represent unknown numbers rather than variables in the sense of elementary algebra. Modern higher analysis may indeed be founded logically upon the Weierstrassian "static theory of the variable" which represents in a sense of partial return to the ancient Greek view; but the considerations leading to the original development of analytic geometry (as also of the calculus) were definitely phoronomic. During the later Scholastic period, and particularly during the fourteenth century, there arose at Paris and Oxford lively discussions on questions concerning rates of change, both uniform and non-uniform. The ideas of inertia and of acceleration, ubiquitously but erroneously ascribed today to Galileo, were developed at this time. In connection with this work Nicole Oresme made one of the earliest attempts to represent graphically the manner in which one quantity varied with another. For example, the velocity of a freely falling body, which was taken to be proportional to the time of fall, he plotted as a function of the time. Marking off points on a horizontal straight line as time units, the velocity corresponding to any time

was represented as an ordinate at this point, and the resulting graph was an oblique straight line. Oresme attempted to picture also certain functions which were "difformly difform" (that is, the rate of change was not constant), the graphs in these cases being necessarily curvilinear.

In certain quarters Oresme has been vigorously acclaimed as the inventor of analytic geometry. His work undoubtedly represents a very definite step in this direction, for he may be looked upon as the father of systematic graphical representation, but his view is seen to fall short of the conceptions of Descartes and Fermat. In a sense one may say that Oresme showed that a simple proportion can be represented by a straight line and that certain other types of variation are associated with characteristic diagrams. This may be regarded as an advance beyond the coordinate considerations of Apollonius in that the curve is here determined by the coordinate system and by the law of variation; the coordinate frame is not introduced as a device for studying a curve already given. Nevertheless, Oresme was prevented from taking full advantage of his novel idea by deficiencies in geometrical knowledge and algebraic technique. Unfortunately the mathematical interest of the age took the form of abortive speculations (resumed half a millennium later by Cantor) on infinity and the continuum, and the Greek classics were neglected. Consequently there is in the work of Oresme no systematic association of algebra and geometry in which an *equation* in two variables determines a specific curve and conversely.

The shortcomings in algebra which are apparent in the work of Menaechmus, Apollonius, and Oresme were not removed until the sixteenth century. The arithmetic of Diophantus (and, a fortiori, that of the Babylonians) had remained largely unknown to the Latin medieval world, but by the Arabs it had been translated, amplified, and transmitted to Europe. Although Leonardo of Pisa appreciated the significance of this work, it remained latent in Europe for almost three centuries before algebra was systematically developed by

EXAMPLES OF EARLY GRAPHS

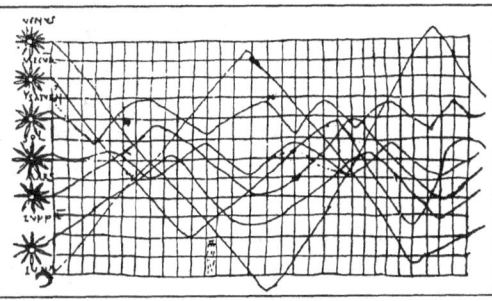

Illustration from a tenth-century manuscript depicting the positions of the "seven heavenly wanders," i.e., five planets, the sun, and the moon, in the night sky during the period of one month. The dependence on a coordinate system is obvious.

Sketches on a fourteenth-century manuscript on kinetics. The sketches relating time and distance were composed by the theologian–natural philosopher, Nicole Oresme, or one of his students.

Chuquet, Pacioli, Cardan, Stifel, and many others. This development happily coincided with a strong revival of interest in ancient geometry, including the *Conics* of Apollonius. Not unnaturally the gap between geometric and algebraic methods became less pronounced than it had been in the works of the classical Greek period. Tartaglia, Cardan, and Bombelli solved equations geometrically and, inversely, algebraic methods were widely advocated by Benedetti, Clavius, Viète, Stevin, Girard, Ghetaldi, Cataldi, Harriot, and Oughtred as a means of simplifying geometrical problems. Viète realized as well the advantage to be gained by Plato's analytic approach. Moreover, he contributed to algebra a significant point of view: the subject was to be looked upon not simply as numerical logistic but as the logic of magnitudes in general—*logistica speciosa*. These magnitudes he represented by letters, vowels for the unknown and consonants for those assumed known. It is to be remarked clearly, however, that these symbols denoted determinate quantities rather than variables. The geometric problems were reducible to equations in a single unknown, and the value of this unknown was determined by solving this equation. In fact, the algebra of the time was predominantly concerned with the solution of such equations. Diophantus had considered equations in several unknowns from the point of view of the theory of numbers; but these did not attract the algebraist of the sixteenth century, for they could not be "solved" in the ordinary sense. When a geometric problem led to an equation in two unknowns, it was abandoned as unsolvable. This is where the genius of Fermat and Descartes appeared. These men saw that lack of a unique solution did not in this case render the problem devoid of interest. They interpreted the one unknown as a variable horizontal line segment (abscissa) with a fixed initial end-point (origin), and at the other extremities of this they erected perpendicular line segments (ordinates) of lengths corresponding to the values determined for the second unknown in terms of the first. The extremities of these ordinates, infinite in number, formed a curve which, for a given equation and coordinate system, was uniquely determined. As Fermat stated it (*Oeuvres*, I, 91; III, 85): "whenever in a final equation two unknown quantities are found, we have a locus, the extremity of one of these [unknown magnitudes] describing a line, straight or curved."

That the crux of the matter lies in the interpretation of problems leading to a final equation in more than one unknown is quite clear also from the work of Descartes: Book I deals with determinate problems; i.e., those which lead to "as many such equations as there are supposed to be unknown lines." (*Geometry*, p. 6.) Book II, on the other hand, deals with plane problems on loci in which "this [final] condition can be expressed by a single equation in two unknown quantities." (*Geometry*, p. 34.) In suggesting the extension of this discovery to three dimensions he adds: "If two conditions for the determination of the point are lacking [i.e., if the final equation has three unknowns], the locus of the point is a surface." (*Geometry*, p. 80.) Fermat stated the principle still more sharply as follows: "There are certain problems which involve only one unknown, and which can be called determinate, to distinguish them from the problems of loci. There are certain others which involve two unknowns and which can never be reduced to a single one: these are the problems of loci. In the first problems we seek a unique point, in the latter a curve. But if the proposed problem involves three unknowns, one has to find, to satisfy the question, not only a point or a curve, but an entire surface." (*Oeuvres*, III, 1161.)

The object of this paper has not been to decide whether or not Descartes and Fermat invented analytic geometry. An answer to this problem depends on one's definition and sense of proportion; but *de gustibus non est disputandem*. The aim has been the simpler one of indicating just what it was that these men did discover in this connection. Neither one of them invented the use of coordinates or of the analytic method. Neither was first in applying algebra to geometry or in graphically representing variables. Moreover, it had long been known that, for a given curve, certain distance relationships are determined which may be interpreted as equations of the curve with respect to coordinate systems. However, there appears to have been no conception before their time of the converse—the fact that in general an arbitrary *given equation* involving *two unknown quantities* can be regarded as determining *per se*, with respect to a coordinate system, a plane curve. This latter recognition, together with its fabrication into a formalized algorithmic procedure, constituted the decisive contribution of Fermat and Descartes. This is the sense in which these men may be regarded as the founders of analytic geometry, much as Newton and Leibniz are generally regarded as the inventors of the calculus through precise formulation of the mutually inverse nature of area and tangent problems in terms of a definitely regularized operational procedure. The publication of Descartes's *Géométrie* in 1637 preceded by 42 years that of Fermat's *Isagoge*, but both men were in independent possession of their methods well before this time—about 1619 for Descartes and 1629 for Fermat. As in the calculus the root idea was due to the two greatest figures of the second half of the seventeenth century, so likewise in analytic geometry the fundamental principle was recognized by the two greatest mathematicians of the first half of this same century.

REFERENCES

1. Apollonius of Perga, *Treatise on Conic Sections* (ed. by T. L. Heath, Cambridge, 1896). [Heath points out that the method of Apollonius does not differ essentially from that of modern analytic geometry except that geometrical operations take the place of algebraical calculations.]

2. Bosmans, Henri, "La première édition de la 'Clavis mathematica' d'Oughtred, son influence sur la 'Géométrie' de Descartes," *Annales de la Société Scientifique de Bruxelles*, XXXV (1910–1911), 24–78. [Oughtred as a link from Viète to Descartes.]

3. Chasles, Michel, *Aperçu historique sur l'oringine et le développement des methodes en géométrie* (Paris, 1875). [Strong claim for Descartes as sole inventor (pp. 94–95).]

4. Coolidge, J. L., "The origin of analytic geometry," *Osiris*, I (1936), 231–250. Appears also as a section in his *History of Geometrical Methods* (Oxford, 1940). [Defends thesis that analytic geometry was an invention of the Greeks, perhaps of Menaechmus.]

5. Descartes, René, *The Geometry of René Descartes* (transl. by D. E. Smith and M. L. Latham, Chicago and London, 1925). [Excellent English edition.]

6. Duhem, Pierre, *Études sur Léonard de Vinci* (3 vols., Paris, 1906–1913). [Fullest available account of the medieval precursors of Galileo and Descartes.]

7. Duhem, Pierre, "Oresme," *Catholic Encyclopedia*, XI (1911), 296–297. [Asserts that Oresme "forestalls Descartes in the invention of analytic geometry."]

8. Fermat, Pierre de, *Oeuvres* (ed. by Paul Tannery and Charles Henry, 4 vols. and supp., Paris, 1891–1922). [See vol. I for Latin of "Introduction to plane and solid loci," vol. III for French translation.]

9. Funkhouser, H. G., "Historical development of the graphical representation of statistical data," *Osiris*, III (1937), 269–404. [Includes reference to Oresme's work.]

10. Gelcich, E., "Eine Studie ueber die Entdeckung der analytischen Geometrie mit Berücksichtigung eines Werkes des Marino Ghetaldi Patrizier Ragusaer aus dem Jahre 1630," *Abhandlungen zur Geschichte der Mathematik*, IV (1882), 191–231. [Excellent critical account of relations between algebra and geometry at that time. Says Ghetaldi lacked principle of coordinates.]

11. Günther, Sigismund, "Le origini ed i gradi di sviluppo del principio delle coordinate," *Bullettino di Bibliografia e di Storia delle Scienze Matematiche e Fisiche*, X (1877), 363–406. [Shows that use of coordinates goes back to Greek times, and that work of Oresme and Kepler resembles analytic geometry in some respects.]

12. Karpinski, L. C., "Is there progress in mathematical discovery and did the Greeks have analytic geometry?", *Isis*, XXVII (1937), 46–52. [Rejects Coolidge's thesis that Greeks had analytic geometry. Emphasizes idea of progress and development of algebraic notations.]

13. Libri, Guillaume, *Histoire des sciences mathématiques in Italie, depuis la renaissance des lettres jusqu'a la fin du dix-septième siècle* (vols. III and IV, Paris, 1840–

1841). [Emphasizes Benedetti (III, 124) and Cataldi (IV, 95).]

14. Loria, Gino, "Da Descartes e Fermat a Monge e Lagrange. Contributo alla storia della geometria analitica," *Reale Accademia dei Linceri*. Atti. *Memorie della classe di scienze fisiche, matematiche e naturali* (5), XIV (1923), 777–845. [Excellent general summary with emphasis on development after Descartes.]

15. Loria, Gino, "Descartes géomètre," *Etudes sur Descartes* (Paris, 1937), pp. 199–220. [Summary and analysis of the Géométrie.]

16. Loria, Gino, *Il passato e il presente delle principali teorie geometriche; storia e bibliografia* (4th ed., Padova, 1931). [Excellent and ample account of the development of geometry in modern times.]

17. Loria, Gino, "Pour une histoire de la géométrie analytique," *Verhandlungen des dritten internation-alen Mathematiker-Kongresses in Heidelberg vom 8. bis 13. August 1904* (Leipzig, 1905). [Critical analysis of contributions of Fermat and Descartes and of subsequent development.]

18. Loria, Gino, "Qu'est-ce que la géométrie analytique?", *L'Enseignement Mathématique*, XIII (1923), 142–147. [Emphasizes Euler's *Introductio in analysin infinitorum* for use of formulas in solving geometrical problems.]

19. Milhaud, Gaston, *Descartes savant* (Paris, 1921). [Excellent analysis of the development of Descartes' scientific and mathematical thought.]

20. Morley, F. V., "Thomas Hariot," *Scientific Monthly*, XIV (1922), 60–66. [Claims analytic geometry for Harriot. This claim is no longer substantiated.]

20a. Müller, Felix, "Zur Literatur der analytischen Geometrie und Infinitesimalrechnung von Euler," *Jahresbericht der Mathematiker-Vereinigung*, XIII (1904), 247–253. [Emphasizes the use of analytic methods between Fermat and Euler.]

21. Ritter, F., "Première série de notes sur la logistique specieuse par François Viète," *Bullettino di Biblio-grafia e di Storia delle Scienze Matematiche e Fisiche*, I (1868), 245–276. [On Viète's combination of algebra and geometry.]

22. Saltykow, N., "'La géométrie' de Descartes. 300ᵉ anniversaire de géométrie analytique," *Bulletin des Sciences Mathématiques* (2), LXII (1938), 83–96, 110–123. [Emphasizes originality of Descartes in combining many elements previously known.]

23. Strong, E. W., *Procedures and Metaphysics. A Study in the Philosophy of Mathematical-Physical Science*

in the Sixteenth and Seventeenth Centuries (Berkeley, Calif., 1936). [A critical discussion of tendencies, some of which concern analytic geometry. Work of Cardan, Tartaglia, Benedetti, Stevin, Viète discussed.]

24. Tropfke, Johannes, *Geschichte der Elementarmathematik* (vol. VI, Berlin and Leipzig, 1924). [One of the fullest available histories of analytic geometry is included in pp. 92–169.]

25. Wieleitner, Heinrich, "Der 'Tractatus de latitudinibus formarum' des Oresme," *Bibliotheca Mathematica* (3), XIII (1912–1913), 115–145. [Oresme's graphical representation.]

26. Wieleitner, Heinrich, "Marino Ghetaldi und die Anfänge der Koordinatengeometrie," *Bibliotheca Mathematica* (3), XIII (1912–1913), 242–247. [Sees combination of algebra and geometry, but not analytic geometry, in Ghetaldi.]

27. Wieleitner, Heinrich, "Ueber den Funktionsbegriff und die graphische Darstellung bei Oresme," *Bibliotheca Mathematica* (3), XIV (1914), 193–243. [Wieleitner rejects Oresme as the founder of analytic geometry but sees a likelihood that he influenced Descartes. (See p. 241.)]

28. Zeuthen, H. G., *Geschichte der Mathematik im XVI und XVII. Jahrhundert* (German ed. by Raphael Meyer, Leipzig, 1903). [A brief summary of the geometry of Fermat and Descartes is given on pp. 192–233.]

See also the standard histories of mathematics by Archibald, Ball, Bell, Cajori, Cantor, Hankel, Heath, Kaestner, Marie, Montucla, Smith, Wieleitner, Zeuthen; also notes in *Encyclopédie des sciences mathématiques.*

A further contribution by Gino Loria, "Perfectionnements, évolution, métamorphoses du concepte de coordonnées. Contribution à l'histoire de la géométrie analytique," was scheduled for publication in volume VIII of *Osiris,* but the appearance of this volume has been held up by the German invasion of Belgium.

22

The Young Pascal

$\mathcal{V}$ERSATILITY is not inevitably the companion of genius. It is not altogether common to find a man who is at the same time a clever experimental physicist, a creative mathematician, an inventor with an eye to money-making, a gifted writer whose artistry places him among the foremost French stylists, and a religious philosopher of singular originality and ardor. Blaise Pascal was such a man. He could write an important treatise on the vacuum as well as produce those incomparable examples of controversial literature, the famous *Provincial Letters*. He invented the first adding machine of practical consequence and tried (in vain) to realize a profit from its sale. In the fragments of his projected apology for the Christian Religion, left unfinished by his death at the early age of thirty-nine, are to be found many evidences of his mathematical genius as well as of a remarkable piety and zeal. A clear and complete picture of his early life and education would not only be of rare interest, but it could not fail to contain many suggestions of value to the modern teacher. It is indeed a pity that such incomplete information is available. The story is soon told, but it is well worth the telling, and perhaps it holds some inspiration or lesson for our own times.

Blaise Pascal belonged to a family of provincial officials. Although it could boast a title of nobility granted by Louis XI in recognition of the faithful services of one Etienne Pascal, an important fiscal

Reprinted from *Mathematics Teacher* 30 (Apr., 1937): 180–85; with permission of the National Council of Teachers of Mathematics.

officer in the King's entourage, it was of parliamentary rather than noble condition, and the Province of Auvergne provided places in its local magistracy and revenue offices for many of the younger members of the family. The Pascals were well-to-do, substantial people of considerable local importance, and had for generations audited the accounts of the province or presided over its courts.

Another Etienne Pascal was President of the *Cour des Aides*, a court having jurisdiction in all matters pertaining to indirect taxation, when his son, Blaise, was born at Clermont-Ferrand on June 27th, 1623. This strong-minded and talented magistrate had married in 1618 Antoinette Begon, the pious daughter of a family of the merchant class. A son born in 1619 lived only long enough to be baptized, but 1620 saw the arrival of a daughter, Gilberte, who was to become the biographer of her famous younger brother and sister, Jacqueline, born in 1625.

This was a period of scientific discovery and advance. The first half of the seventeenth century was to see the death of Francis Bacon, the work of Galileo, Kepler, Descartes, Roberval, and Fermat, the birth of Newton, Huygens, and Leibniz. We shall find Etienne Pascal and his son taking their parts in the work of the age. But in spite of the spirit of true scientific thought, there was still a wide-spread faith in astrology, "spells," witchcraft, signs, and omens. Even Kepler had to make astrological predictions to help to earn his living, and it was only by exerting considerable influence that he was able to save his mother from conviction of witchcraft. A curious story is told of the infant

Blaise Pascal by his niece, Marguerite Perier,[1] which illustrates how even a man of his father's sagacity was forced to take account of the prevailing superstitions.

According to this story, when Blaise was between one and two years of age, he fell into a strange sort of languor attended by two surprising symptoms. The sight of water threw him into convulsions, and, although he enjoyed the caresses of his mother or of his father separately, their approach together sent him into transports of childish rage. This illness lasted for over a year during which time he grew so much worse that his parents began to despair of his life. Everybody said that this malady was the result of a "spell" cast over the child by a woman who had received charity from his mother. Although his father was skeptical, he seems to have sent for the women and to have wrung from her a confession of the sorcery. She was able to explain how the spell could be removed, but it would require the sacrifice of the life of another to whom the enchantment could be transferred. Fortunately an animal would do. The anxious father offered his horse. But, the witch remarked, the thing could be accomplished at much less sacrifice: a cat would suffice. So a cat was produced. When thrown out a window only six feet from the ground, it instantly died. Next morning a poultice was made from three leaves each of three different herbs gathered before sunrise by a girl less than seven years of age. Upon application of this poultice to his body the child fell into a coma from which he awoke—just at midnight— completely cured.

Unlikely as this incident appears, it may well symbolize the spirit of the times, while the illness itself was the forerunner of that ill health which persisted throughout so much of Pascal's life.

When Blaise was but five years old his mother died. The father, a man of about forty, thenceforth lavished his care and affection upon his three children. Two years later, in 1631, he sold his magistracy at Clermont, and moved with his family to Paris where he planned to devote his time to the education of his children, especially his son. He would permit no one to help or to interfere with his original plan of teaching the boy. Although the Jesuit schools were then flourishing and providing what was regarded as a well balanced and methodical course of instruction, Etienne Pascal would have none of it. He preferred to keep Blaise at home and to teach him in his own way. He failed to count upon his own insufficiency in many branches of knowledge. He knew the Law as a former official, could use Latin about as well as most educated men of the day, was acquainted with mathematics and physics, and was a fervent believer in experimental science. He made up for the gaps in his knowledge and for his lack of experience by a method of teaching which was of his own personal invention. His basic principle was that the child's lesson should always be entirely within his grasp, easily and completely understood. A glimpse of the whole subject under consideration was to suggest simple and easily discerned general laws and to arouse curiosity. From these general principles were to be deduced explanations of particular facts observed or questioned by the pupil. Etienne Pascal would make no attempt to force or strain an inquiring mind.

That the youthful Blaise had an inquiring mind—not easily restrained by his father's notion that the child should be "held well beyond his work"—is evident from the biography written by his sister Gilberte.[2] He wanted to know the reasons for everything, and if good ones were not given he would seek them for himself and not give up until he had a satisfactory explanation. For instance, we are told that when he was eleven years old he noticed that a china dish struck by a knife produced a loud sound which ceased when the hand touched the dish. He was not content until, after several experiments, he had discovered the reason. He then wrote, no doubt with parental encouragement, a little essay on sound.

There was an incessant inquiry into the origin and nature of things in general, into their properties and their uses. For example, what is grammar?

How does it happen that all languages are communicable from one country to another? What of the extraordinary effects of nature such as gunpowder exploding in a cannon? It was expected that these inquiries would last until Blaise was twelve years old when he was to be put to the study of Latin and Greek. He was also to acquaint himself with Spanish and Italian, and always by methods and according to rules imposed by his father. When he should have mastered languages, he was to take up mathematics. According to this plan he would have been about fifteen or sixteen years old when introduced to geometry, and should by that time have reached a maturity and orderliness of mind properly fitting him to appreciate this absorbing subject. The whole scheme was intended to accustom the child to seek out knowledge for himself while guided in broad paths of learning, to render account of his work to himself, and never to take preconceived notions and hypotheses for truths.

While this method of education suited the boy's nature admirably, while it developed his good qualities and nourished his originality, it overlooked many important matters. History was neglected. Every idea was a personal discovery of the pupil and appeared to him as his own exclusive property. Such a lack of historical perspective encourages a man to exaggerate the consequences of his own ideas and to disdain the work of others to whom he feels he owes nothing. Who can say that Pascal did not at times exhibit evidences of a lofty egotism? "He who reads the *Thoughts* of Pascal should give a thought to the president of the *Cour des Aides* of Clermont-Ferrand who taught his son so well while teaching him so badly."[3]

That the method had surprising effects upon the lad is clear if we can believe an often repeated story[4] told by his sister Gilberte. One day his father happened to enter the play room unobserved. What was his surprise to find the boy of twelve engaged in drawing figures upon the stone floor with a piece of charcoal. Although he had not been allowed to study Euclid, when questioned the boy explained that he was trying to prove that the sum of the angles of a triangle is two right angles. His father then drew from him step by step the method by which he had used his "bars" (lines) and "rounds" (circles) to arrive at this thirty-second proposition of Euclid. Overcome with emotion, Etienne Pascal withdrew, and soon Blaise was given mathematical books to study. He had, no doubt, overheard in the conversation of his father's friends some references to geometry, and, not content to regard lessons in geometry as simply a promised reward for proficiency in Latin and Greek, had set out to investigate the subject for himself. This story may be largely the product of an adoring sister's imagination, and yet there must be at least some element of truth in it. She was older than her brother and was taking the place of a mother and a housewife. A young woman matured beyond her years by this experience, while perhaps possessing strong family partiality, must have seen clearly the remarkable ability of the boy. Whether in later years she dramatized this ability in some such imaginary incident, or whether the story is strictly accurate in all details, unlikely as this may seem, cannot be definitely said. However, there can be no doubt that young Blaise showed such proficiency in mathematics that his father, perhaps surprised and gratified at this result of his educational method, felt constrained to revise his plan to withhold mathematical studies until languages had been mastered.

Etienne Pascal had found his level in Parisian society. He was admitted to a circle of scientific men such as Roberval, Carcavi, Le Pailleur, Desargues, and gifted amateurs of high spirit. The central figure of the little group was Father Mersenne, a clever and agreeable Minimist friar who was in close touch with Galileo, Descartes, Torricelli, and other scientific men of the day. The evident genius of young Blaise prompted his father to introduce him to this circle which was the immediate ancestor of the French Academy of Sciences. Here, although he was but in his early teens, he took part in work and discussions. And here, too, he found a school of manners. The right to speak depended upon the amount of information. Older men who knew little of a subject had to be content to listen to younger men who knew much. Such experiences led him to attempt to judge everyone by purely intellectual standards. It did not take him many years to discover that the world does not employ such a standard of judgment, for he observes in the *Thoughts*, "Rank is a great advantage, for it gives to a man of eighteen years of age a degree of acceptance and respect which another man can scarcely obtain by merit at fifty. Here is a gain, then, of thirty years without difficulty."[5]

It must not be supposed that this gifted youth led a life of solemn study with no amusement. While his father was a former magistrate and a *savant*, he was not solely a meditative philosopher. He was an intimate friend of the versatile and volatile Le Pailleur who ran about France and England, changed his religion from Protestant to Catholic and back again, taught himself mathematics, sang songs—he is said to have sung eighty-eight songs during one gay evening—danced, and perpetrated practical jokes upon unsuspecting friends. The mere fact that a man of this character was an intimate friend of the family is enough to dispel the picture of an austere cultivation of the intellectual at the expense of all other powers. A rare cordiality and familiarity characterized the family life. President Pascal was a most affectionate father. Gilberte was a sagacious and practical young housekeeper who, when away from her household duties, took a quiet place. Jacqueline, though much younger, was much more noticed.

The Pascals had visited Auvergne only once, in 1636, since their establishment in Paris. Most of their interests were to be found in the capital where their fortune had been invested in municipal bonds. But unfortunately Richelieu's policy of opposition to the House of Hapsburg had involved a considerable drain on French finances, and in 1638 in order to meet certain exigencies, the interest rate on the Parisian bonds was arbitrarily cut. This of course embarrassed and angered Etienne Pascal who associated himself with a group of protesting investors. Unhappily there was some show of violence, and two of the leaders were arrested and summarily removed to the Bastille. Pascal was able to escape and to take refuge in Auvergne. He risked return only when Jacqueline was dangerously ill with the smallpox. In the meantime the pathetic situation of the three children had struck the fancy of the fashionable world. In February, 1639, Jacqueline was encouraged to take part in the play, *L'Amour Tyrannique*, which was performed before the Cardinal. She so charmed Richelieu that his attention could be favorably attracted to the gifted son and exiled father, and Jacqueline was instructed to tell her father to return.

The evident ability and well-known integrity of President Pascal so appealed to the Cardinal that he resolved to send the former magistrate to Rouen to act as Intendant. Like most people, the French did not like to pay taxes. In 1639 the national treasury was seriously depleted, and in spite of the most determined efforts of the Minister, the provincial parlements failed to squeeze any more money out of the people. The local officials sympathized with their neighbors and did nothing. The Cardinal, resolving to put an end to such nonsense, created the office of Intendant whose occupant was to act as a sort of vice-regent in the matter of tax collections. His power was as great as his popularity was minute. Normandy was one of the most unruly of the provinces and needed a firm hand to bring it up to the mark, and the capable and courageous Etienne Pascal was the man for the job. Accordingly he arranged to remove his family to Rouen in the autumn of 1639.

Meanwhile Blaise had been thinking about mathematics and joining in the discussions of Mersenne's circle of friends. He had undoubtedly studied carefully many of the mathematical books in his father's library, and had also made himself familiar with the new geometrical methods of

Desargues. His father's educational method had encouraged original work on his part, and at the age of sixteen he was to be found engaged in the production of a new theory of the conic sections, a forerunner of which appeared in 1640 as *Essay pour les coniques*. In this little handbill was announced a lemma which has become famous as "Pascal's Theorem" and amounts to the statement that the intersections of the three pairs of opposite sides of a hexagon inscribed in a conic are collinear. The youthful investigator acknowledged his indebtedness to the methods of Desargues who had endeavored to reduce the properties of the conic sections to a small number of propositions. The improvement over the old method of considering each conic as a separate curve was in treating these curves simply as various perspectives of a circle. That Pascal had carried this new method much farther than had its inventor was at once acknowledged by Desargues who gave the "mystic hexagram" the name *La Pascale*. Great praise immediately fell to the lot of the sixteen year old mathematician. Everyone was enthusiastic—everyone but Descartes who grumbled to Mersenne, "It is just as I thought! I had not read half the little essay on Conics by M. Pascal's son before I saw that he had taken most of his ideas from M. des Argues, and this was confirmed soon afterwards by his own confession."[6] But there can be no doubt of Pascal's originality. The *Essay pour les coniques* was the announcement of a treatise which he was preparing. It concludes with the modest statement

> We have several other problems and theorems and several consequences deducible from the preceding, but the mistrust which I have of my slight experience and capacity does not permit me to advance more until my present effort has passed the examination of able men who may oblige me by looking at it. Afterwards, if they think it has sufficient merit to be continued, we shall endeavor to push our studies as far as God will give the power to conduct them.[7]

Mersenne reported that in this treatise Pascal had deduced the properties of the conic sections from his "mystic hexagram" in four hundred corollaries.

While the work was never published, Leibniz knew of it and gave a summary of it in his letter of August 30th, 1676 to Etienne Perier, Pascal's nephew.

This discovery of Pascal indicates clearly his view of mathematics. It rests upon the idea that the properties of a complex figure may be considered as a modification of those of a more simple figure. The effort of his father to have him always take a comprehensive and unifying view of everything helped to save him from the all too common error of the student who so concentrates upon some small details that he misses the simplicity and unity of his subject.

The removal to Rouen when Blaise was seventeen soon put a stop to these extended researches in pure theory. There his father plunged into the business of straightening out the chaotic finances of Normandy, and his son was expected to help him as much as possible. But his native genius, combined with the encouragement which his education had given to original work of all sorts, soon (1643) led to the idea of an "arithmetical machine." Before this adding machine could be finally perfected, more than fifty models had to be constructed, and it was not until some years later that a patent was secured which forbade the copying or construction of the machine by anyone else.

The achievements of Pascal in the fields of mathematics, physics, literature, and philosophy, and the story of his later life are too well known to be repeated here. But it is well to pause to consider the remarkable effect which the manner of his education must have had upon him. It fostered original investigation—even of simple things well known to the world of his day—above all else. We see its flowering, if not in the apocryphal "discovery" of geometry, certainly in the theorem of the hexagon, at an early age. Undoubtedly his genius would have asserted itself under any circumstances, but it was surely awakened and nourished by that schooling so well adapted to his needs. We should be fortunate indeed if today we could make available such an opportunity for every talented young man and woman.

NOTES

1. Marguerite Perier, "Mémoire de la vie de M. Pascal, écrit par mademoiselle Perier, sa nièce" in *Oeuvres de Pascal* ed. by Brunschvicg and Boutroux, Paris, 1908, 14 vols., vol. I, pp. 125–36.

2. Gilberte Perier, "La vie de Monsieur Paschal, escrite par Madame Perier, sa soeur, femme de Monsieur Perier, conseiller de la Cour des Aides de Clermont" in *Oeuvres de Pascal*, vol. I, pp. 50–114. This biography, written soon after his death, is the chief source of information concerning his early life.

3. Fortunat Strowski, *Pascal et son Temps, deuxième partie, l'histoire de Pascal*, Paris, 1907, p. 9.

4. It appears on pp. 53–56 in vol. I of *Oeuvres de Pascal.*

5. Blaise Pascal, *Oeuvres de Pascal*, vol. XIII, pp. 240–41.

6. René Descartes, Letter of April 1, 1640, to Mersenne in *Oeuvres de Descartes* ed. Charles Adam and Paul Tannery, Paris, 1899, 12 vols., vol. III, p. 47.

7. Blaise Pascal, *Oeuvres de Pascal*, vol. I, pp. 259–60.

HISTORICAL EXHIBIT 10

Roberval's Quadrature of the Cycloid

As a circle rolls along a horizontal plane, a fixed point on its circumference traces out a curve known as a cycloid. This name was given to the curve by Galileo who was fascinated by its properties.

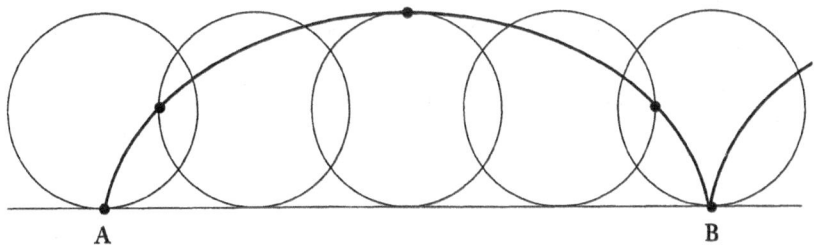

The first mathematician known to study the properties of this mechanical curve was the Frenchman Charles Bouvelles (ca. 1470–1553). Galileo admired the graceful arches of the cycloid and sought to incorporate them into bridge construction. He attempted to approximate the area contained within one arch by balancing a cycloidal template against circular templates of the generating circle. Through this experimentation he concluded that the cycloidal area was about three times that of the generating circle. Further investigation of this question was needed. In 1630, Père Marin Mersenne (1588–1648) suggested to his community of mathematical correspondences which included such notables as Descartes and Fermat that finding the quadrature of the cycloid would provide a good test for the newly-devised techniques of infinitesimals. This challenge was undertaken by Gilles Personne de Roberval (1602–1675), a profesor of mathematics at the Royal College. Roberval's method is outlined below:

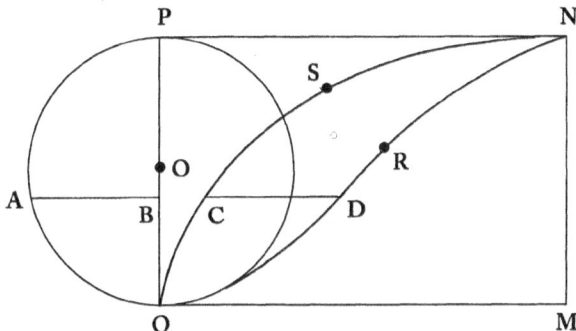

Let *QMNS* be half an arch of the cycloid generated by circle *O*. The area of *QMNP* equals twice the area of circle *O*. Construct infinitesimals (line segments) parallel to *QM* with lengths determined by the horizontal distance between diameter *PQ* and the circumfer-

(continued)

ence of a semi-circle, that is, *AB*. For each line in the semi-circle, a corresponding line of the same length is constructed from the cycloid as shown, that is, *CD*. The sequence of such infinitesimals with the cycloid, by their end points determine a curve *NRS* called the companion to the cycloid.

The area contained between the cycloid and its companion is composed of infinitesimals corresponding to those of the semi-circle *PAQ*, therefore this area is equal to that of the semi-circle or one-half the area of circle *O*. Now, by Cavalieri's Theorem [If two areas are everywhere of the same width, then the areas are equal], *NRQ* bisects *QMNP*; therefore, the area under *NRQ* equals the area of circle *O*. Adding together the area of these two regions, *QMNR* and *QRNS*, we find that the area under half the arch of the cycloid is equal to ³/₂ area of circle *O*. Thus it can be concluded that the area contained in an arch of a cycloid is three times the area of its generating circle.

23

Isaac Newton:
Man, Myth, and Mathematics

V. FREDERICK RICKEY

*T*HREE hundred years ago, in 1687, the most famous scientific work of all time, the *Philosophiae Naturalis Principia Mathematica* of Isaac Newton, was published. Fifty years earlier, in 1637, a work which had considerable influence on Newton, the *Discours de la Méthode*, with its famous appendix, *La Géométrie*, was published by René Descartes. It is fitting that we celebrate these anniversaries by sketching the lives and outlining the works of Newton and Descartes.

In the past several decades, historians of science have arranged the chaotic bulk of Newton manuscripts into a coherent whole and presented it to us in numerous high quality books and papers. Foremost among these historians is Derek T. Whiteside, of Cambridge, whose eight magnificent volumes overflowing with erudite commentary have brought Newton to life again.

> By unanimous agreement, the *Mathematical Papers* [of Isaac Newton] is the premier edition of scientific papers. It establishes a new criterion of excellence. Every further edition of scientific papers must now measure itself by its standard. [26, p. 87]

Other purposes of this article are to dispel some myths about Newton—for much of what we previously "knew" about him is myth—and to encourage the reader to look inside these volumes and to read Newton's own words, for that is the only way to appreciate the majesty of his intellect.

Newton's Education and Public Life

Isaac Newton was born prematurely on Christmas Day 1642 (O.S.), the "same" year Galileo (1564–1642) died, in the family manor house at Woolsthorpe, some 90 km NNW of Cambridge. His illiterate father—a "wild, extravagant, and weak man"—had died the previous October. His barely literate mother, Hanna, married the Reverend Barnabas Smith three years later, leaving Newton to be raised by his aged grandmother Ayscough.

Newton attended local schools and then, at age 12, traveled 11 km north to the town of Grantham, where he lived with the local apothecary and his books while attending grammar school. The town library had two or three hundred books, some 85 of which are still chained to the walls. Of course he studied Latin, also some Greek and Hebrew. Four years later, in 1658, he returned home to help his now twice-widowed mother manage the farm. Recognizing that Newton was an absent-minded farmer, his uncle William Ayscough (M.A. Cambridge, 1637) and former Grantham schoolmaster, Henry Stokes, persuaded his mother to send him back to Grantham to prepare for Cambridge. Judging by a mathematical copybook in use at Grantham in the 1650s, Stokes was a most unusual schoolmaster. The copybook con-

Reprinted from *College Mathematics Journal* 60 (Nov., 1987): 362–89; with permission of the Mathematical Association of America.

tained arithmetic through the extraction of cube roots, surveying, elementary mensuration, plane trigonometry, and elaborate geometric constructions, including the Archimedean bounds for π. This went far beyond anything taught in the universities of the period; consequently, contrary to tradition, Newton had a superior knowledge of mathematics before he went to Cambridge [33, pp. 110–111; 34, p. 101; updating 20, I, p. 3].

In 1661, eighteen-year-old Newton matriculated at Trinity College, the foremost college at Cambridge, as a subsizar (someone who earned his way by performing simple domestic services). This position reflected his wealthy mother's reluctance to send him to the university. At that time, Cambridge was little more than a degree mill. Lectures were seldom given. Fellows tutored primarily to augment their income. Although Newton did not finish any of the books from the established curriculum, which consisted mostly of Aristotelian philosophy, he did learn the patterns of rigorous thought from Aristotle's sophisticated philosophical system. A chance encounter with astrology in 1663 led him to the more enlightened "brisk part of the University" that was interested in the work of Descartes [28, p. 90]. The laxity of the university allowed him to spend the last year and a half of his undergraduate studies in the pursuit of mathematics. In 1665, Newton received his B.A. "largely because the university no longer believed in its own curriculum with enough conviction to enforce it." [28, p. 141].

In the summer of 1665, virtually everyone left the university because of the bubonic plague. The next March the university invited its students and Fellows to return for there had been no deaths in six weeks, but by June it was clear that the plague had not left, so the students who had returned left again. The university was able to resume again in the spring of 1667. Newton had left by August 1665 for Woolsthorpe. He returned on 20 March 1666, probably left again in June, but not until he had written his famous May 1666 tract on the calculus. He did not return to Cambridge until late April 1667, having revised the May tract into the October 1666 tract while back on the farm. "For whatever it is worth, the papers do not indicate that anything special happened at Woolsthorpe." [27, p. 116]. Much has been written about these plague years as the *anni mirabiles* of Newton, but the record clearly shows that he wrote the bulk of his mathematical manuscripts on the calculus while he was at Cambridge.

Myth: At the Woolsthorpe farm, during the plague years, Newton invented the calculus so that he could apply it to celestial mechanics.

The primary source for the myth [27, p. 110] of Newton's miracle years is this 1718 (unsent?) letter from Newton to Pierre DesMaizeaux:

In the beginning of the year 1665 I found the Method of approximating series & the Rule for reducing any dignity [= power] of any Binomial into such a series. The same year in May I found the method of Tangents . . . , & in November had the direct method of fluxions & the next year in January had the Theory of Colours & in May following I had entrance into y^c inverse method of fluxions. And the same year I began to think of gravity extending to y^c orb of the Moon & (having found out how to estimate the force with wch globe revolving within a sphere presses the surface of the sphere) from Keplers rule . . . I deduced that the forces wch keep the Planets in their Orbs must [be] reciprocally as the squares of their distances from the centers about wch they revolve: & thereby compared the force requisite to keep the Moon in her Orb with the force of gravity at the surface of the earth, & found them answer pretty nearly. All this was in the two plague years of 1665 & 1666. For in those days I was in the prime of my age for invention & minded Mathematicks & Philosophy [= Science] more then at any time since. [27, p. 109]

LUCASIAN PROFESSOR

At Trinity College, Newton became a Minor Fellow in 1667 and a Major Fellow in 1668. On 29 October 1669, at the age of 26, Newton became the second Lucasian Professor of Mathematics at Cambridge, succeeding Isaac Barrow (1630–1677). This post gave him security, intellectual independence, and a good salary. According to the Lucasian statutes, Newton was to lecture once a week during each of the three terms and to deposit ten of the lectures in the library. Even though this position had been designed by its founder Henry Lucas as a teaching post, not a research position [20, V, xiv], Barrow had already turned the position into a sinecure and Newton did not work much harder at the teaching aspects of the post. He deposited 3–10 lectures per year for the first seventeen years as Lucasian Professor, and none thereafter.

As a teacher, Newton left no mark whatsoever. Years later, when he was duly famous, one would expect that many people would have claimed to have attended his lectures, yet we know of only three. Perhaps the situation is best summed up by Newton's amanuensis (a human wordprocessor), the unrelated Humphrey Newton:

> He seldom left his chamber except at term time, when he read in the schools as being Lucasianus Professor, where so few went to hear him, and fewer that understood him, that ofttimes he did in a manner, for want of hearers, read to the walls. [6, X, 44]

LONDON AND BEYOND

In 1696, Newton accepted the post of Warden of the Mint (moving to London in March or April of 1696) and four years later became Master. In 1701, Newton resigned the Lucasian professorship. In 1703, he was elected President of the Royal Society, which he ruled with an iron hand until his death. In 1705, Newton was knighted by Queen Anne—not for his scientific advances, but for the service he had rendered the Crown by running (unsuccessfully) for Parliament in 1705 [28, p. 625]. For the rest of his life, Newton looked after the Mint and the Royal Society, twice revised his *Principia* (1713 and 1726), engaged in the infamous priority dispute with Leibniz, and toiled on secret research in religion and church history. His creative scientific life essentially ended when he left Cambridge.

Newton died 20 March 1727, at the age of 84, having been ill with gout and inflamed lungs

for some time. He was buried in Westminister Abbey.

NEWTON'S NACHLASS

At the time of his death Newton was wealthy. Income from the Lucasian Chair and farm rents brought £250 per year, sufficient for a handsome living for a bachelor don. When he became Master of the Mint, his salary jumped to £600 and he also received the perquisite of a commission on the amount of coinage. This amounted to some £1500 pounds per year, thus bringing his income to over £2000 pounds per year, a very substantial figure at that time. On his death his estate was valued at £30,000.

Newton left his library of some two thousand volumes to his nieces and nephews. The books were quickly sold to the Warden of Fleet Prison for £300 for his son Charles Huggins who was a cleric near Oxford. On Huggins's death in 1750 they were sold to his successor, James Musgrave, for £400. They remained in the Musgrave family until 1920, when some of them were sold at auction as part of a "Library of miscellaneous literature," fetching only £170. Although the family

FIGURE I

Isaac Newton

didn't know what they had sold, the book dealers knew what they had bought. Newton's annotated copy of Barrow's *Euclid*, which sold for five shillings, was soon in a bookseller's catalogue for £500. In 1927, the remaining 858 volumes were offered for £30,000 but remained unsold until 1943 when they were purchased for £5,500 and donated to the Wren Library at Trinity College. Of the thousand or so that were dispersed in 1920, some still show up unrecognized in bookshops. As recently as 1975, one was purchased in a Cambridge bookshop for £4. The books are easily identified by Newton's peculiar method of dog-earing by folding a page down to point to the precise word that interested him.

From Newton's library, 1736 books have now been located. Since his was a working library, a subject classification of the nonduplicates provides some information about Newton's interests. (For additional details, see [10, p. 59], from which the table below is condensed.) Newton also had access to the library of Barrow until Barrow's death in 1677, and to the Cambridge libraries until he moved to London in 1696.

Whiteside has tracked down every available scrap of material on Newton's mathematics and published it in *The Mathematical Papers of Isaac Newton* [20]. To really appreciate Newton's mathematical genius, one must grapple with his mathematics as he wrote it. The best place to gain an overview for this project is in Whiteside's wonderful introductions to these volumes and to the various papers in them. They have been used extensively in preparing this paper.

This biographical sketch has been intentionally kept short. For further details about Newton and his work, see the article by I. B. Cohen in the *Dictionary of Scientific Biography* (DSB) [6, X, pp. 42–103]. This is the single most authoritative reference work about the lives and contributions of deceased scientists. To avoid frequent references to it, we give dates after the first occurrence of an individual's name if the DSB contains an article about him. Two excellent biographies of Newton are Westfall's full scientific biography, *Never at Rest* [28], and Manuel's psychobiography *A Portrait of Isaac Newton* [15], some conclusions of which must be taken with care. For mathematical details, consult the many papers of Whiteside, only a few of which are cited here.

Newton's Mathematical Readings

The year 1664 was a crucial period in Newton's development as a mathematician and scientist, for it was then that he began to extend his readings beyond the traditional Aristotelian texts of the moribund curriculum to the new Cartesian ideas. (For details of Newton's nonmathematical readings, see McGuire [16].) According to Abraham DeMoivre (1667–1754), the expatriate French intimate of Newton during Newton's last years, the immediate impulse for Newton taking up mathematics was:

> In 63 [Newton] being at Sturbridge [international trade] fair bought a book of Astrology, out of a curiosity to see what there was in it. Read in it till he came to a figure of the heavens which he could not understand for want of being acquainted with Trigonometry.
> Bought a book of Trigonometry, but was not able to understand the Demonstrations.
> Got Euclid to fit himself for understanding the ground of Trigonometry.
> Read only the titles of the propositions, which

Subject	Mathematics	Physics and Astronomy	Alchemy	Theology
Number of titles (%)	126 (7.2)	85 (4.9)	169 (9.6)	477 (27.2)

Subject	History	Other Science	Other
Number of titles (%)	143 (8.2)	158 (9.0)	594 (33.9)

he found so easy to understand that he wondered how any body would amuse themselves to write any demonstrations of them. Began to change his mind when he read that Parallelograms upon the same base & between the same parallels are equal, & that other proposition that in a right angled Triangle the square of the Hypothenuse is equal to the squares of the two other sides.

Began again to read Euclid with more attention than he had done before & went through it.

Read Oughtreds [Clavis] which he understood tho not entirely, he having some difficulties about what the Author called Scala secundi & tertii gradus, relating to the solution of quadratick [&] Cubick Equations. Took Descartes's Geometry in hand, tho he had been told it would be very difficult, read some ten pages in it, then stopt, began again, went a little farther than the first time, stopt again, went back again to the beginning, read on till by degrees he made himself master of the whole, to that degree that he understood Descartes's Geometry better than he had done Euclid.

Read Euclid again & then Descartes's Geometry for a second time. Read next Dr Wallis's Arithmetica Infinitorum, & on the occasion of a certain interpolation for the quadrature of the circle, found that admirable Theorem for raising a Binomial to a power given. But before that time, a little after reading Descartes Geometry, wrote many things concerning the vertices Axes [&] diameters of curves, which afterwards gave rise to that excellent tract de Curvis secundi generis.

In 65 & 66 began to find the method of Fluxions, and writt several curious problems relating to that method bearing that date which were seen by me above 25 years ago. [20, I, pp. 5–6]

These words of DeMoivre, which agree with the report of Conduitt [20, I, pp. 15–19], certainly have an air of authenticity to them, and we know, based on extant manuscripts, that they are substantially correct (modulo Stokes's copybook). In the years 1664–1665, Newton made detailed notes on the following contemporary high level books, which influenced him at the very beginning of his mathematical studies.

- Barrow's *Euclidis Elementorum* (1655)
- Oughtred's *Clavis Mathematicae* (1631) in the 1652 edition

- *Geometria, à Renato des Cartes*, 1659–1661 edition of Schooten
- Schooten's *Exercitationum Mathematicarum Libri Quinque* (1657)
- Viète's *Opera Mathematica*, 1646 edition of Schooten
- Wallis's *Arithmetica Infinitorum* (1655)
- Wallis's *Tractatus Duo* (1659).

Let us look carefully at each of them to see what Newton learned.

EUCLID (FL. CA. 295 B.C.)

As DeMoivre indicated, Newton read Euclid as a student, although he did not develop any deep knowledge of the work then. Recall the story [28, p. 102] that Barrow examined Newton on Euclid and found him wanting. Newton was mainly influenced by books II (geometrical algebra), V (proportion), VII (number theory), and X (irrationals). The primary thing that he learned from Euclid was the traditional forms of mathematical proof [20, I, p. 12].

WILLIAM OUGHTRED (1575–1660)

At age fifteen, Oughtred went to Cambridge where he studied mathematics diligently on his own, for there was then hardly anyone there to teach him. He graduated B.A. in 1596 and M.A. in 1600. In 1603, he became a (pitiful) preacher and soon settled in as rector at Albury where he remained until his death.

It was as a teacher that he was renowned. He taught privately and for free. People came from the continent to talk to him, so wide was his reputation in mathematics. To instruct a young Earl, Oughtred wrote a little book of 88 pages that contained the essentials of arithmetic and algebra. *Clavis Mathematicae* (*Key to Mathematics*) published in 1631, was "a guide for mountain-climbers, and woe unto him who lacked nerve." [2, p. 29]. The style was obscure, the rules so involved they were difficult to comprehend. Oughtred carried symbolism to excess, a habit acquired by his most famous pupil, John Wallis. Nonetheless, *Clavis* established Oughtred as a capable mathematician and exerted a considerable effect in England, for it was a widely studied book in higher mathematics [32, p. 73].

Oughtred's Clavis, in the 1652 edition, was one of the first mathematical books that Newton read. From it he learned a very important lesson: Oughtred taught that *algebra was a tool for discovery* that did not need to be backed up by geometry [13, p. 408]. Newton held Oughtred in high regard, describing him as "a Man whose judgment (if any man's) may be safely relyed upon." [19, III, p. 364]

René Descartes (1596–1650)

René du Perron Descartes was born 31 March 1596 in La Haye (now La Haye-Descartes), France, a small town 250 km SSW of Paris. At the age of eight, he enrolled in the new Jesuit *collège* at La Flèche. There Descartes received a modern education in mathematics and physics—including the recent telescopic discoveries of Galileo—as well as more traditional schooling in the humanities, philosophy, and the classics. It was there, because of his then delicate health, that he developed the habit of lying abed in the morning in contemplation. Descartes retained an admiration for his teachers at La Flèche but later claimed that he found little of substance in the course of instruction and that only mathematics had given him any certain knowledge.

Descartes graduated in law from the University of Poitier in 1616, at age 20, but never practiced law as his father wished. By this time, his health improved and he enjoyed moderately good health for the rest of his life. Because he decided that he could not believe in what he had learned at school, he began a ten year period of wandering about Europe, spending part of the time as a gentleman soldier. It was during this period that Descartes had his first ideas about the "marvelous science" that was to become analytic geometry.

Although we have little detail about this period of his life, we do know that he hoped to learn

FIGURE 2

DISCOURS
DE LA METHODE
Pour bien conduire sa raison, & chercher la verité dans les sciences.
PLUS
LA DIOPTRIQVE.
LES METEORES.
ET
LA GEOMETRIE.
Qui sont des essais de cete METHODE.

A LEYDE
De l'Imprimerie de IAN MAIRE.
CIƆ IƆ C XXXVII.
Avec Privilege.

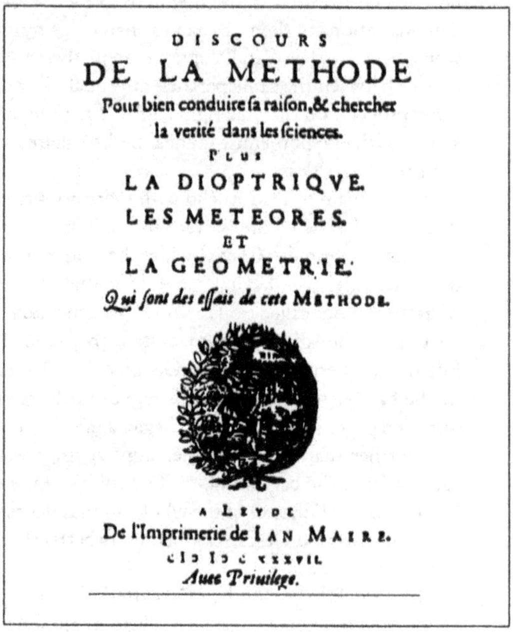

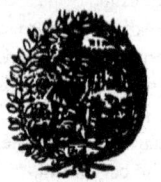

FIGURE 3

Pólya was very much influenced by Descartes [22, I, p. 56].

DESCARTES'S RULES:

The first was never to accept anything as true that I did not know evidently to be such; that is to say, carefully to avoid haste and bias, and to include nothing more in my judgments than that which presented itself to my mind so clearly and so distinctly that I had no occasion to place it in doubt.

The second was to divide each of the difficulties that I examined into as many parts as possible, and according as such division would be required for the better solution of the problems.

The third was to direct my thinking in an orderly way, by beginning with the objects that were simplest and easiest to understand, in order to climb little by little, gradually, to the knowledge of the most complex; and even for this purpose assuming an order among those objects which do not naturally precede each other.

And the last was at all times to make enumerations so complete, and reviews so general, that I would be sure of omitting nothing. [4, p. 16]

FIGURE 4

René Descartes

from "the book of the world." Descartes reached two conclusions. First, if he was to discover true knowledge he must carry out the whole program himself, just as a perfect work of art is the work of one master. Second, he must begin by methodically doubting everything taught in philosophy and looking for self-evident, certain principles from which to reconstruct all science.

In November 1628, Descartes had a public encounter with Chandoux, who felt that science was founded only on probability. By using his method to distinguish between true scientific knowledge and mere probability, Descartes easily demolished Chandoux. Among those present was the influential Cardinal de Bérulle, who charged Descartes to devote his life to working out the application of "his manner of philosophizing . . . to medicine and mechanics." To execute this design, Descartes moved to the Netherlands in 1628, where he lived for the next twenty years.

In Holland, Descartes worked at his system and, by 1634, had completed a scientific work

entitled *Le Monde.* He immediately suppressed the book when he heard about the recent condemnation of Galileo by the inquisition. He learned this from Marin Mersenne (1588–1648), a fellow student at LaFlèche and later the hub of the scientific correspondence network in Europe. This reveals Descartes's spirit of caution and conciliation toward authority (he was a lifelong devout Catholic). Later he took care to present his less orthodox views more obliquely.

Three hundred and fifty years ago, in 1637, the *Discours de la Méthode* [Figure 2], with appendices *La Dioptrique, Les Meteores,* and *La Géométrie,* appeared anonymously in Leyden, although it was soon widely known that Descartes was the author. The opening *Discours* is notable for its autobiographical tone, compressed presentation, and elegant French style. It was written in French since he intended—as did Galileo—to aim over the heads of the academic community to reach the educated people. Today, it is this opening *Discours,* with its problem-solving techniques [Figure 3], that is read.

In 1644, Descartes published *Principia Philosophiae,* a work in which he presented his views on cosmology. He expounded a mechanical philosophy in which a body could influence only those other bodies that it touched. Thus, for example, Descartes imagined space filled with "vortices" that moved the planets. This world view quickly became dominant in Europe. After the publication of Newton's *Philosophiae Naturalis Principia Mathematica,* the two scientific outlooks competed until well into the eighteenth century. Significantly—and this is reflected in the titles—Newton made mathematics indispensable for understanding the universe.

Queen Christiana of Sweden, ambitious patron of the arts and collector of learned men for her court, had seen the works of Descartes and pleaded with him to join her and teach her philosophy. She sent a man-of-war to fetch him but he was loath to go, in his words, to the "land of bears between rock and ice." But go he did. Being more of an athlete than a scholar, the 23-year-old

Queen wanted her lessons at five in the morning in a cold library with windows thrown wide open. This harsh land, where "men's thoughts freeze during the winter months," was too much for Descartes. A few months later he caught pneumonia and died on 11 February 1650.

CONTENTS OF THE *GEOMETRY*

The *Geometry* of Descartes is available to us in two English editions, the well known Smith-Latham translation [3] and the only complete English translation of the whole *Discours de la Méthode* by Olscamp [4]. The latter should be consulted since the appendix on *Optics* contains much interesting material on the conics.

In the first book of the *Geometry*, Descartes gave new geometric solutions of quadratic equations. For example [Figure 5], to solve the equation $z^2 = az - b^2$ (where a and b are both positive), Descartes drew the base line LM of length b and a perpendicular line LN of length $a/2$. Then he drew the circle with center N and radius NL. This circle cut the line perpendicular to LM at M in two points. The line segments MR and MQ are the solutions of the equation, as the reader can easily check. Descartes was aware that if the circle misses (only touches) the perpendicular to LM at M, then there is no (only one) solution to the equation.

Observe [Figure 5] that we have adopted Descartes's notation. In fact, his *Geometry* is the oldest mathematics text that we can read without having great difficulties with the algebraic notation. Descartes introduced the use of x, y, z for variables and a, b, c for constants, and he also introduced the exponential notation (except that he sometimes writes "aa" for our "a^2"). The only significant difference is that Descartes uses the symbol ∞ for equality.

Another problem Descartes dealt with in the first book was the problem of Pappus (fl. A.D. 300–350), which he mistakenly believed was still open. The problem asks for the locus of points such that the product of the distances (measured at fixed angles) to half of a fixed set of lines is equal to the

FIGURE 5

From p. 303 of Descartes's Géométrie

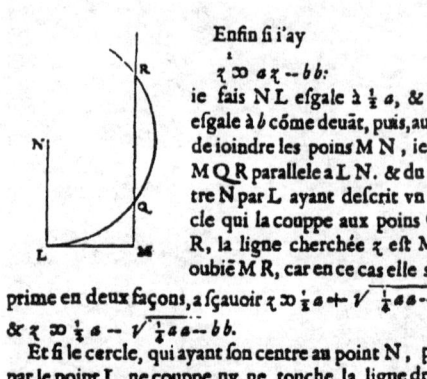

product of the distances to the other half (times a constant if the number of lines is odd). If there are three or four lines, Descartes showed that the locus is a conic. As an example with five lines, Descartes considered one horizontal line and four equally spaced vertical lines (Figure 6).

He set the product of the distances to the first, third, and fourth vertical lines equal to the product

FIGURE 6

Cartesian Parabola

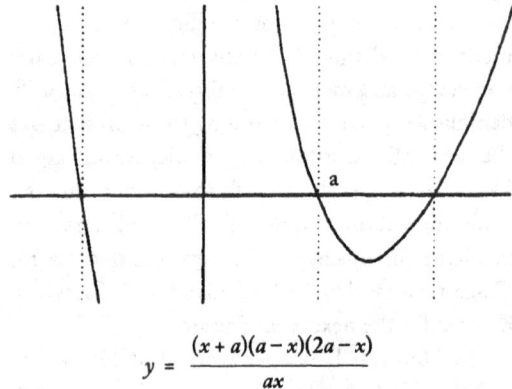

$$y = \frac{(x+a)(a-x)(2a-x)}{ax}$$

of the constant distance *a* between the lines, the distances to the second vertical line and the horizontal line, and obtained the equation $axy = (x + a)(a − x)(2a − x)$. Newton later called this curve the Cartesian Parabola. Since there were very few curves in Descartes's day, each received its own fancy name. This curve was only the second cubic (that is, a polynomial in two variables of degree three) ever discussed. The first was the Cissoid of Diocles (fl. ca. 190 B.C.). Descartes used his new curve extensively in his third book to solve equations of the fifth and sixth degrees as intersections of it and a circle.

GEOMETRICAL VS. MECHANICAL CURVES

The second book of Descartes's *Geometry* begins with a discussion of those curves which Descartes believed should be admitted into geometry. He does not consider the equation to be a sufficient representation of a curve, for equations are clearly algebraic objects. This forced him to always define curves by giving some geometric criterion. Later he derived the equation.

Descartes made a strict distinction between the curves that he called "geometrical" and those which he called "mechanical," but his explanation was none too clear. It has turned out that Descartes's geometrical (mechanical) curves are just the graphs of our algebraic (transcendental) functions. See Bos [1] for a full discussion. Descartes said that a curve is geometrical if it "can be conceived of as described by a continuous motion" [3, p. 43]. This excludes the spiral and the quadratrix because "they must be conceived of as described by two separate movements whose relation does not admit of exact determination" [3, p. 44]. Descartes allowed the use of a loop of thread to trace out a geometrical curve, as long as the shape of the string remained polygonal [3, p. 91]. Thus, the ellipse is a geometrical curve since it can be traced out using the familiar gardener's construction using string and pegs. In *La Dioptrique*, Descartes showed how to construct the hyperbola using straightedge and string [4, p. 135]. However, the curve generated by the moving end of a piece of thread as it unwinds from a spool is a mechanical curve, for the thread was curved while wound around the spool and straight after it unwinds.

> On the other hand, geometry should not include lines that are like strings, in that they are sometimes straight and sometimes curved, since the ratios between straight and curved lines are not known, and I believe cannot be discovered by human minds, and therefore no conclusion based upon such ratios can be accepted as rigorous and exact. [3, p. 91]

That straight and curved lines cannot be compared is an old dictum of Aristotle. Descartes's adoption of it was important for it set up the question of rectification of curves—that is, the problem of finding arc length of curves.

Let us now consider Descartes's argument for the Cartesian Parabola being a geometrical curve. He gave the following definition of a geometrical curve, then found its equation. Since its equation is the same as that of the Cartesian Parabola, the Cartesian Parabola is a geometric curve.

> I shall consider next the curve *CEG* [Figure 7], which I imagine to be described by the intersection of the parabola *CKN* (which is made to move so that its axis *KL* always lies along the straight line *AB*) with the ruler *GL* (which rotates about the point *G* in such a way that it constantly lies in the plane of the parabola and passes through the point *L*). [3, p. 84]

FIGURE 7

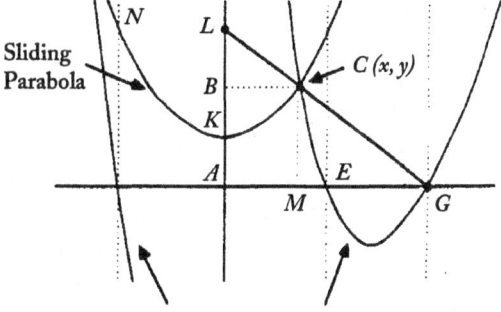

Cartesian Parabola

If we let *AB* be the *y*-axis and *AG* be the *x*-axis (Descartes used the opposite convention), then the Cartesian Parabola is the locus of all points *C(x,y)* of intersection of the parabola that slides up and down the *y*-axis and the ruler that pivots at the fixed point *G(2a, 0)* and passes through the point *L* moving along the *y*-axis with the parabola. The parabola has equation $x^2 = az$, where $a = KL$ and $z = BK$ (the focus of the parabola is one-fourth of the way from *K* to *L*). Descartes found the equation of the curve using classical geometry: Since the triangles *GMC* and *CBL* are similar, *GM/MC* = *CB/BL*, that is, $(2a-x)/y = x/BL$. Thus, we have

$$BK = a - BL = a - \frac{xy}{2a-x}.$$

But the equation of the parabola *CKN* can be written $BK = x^2/a$. Equating these expressions for *BK*, and simplifying, we obtain,

$$x^3 - 2ax^2 - a^2x + 2a^3 = axy,$$

which is the equation of the Cartesian parabola. (Note that the name comes from the fact that a parabola is sliding up and down the line.)

DESCARTES'S SUBNORMAL METHOD

In our calculus classes, one important problem is to find an equation of the tangent line to a curve at a given point on the curve. Problems were not phrased this way in the seventeenth century, because equations of lines was not a well developed topic. They asked (equivalently) for the subnormal for a given point on the curve, that is, the length of the segment on the *x*-axis between the abscissa of a point on the curve and the *x*-intercept of the normal line at that point. The subtangent was defined analogously.

Descartes presented a method for finding the subnormal [Figure 8]. If we can find a circle, with center *P* on the *x*-axis, that cuts a curve in precisely one point *C*, then the radius at that point is normal to the curve. But if the center of the circle through the point *C* be moved "ever so little" along the *x*-axis, the circle will cut the curve at two points.

FIGURE 8

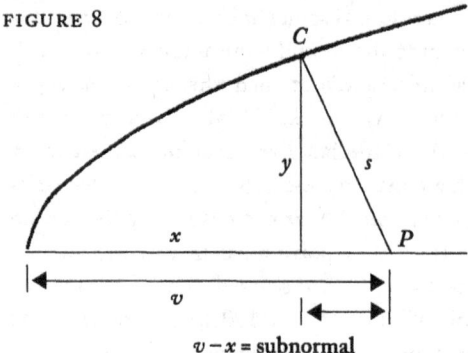

$v - x = \text{subnormal}$

This idea provided a means of finding the subnormal for any point (x_0, y_0) on the curve. Starting with the equation of the curve and the equation of a variable circle with center $P = (v, 0)$, find the equation giving their intersection. Then choose *P* so that the intersection equation has a double root.

Let us consider the case of the parabola $y^2 = kx$. The circle having center $P = (v, 0)$ and radius *s* that passes through the point (x_0, y_0) has equation $(v - x_0)^2 + y_0^2 = s^2$. Since (x_0, y_0) is on the parabola, $y_0^2 = kx_0$, and we obtain

$$x_0^2 + (k - 2v)x_0 + (v^2 - s^2) = 0.$$

This equation will have a double root if and only if the discriminant is zero; in which case, $x_0 = -(k - 2v)/2$, or

$$v - x_0 = k/2.$$

This looks mysterious today, but any mathematically literate contemporary of Descartes would know that the parabola has constant subnormal. Perhaps we should check this result using the new calculus: If $y^2 = kx$, then $2yy' = k$. So $y' = k/(2y)$. Thus, the normal line at (x_0, y_0) has slope $-y_0/(k/2)$. To plot the normal line, we go down y_0 from the point (x_0, y_0) to land on the *x*-axis, and then go right the constant distance $k/2$. Thus, the subnormal for any point on this parabola does indeed have constant length, $k/2$.

Descartes was justly proud of this work, for he wrote:

I have given a general method of drawing a straight line making right angles with a curve at an arbitrarily chosen point upon it. And I dare say that this is not only the most useful and most general problem in geometry that I know, but even that I have ever desired to know. [3, p. 95]

There is one final quotation from Descartes that is important here, for it deceived Newton—in a positive way:

> When the relation between all points of a curve and all points of a straight line is known [that is, when we have the equation of the curve] . . . it is easy to find . . . its diameters, axes, center and other lines [e.g., tangent and normal lines] or points which have especial significance for this curve . . . By this method alone it is then possible to find out all that can be determined about the magnitude of their areas, and there is no need for further explanation from me. [3, p. 92]

Newton believed Descartes's claim, that from the equation of a curve one can tell everything about it. This encouraged Newton to develop the variety of ad hoc techniques which he learned from the works of Descartes and Wallis into algorithms for solving problems about all curves. This was just one of the motivations that Newton had for inventing the calculus.

For further information about Descartes, see the DSB article by Crombie, Mahoney, and Brown [6, IVC, pp. 51–65]. The book by Scott [23] contains a detailed discussion of his mathematical work. Bos [1] gives an interesting study of Descartes's concept of curve. Of course, one should read the *Geometry* itself [3], [4].

FRANS VAN SCHOOTEN (1615–1660)

Schooten enrolled at the University of Leiden at age 16, where he was carefully trained by his father in the Dutch school of algebra. He met Descartes when the latter was in Leiden to supervise the printing of the *Discours de la Méthode* (1637). Schooten recognized the value of the work but had difficulty mastering its contents. So he went to Paris for further study, where he was cordially welcomed by Mersenne.

While in Paris, Schooten read the manuscripts of Pierre de Fermat (1601–1665) and Françoise Viète (1540–1603), and under commission of the famous Leiden publishing house of Elsevier, gathered all the printed works of Viète. This included Viète's most famous work, *In Artem Analyticam Isagoge* (*Introduction to the Analytic Art*) of 1591, which dealt mainly with the theory of equations. Because of this work, Viète is known as the father of algebra. Conscious of the great importance of the scattered works of Viète on algebra, geometry, and analysis, which had been published separately from 1579 to 1615, Schooten republished them with commentary as *Francisci Vietae Opera Mathematica* (1646). The work quickly became an indispensable collection of mathematical source materials, and Newton carefully studied a copy from the Cambridge libraries [20, I, p. 21]

Schooten returned to Leiden in 1643 and began working on a Latin translation of Descartes's *Géométrie*, which he published in 1649. Descartes had been dissatisfied with the form and argument of his *Géométrie* from the very day of its publication, and therefore encouraged the writing of commentaries clarifying its obscurities and developing its approach. Because of its valuable commentary and excellent figures, Schooten's edition was enthusiastically received. This success led him to prepare a much enlarged second edition that appeared in two volumes (1659–1661). It contained about 800 pages of commentary and new work, in addition to the 100 page translation of Descartes's *Géométrie*, and included [20, I, pp. 19–20]:

- Schooten's extremely valuable commentaries. Many of these details were derived directly from Descartes's own criticisms made in correspondence with Schooten.

- Florian Debeaune's (1601–1652) *Notae Breves*, a work which Descartes welcomed as a perceptive exposition of the more elementary aspects of his work. Debeaune posed the first inverse tangent problem.

- Jan Hudde's (1628–1704) studies on equations and extreme values. His rule for locating double roots of equations was useful in applying Descartes's tangent method. It was an important precursor of the derivative.

- Jan de Witt's (1629–1695) excellent tract on conic sections.

- An example of Fermat's extreme value and tangent method.

- Christiaan Huygens's (1629–1695) first publication, an improved method for finding the tangent to the conchoid.

- Hendrik van Heuraet's (1633–ca. 1660) rectification method, of which we shall say more below.

All of this shows the great effort that Schooten devoted to the training of his students and to the dissemination of their findings. Much of their work is available only in correspondence, careful studies of which are currently being made. It was from these editions of Schooten that mathematicians learned of the work of Descartes. It was the second Latin edition that Newton borrowed and annotated in the summer and autumn of 1664 (the copy he bought the following winter may have been the 1649 edition). It had an immense impact on his mathematical development; for after mastering it, he was current with research in the new analysis.

JOHN WALLIS (1616–1703)

Before attending Emanuel College, Cambridge, the only mathematics Wallis knew was what he learned from his brother who was preparing for a trade. At Cambridge, mathematics "were scarce looked upon." He took his M.A. in 1640 and was ordained. In 1649, he was appointed Savilian professor of Geometry at Oxford, an appointment that must have surprised those who thought the only mathematics he had done was to decode a few messages for the Parliamentarians.

This is not quite true, but, wrote Wallis "I had not then [in 1648] seen Descartes' Geometry." [20, III, p. xv]. In 1647 or 1648, he chanced upon Oughtred's *Clavis*, mastering it in a few weeks, and then rediscovered Cardano's formula for the cubic. In 1648, at the request of Cambridge professor of mathematics John Smith, he reworked Descartes's treatment of the fourth degree equation by factoring it into two quadratics. As soon as he was appointed Savilian professor at Oxford, he took up the study of mathematics, with rare energy and perseverance, and soon became one of the best mathematicians in Europe. He held the post for 50 years.

Wallis's *Operum Mathematicorum Pars Altera* (Oxford, 1656) was a fat and rather motley two-part collection of his early mathematical lectures, commentaries, and researches [20, I, p. 23]. It contained his *De Sectionibus Conicis* (dated 1655), a treatise of 110 pages that was the first elementary text on the conics treated from the Cartesian viewpoint. In an appendix, Wallis tried to extend the approach to higher plane curves, especially the cubical parabola $a^2y = x^3$, where the constant a^2 was used to preserve dimensionality. He successfully found the subtangent, but had trouble with the graph because he did not feel comfortable with negative numbers. He also introduced the semi-cubical parabola $ay^2 = x^3$, a curve that played a very important role in the development of the calculus [30, p. 295–298]. Quite suddenly the mathematical world had been presented with a powerful analytic geometry, only to find that there were few curves on which to practice it. The new perspective of Wallis—which took some time to be adopted by the mathematical community—was that any algebraic equation in two variables defines a curve [13, p. 238].

Together with his conic sections, Wallis published the work on which his fame rests, *Arithmetica Infinitorum* (dated 1656; printed 1655). This volume developed from his study of the *Opera Geometrica* (1644) of Torricelli (1608–1647). Wallis tried to apply these methods to the quadrature of the circle, but not even the study of the voluminous *Opus Geometricum* (1647) of Gregorius Saint Vincent (1584–1667), helped. Out of the project of squaring the circle, he did get his famous infinite product for $\pi/4$.

The *Arithmetica Infinitorum* exerted a singularly important influence on Newton when he studied it in the winter of 1664–1665. From it, Newton learned of the problem of quadratures, or, as we now say, finding areas under curves. Newton probably also read Wallis's *Tractatus Duo* (1659) that presented his research on the cycloid, cissoid, and other geometrical figures.

RECTIFICATION OF CURVES

By 1638, Descartes suspected that the logarithmic spiral might be rectifiable; that is, the length of an arc of the curve could be computed. Even if correct, this would not cause him any difficulties because the spiral is a mechanical curve, and Descartes only accepted Aristotle's dictum that straight lines and curved lines could not be compared for geometrical curves. In 1657, Huygens found the length of an arc of a parabola; but he used a mechanical curve in his solution, and thus Descartes's version of Aristotle's dictum was still intact. Also Huygen's method did not generalize.

The first geometrical curve to be rectified in a geometric way was Wallis's semi-cubical parabola $ay^2 = x^3$. As often happens, several people solved the problem simultaneously: William Neil in 1657, Hendrick van Heureat in 1659 [14], and Pierre de Fermat in 1660. Of course, a priority dispute erupted. Heureat's solution was the most influential because it was published in Schooten's second Latin edition of Descartes's *Geometry*. The proof used the new classical geometry of the seventeenth century and was fairly intricate (for details, see [8] or [13]). The method of proof was to replace the problem of rectification of the semi-cubical parabola by a simpler problem, the quadrature of an ordinary parabola.

This transformation of the problem to a simpler one shows up even when we do the problem today with the calculus, but it is so slick that it is easy to miss what happens. Starting with $y^2 = x^3$ (it is no accident that we still do this first today), we obtain $(y')^2 = 9x/4$. Thus, the arc length from, say, $(0, 0)$ to $(4, 8)$, is

$$L = \int_0^4 \sqrt{1 + (9x/4)} \, dx.$$

The substitution $u = 1 + (9x/4)$ transforms this into

$$L = (4/9) \int_1^{10} \sqrt{u} \, du.$$

The first of these integrals represents an arc length, whereas the second stands for the area under a parabola. Today, we just look at these as two simple integration problems, but in the old days B(efore) C(alculus), these were viewed as two separate kinds of problems.

Heuraet's method was entirely general. When Newton saw the proof, he realized the value of transforming one type of problem into another. This is one of the roots of the Fundamental Theorem of Calculus. It is the biggest swap of all—we trade integration for anti-differentiation. This is precisely what Newton did soon after he read Heureat's proof. (For a full history of the rectification problem, see Hofmann [11, Ch. 8].)

CONCLUDING REMARKS ABOUT NEWTON'S READINGS

In order to do creative work, a mathematician "needs an adequate notation, a competent knowledge of mathematical structure and the nature of axiomatic proof, an excellent grasp of the hard core of existing mathematics and some sense of promising line for future advance." [20, I, p. 11]. The works that Newton chose to read in 1664 and 1665 magnificently met these needs. He took his arithmetic symbolism from Oughtred, his geometrical form from Descartes. Of course, he grafted on new modifications of his own while creating the calculus. He learned elementary scholastic logic in grammar school and traditional forms of mathematical proof from Euclid. He learned the new analytic geometry of the seventeenth century from Schooten and de Witt, topics in algebra and the theory of equations from Viète, Oughtred, Schooten, and Wallis. Most importantly, he learned of the twin problems of infinitesimal analysis: From Descartes, the method of tangents; from Wallis, quadrature. There were plenty of open problems for Newton to attack. Without doubt, the two strongest influences on Newton were Descartes and Wallis. [20, I, pp. 11–13]

It is of as much interest to note *what Newton did not read.* We miss the names of Napier, Briggs, Harriot, Desargues, Pascal, Fermat, Stevin, Kepler, Cavalieri, and Torricelli. Among the Greeks there is only Euclid, not Apollonius nor Archimedes. In fact, Newton seemed to dislike the method of exhaustion. There is great significance in this lack of knowledge of ancient mathematics and of the new classical (as opposed to analytic) geometry of the seventeenth century. He was not hampered by its knowledge. Had Newton gained a deep knowledge of classical geometry and the new classical geometry of his century, I conjecture it would have hindered his invention of the calculus (and similarly for Leibniz who was also ignorant of classical geometry).

As Westfall points out [28, p. 100] about Newton's readings: "In roughly a year, without benefit of instruction, he mastered the entire achievement of seventeenth-century analysis and began to break new ground." In fact, by mid-1665, Newton's urge to learn from others seems to have abated [20, I, p. 15].

Newton's Works

Newton was an extraordinary scientist because he made so many fundamental contributions to different fields:

- Mathematics, both pure and applied
- Optics and the theory of light and color
- Design of scientific instruments

- Synthesis and codification of dynamics
- Invention of the concept and law of universal gravity.

In addition, we now know, and are willing to admit, that he spent immense amounts of time working on:

- Alchemy
- Chronology, church history, and interpretation of the Scriptures.

The range and depth of Newton's intellectual pursuits never ceases to amaze us.

As a first step in understanding Newton's contributions, consider the chart below that indicates when Newton was involved in various research areas. One might think that Newton thought about everything all of the time, but the manuscript record shows that he worked on only a few areas at any one time, and these were not necessarily—in his mind at least—disjoint.

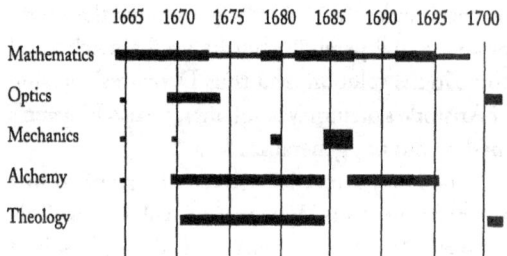

FIGURE 9

Newton's areas of activity

We begin with a synopsis of Newton's mathematics as presented in Whiteside's edition of Newton's *Papers* [20]. This will be followed by a thumbnail sketch of each of these areas of Newton's intellectual efforts. Since it is impossible to discuss all of his contributions here, only a few examples of Newton's mathematical work will be discussed in detail. These were chosen with the teacher in mind, to provide examples that can be used in the classroom.

Volume I. (1664–1666). The volume begins with Newton's annotations on the works of Oughtred, Descartes, Schooten, Viète, and Wallis. The bulk consists of research on analytic geometry and the calculus. Newton turns Descartes's subnormal technique into the notion of curvature, and Hudde's rule for double roots into fluxions (differentiation). We see the calculus become an algorithm in mid-1665. This early work on the calculus was summarized in the October 1666 tract on fluxions. In a schematic diagram, Whiteside [20, I, p. 154] shows how all of these ideas came together to give birth to the calculus. The volume ends with miscellaneous work on trigonometry, the theory of equations, and geometrical optics.

Volume II. (1667–1670). Work on classification of cubics begins here and was published as an appendix to his *Optics* (1704). In this volume, we see Newton struggling with the graphs. The most important work on the calculus is the hastily composed 1669 tract *De Analysi* that summarizes all of his work thus far. He gave a copy of this to Barrow in 1669 to assert his priority over Nicolaus Mercator (1619–1687) whose *Logarithmotechnia* (1668), with its infinite series for the logarithm, had just appeared. Half the volume consists of his annotations on the *Algebra* of Kinckhuysen. One piece of Newton's advice here is too good not to pass on to our students:

> After the novice has exercised himself some little while in algebraic computation . . . I judge it not unfitting that he test his intellectual powers in reducing easier problems to an equation, even though perhaps he may not yet have attained their resolution. Indeed, when he is moderately well versed in this subject . . . then will he with greater profit and enjoyment contemplate the nature and properties of equations and learn their algebraic, geometrical and arithmetical resolutions. [pp. 423–425]

Volume III. (1670–1673). Although Barrow encouraged Newton to revise *De Analysi* for publication, the booksellers were uninterested. But he did combine the two earlier works on the calculus and many new results in a 1671 tract, with an important foundational change: he postulated a fluent variable of time for his fluxions; that is, all his derivatives are time derivatives. Also, here is an investigation of Huygen's pendulum clock and more research on geometric optics.

Volume IV. (1674–1684). Research in theology and alchemy kept him busy (Figure 9), though his work on mathematics never entirely stopped. This volume contains some of Newton's research on algebra, number theory, trigonometry, and analytical geometry. In the middle of this period, he became fascinated with the classical geometry of the Greeks. Only at the end of this period did Newton show great interest in fluxions and infinite series.

Volume V. (1683–1684). The bulk of this volume consists of Newton's ninety-seven self-styled "lectures," deposited as his Lucasian lectures on algebra for the period 1673–1683. The *Arithmetica Universalis* given here is an incomplete revision of the algebra lectures. Its published version was his most read work, not the papers on calculus.

Volume VI. (1684–1691). Halley's visit in August 1684 turned Newton's interest to the geometry and dynamics of motion, the subject of this entire volume. The work dates from the period 1684–1686, and is arguably as creative as the miracle years of 1664–1666.

Volume VII. (1691–1695). In the early winter of 1691–1692, Newton wrote *De Quadratura Curvarum*, on the quadrature of curves. He also dealt with classical geometry (1693), higher plane curves, and finite-difference approximations (1695). As always, Whiteside has "taken care to preserve all the significant idiosyncrasies, contractions, superscripts and archaic spellings" of the "ink-blobbed, much-cancelled and often rudely scrawled manuscripts." [p. ix]

Volume VIII. (1697–1722). Most mathematicians will find this the most interesting volume after the first, for it contains Newton's solution (simply stated without proof) of the brachistochrone problem as well as documents related to the priority dispute. (To see that this dispute involved much more than mathematics, read Hall's *Philosophers at War* [9].)

We calculus teachers should refrain from telling our students that Newton invented the calculus because he was motivated by physical considerations. Although applications are an excellent reason for studying the calculus, in Newton's case the record is clear: first mathematics, then applications.

THE BINOMIAL THEOREM

On the frontispiece of the first volume of Newton's *Papers* we see the manuscript where he took up the age old problem of squaring the circle, or (to make the activity sound more respectable) the quadrature of the circle. He became interested in this problem after reading Wallis's *Arithmetica Infinitorum*. Newton learned there how to evaluate the integrals (here expressed in Leibniz's notation) $\int_0^x (1-x^2)^{n/2}dx$, where n is an even integer. Newton tabulated the values of these integrals in his attempt to find the area of a circle ($n = 1$). To see how he did this consider the case when $n = 6$:

$$\int_0^x (1-x^2)^{6/2}dx = 1(x) + 3(-x^3/3) + 3(x^5/5) + 1(-x^7/7).$$

The factors in parentheses are recorded in the rightmost column of the table below. The coefficients, 1, 3, 3, 1, are recorded in the column labeled $n = 6$. In general, to evaluate $\int_0^x (1-x^2)^{n/2}dx$, sum the products of the values in the nth-column by the corresponding terms in the rightmost column.

$n=0$	$n=2$	$n=4$	$n=6$	$n=8$	...	times
1	1	1	1	1	...	x
	1	2	3	4	...	$-x^3/3$
		1	3	6	...	$x^5/5$
			1	4	...	$-x^7/7$
				1	...	$x^9/9$
					⋮	⋮

Wallis had also tabulated these integrals, but since he used 1 rather than x as an upper limit, he did not see the pattern. But Newton recognized it as "Oughtreds Analyticall table," from his readings of Oughtred's *Clavis* [20, I, p. 452]. We, of course, now call this Pascal's triangle. Newton knew that each number in the table is the sum of the number to its left and the one above that, so he decided to extend the pattern backwards for all even values of n. Thus he obtained:

... $n=-2$	$n=0$	$n=2$	$n=4$	$n=6$	$n=8$	...	times
1	1	1	1	1	1	...	x
-1	0	1	2	3	4	...	$-x^3/3$
1	0	0	1	3	6	...	$x^5/5$
-1	0	0	0	1	4	...	$-x^7/7$
1	0	0	0	0	1	...	$x^9/9$
⋮	⋮	⋮	⋮	⋮	⋮	⋮	⋮

To extend this table to odd values of n, Newton used a complicated proportionality argument (see [31] for details). Later, in a letter to Leibniz [19, I, pp. 130–131], Newton provided an easier explanation for the extension. When n is even, say, $n = 2m$, the kth entry in the nth column is given by the binomial coefficient ( ) $= m!/k!(m - k)!$. Newton ignored the restriction that n must be even and used the formula for binomial coefficients when n was odd. For example, the fourth entry in the $n = 1$ column is given by

$$\binom{1/2}{4-1} = \frac{(1/2)(1/2 - 1)(1/2 - 2)}{(1)(2)(3)}$$

Thus, he obtained the results shown at the top of page 499.

Now, from the $n = 1$ column, Newton was able to draw the conclusion that he sought:

$$\int_0^x (1-x^2)^{1/2}dx = x + (1/2)(-x^3/3) + (-1/8)(x^5/5)$$
$$+ (1/16)(-x^7/7) + \cdots.$$

For $x = 1$, this gives an infinite series for the area of (a quadrant of) a circle. From this, Newton jumped to the conclusion that a similar "interpolation" could be done on curves (we would say, on func-

... $n=-3$	$n=-2$	$n=-1$	$n=0$	$n=1$	$n=2$	$n=3$	$n=4$	$n=5$	$n=6$	$n=7$	$n=8$	...	times
1	1	1	1	1	1	1	1	1	1	1	1	...	x
$-\dfrac{3}{2}$	-1	$-\dfrac{1}{2}$	0	$\dfrac{1}{2}$	1	$\dfrac{3}{2}$	2	$\dfrac{5}{2}$	3	$\dfrac{7}{2}$	4	...	$-\dfrac{x^3}{3}$
$\dfrac{15}{8}$	1	$\dfrac{3}{8}$	0	$\dfrac{1}{8}$	0	$\dfrac{3}{8}$	1	$\dfrac{15}{8}$	3	$\dfrac{35}{8}$	6	...	$\dfrac{x^5}{5}$
$-\dfrac{35}{16}$	-1	$\dfrac{5}{16}$	0	$\dfrac{1}{16}$	0	$\dfrac{1}{16}$	0	$\dfrac{5}{16}$	1	$\dfrac{35}{16}$	4	...	$-\dfrac{x^7}{7}$
$\dfrac{315}{128}$	1	$\dfrac{35}{128}$	0	$\dfrac{-5}{128}$	0	$\dfrac{3}{128}$	0	$\dfrac{-5}{128}$	0	$\dfrac{35}{128}$	1	...	$\dfrac{x^9}{9}$
$\vdots$	$\vdots$	$\vdots$	$\vdots$	$\vdots$	$\vdots$	$\vdots$	$\vdots$	$\vdots$	$\vdots$	$\vdots$	$\vdots$		$\vdots$

tions) as well as on their quadratures (integrals), and then guessed the Binomial Theorem for fractional exponents. He checked this result several ways. First, he formally used the square root algorithm to obtain the series

$$(1 - x^2)^{1/2} = 1 - (1/2)x^2 - (1/8)x^4 - (1/16)x^6 - \cdots .$$

Then he checked that it agreed with the Binomial Theorem. Next, he squared both sides of the above equation to see that an equality resulted. As a further check, he used formal long division to obtain an infinite series for $(1 + x)^{-1}$. Note the wonderful research techniques he is using. Nonetheless,

> The paradox remains that such Wallisian interpolation procedures, however plausible, are in no way a

proof, and that a central tenet of Newton's mathematical method lacked any sort of rigorous justification . . . Of course, the binomial theorem worked marvelously, and that was enough for the 17th century mathematician. [31, p. 180]

Newton became tremendously excited with his new tool, the Binomial Theorem, which became a mainstay of his newly developing calculus. He also did such bizarre computations as approximating $\log(1.2)$ to 57 decimal places.

The Binomial Theorem was Newton's first mathematical publication. It appeared in Wallis's *Treatise of Algebra* (Figure 10) in a summary of Newton's two famous letters to Leibniz in 1676 [24, pp. 330–331]. These letters are readily available, with ample commentary, in Newton's *Correspondence* [19, II, pp. 20–47 and 110–161].

FIGURE 10   *First publication of the Binomial Theorem, 1685*

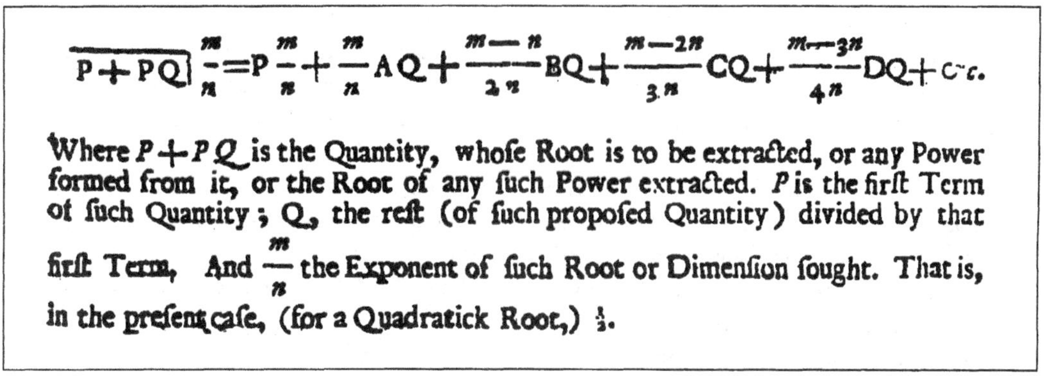

$$\overline{P+PQ}\Big|^{\tfrac{m}{n}} = P^{\tfrac{m}{n}} + \tfrac{m}{n}AQ + \tfrac{m-n}{2n}BQ + \tfrac{m-2n}{3n}CQ + \tfrac{m-3n}{4n}DQ + \&c.$$

Where $P + PQ$ is the Quantity, whose Root is to be extracted, or any Power formed from it, or the Root of any such Power extracted. P is the first Term of such Quantity; Q, the rest (of such proposed Quantity) divided by that first Term, And $\dfrac{m}{n}$ the Exponent of such Root or Dimension sought. That is, in the present case, (for a Quadratick Root,) $\tfrac{1}{2}$.

OPTICS

Newton's earliest work on optics was done at Cambridge and the experiments continued at Woolsthorpe during the plague, but was not put in near final form until he was preparing his Lucasian lectures for 1670–1672. It had long been known (see, for example, Descartes [4, p. 335]) that when light passed through a prism it was dispersed into a colorful spectrum. Newton was able to give a quantitative analysis of this behavior and to devise a new theory of light. In February 1671/72 (the slash date was used because England had not yet adopted the Gregorian calendar), this resulted in Newton's first publication in optics, the lengthy title of which also provides an abstract:

> A Letter of Mr. Isaac Newton, Mathematick Professor in the University of Cambridge; containing his New Theory about Light and Colors: Where Light is declared to be not Similar or Homogeneal, but consisting of difform rays, some of which are more refrangible than others: And Colors are affirm'd to be not Qualifications of Light, deriv'd from Refractions of natural Bodies, (as 'tis generally believed;) but Original and Connate properties, which in divers rays are divers: Where several Observations and Experiments are alleged to prove the said Theory. [18, p. 47].

This work engendered a controversy with Robert Hooke (1635–1703), who claimed to have published the ideas earlier. As a consequence, Newton became extremely reluctant to publish. In fact, the *Optics* was not published until 1704, the year after Hooke's death.

In developing his theory of light, Newton realized that lenses caused chromatic aberration. This set him thinking about telescope design, and he concluded that the problem could be avoided by using mirrors instead of lenses. Consequently he designed a reflecting telescope, built one himself, and then described it in the March 25, 1672 issue of the *Philosophical Transactions*. These first papers of Newton have been photoreproduced by I. B. Cohen [18], along with a valuable introduction by Thomas Kuhn. Rather than describe Newton's theory of light (which has been done by Alan Shapiro in the first volume of *The Optical Papers of Isaac Newton* [21]), we shall briefly discuss telescope design. This provides an interesting classroom example of the reflective properties of the conics.

The first reflective telescope was designed by James Gregory (1638–1675) and published in his *Optica Promota* of 1663, a work which Newton did not read until after he had invented his own telescope. Gregory's telescope consists of a concave primary mirror (on the right in Figure 11a) that is parabolic in shape, and a concave secondary mirror that is elliptical (strictly speaking, the surfaces generated by rotating these conics about the axis of the telescope). The incoming rays of starlight bounce off the parabolic mirror and are reflected through its focus. Beyond that focus is an elliptical mirror that shares a focus with the parabola and has its other focus behind a small hole in the primary mirror. Thus, after the reflected rays of starlight pass through the common focus of the parabola and ellipse, they are reflected off the elliptical secondary mirror and converge at the second focus of the ellipse. Gregory tried to have a telescope built to his design, but the opticians were unable to polish the mirrors properly.

In 1668, Newton placed a flat secondary mirror between the primary parabolic mirror and its focus [Figure 11b]. The eyepiece was located at the side of the telescope. Incoming rays of starlight reflect off the parabolic mirror and head for its focus F. Before they get there, they are reflected off the flat mirror. Then they converge toward F', the point symmetric to F with respect to the plane of the flat mirror. This invention remained unknown until Newton made another one (casting and polishing the mirrors himself) and presented it to the Royal Society of London on 11 January 1672. This so impressed the members that they elected him a Fellow of the Royal Society at that very same meeting.

Later in 1672, another telescope design [Figure 11c] was published by Guillaume Cassegrain (fl. ca. 1672) in France and abstracted in the *Trans-*

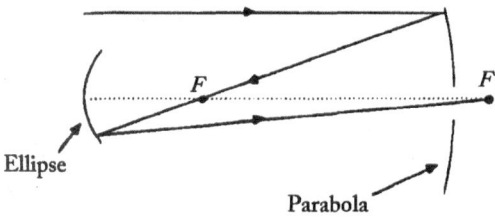

FIGURE IIa   *Gregorian Telescope, 1663*

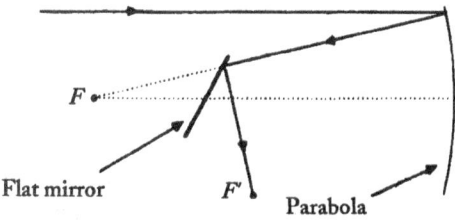

FIGURE IIb   *Newtonian Telescope, 1668*

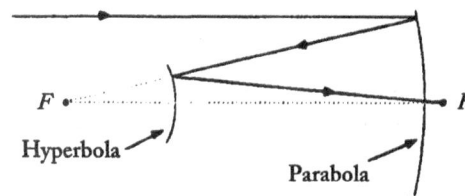

FIGURE IIc   *Cassegrain Telescope, 1672*

actions of the Royal Society. The concave primary mirror is again a parabola with a hole in the center, and the secondary is a convex hyperbolic mirror which shares a focus with the parabola and has its other focus behind the hole in the parabola. Rays of starlight reflected from the primary parabolic mirror head toward the focus of the parabola. Before reaching that focus, they are reflected by the hyperbolic mirror toward the other focus of the hyperbola.

Cassegrain claimed that his design was superior to Newton's. In what was to become his typical style, Newton marshalled his evidence and attacked furiously. He claimed that Cassegrain's idea was not only a minor modification of Gregory's, but also optically inferior. Cassegrain retreated into anonymity. But Newton was wrong

about the superiority of the Gregorian telescope. In 1779, Jesse Ramsden (1735–1800) showed that the combination of a concave and a convex mirror partially corrects the spherical aberrations, whereas in the Gregorian telescope, the aberrations of the two concave mirrors are additive. Today the Cassegrain model is used in most large reflectors.

A mathematical result of Newton's work on optics grew out of the problem of grinding a hyperbolic mirror (although he did not use one, the possibility of a hyperbolic lens was noted by Descartes [4, p. 139]). Newton, and independently Christopher Wren (1632–1723), discovered that the hyperboloid of one sheet was a ruled surface. Newton used this result to show how to make a hyperboloid of one sheet on a lathe by holding the chisel obliquely to the axis of the lathe.

RELIGION.

Despite inheriting his stepfather's theological library and buying several theological books when he came to Cambridge, Newton's serious study of theology began only in the early 1670s. No doubt this came about because the position of Fellow at Trinity required that one had to be ordained in the Anglican Church within seven years of receiving the M.A. In Newton's case, this was by 1675. Not being one to do anything halfway, Newton became engaged in an extensive reading program that took him through all the early Church Fathers. As a result, ordination became impossible for he had become a heretic.

Newton became an Arian or Unitarian—he denied the Trinity—of deep conviction and remained so for the rest of his life. His argument was: "Though Christ was the only begotten son of God, and hence never merely a man, he was not equal to God, not even after God exalted him to sit at his right side as a reward for his obedience unto death." [29, p. 130]. Newton arrived at this position through a careful analysis of Scripture. He believed that a deceitful Roman Church had manipulated the Emperor Theodosius to introduce the false doctrine of the Trinity into the Scriptures in the fourth century. The *Book of Revelation* was crucial to Newton's interpretation. He believed that the Roman Church was the "Great Apostasy" and never ceased to hate and fear it [28, p. 321].

By 1675, Newton was making plans to leave Cambridge for he knew that as a Fellow at the College of the Holy and Undivided Trinity he could not reveal his Arian views. To do so would be socially unacceptable, and he never in his life did so, except obliquely to a few people of similar persuasion. That he read the situation correctly is indicated by the dismissal of William Whiston (1667–1752), Newton's successor in the Lucasian Chair, for the uncompromising expression of Unitarian views. But just at this time, the Crown granted a special dispensation that the occupant of the Lucasian Professorship was not required to be ordained. Thus, Newton could stay at Cambridge. Newton's theological studies continued until work on the *Principia* interrupted [Figure 9]. In London, he was able to take up his theological studies again, and they continued for the rest of his life. (For further details, see [28, pp. 309–334] or [29].)

ALCHEMY

Newton's interest in alchemy has long been embarrassing to some scholars, while others delight in this trace of hermeticism and dub him a mystic. But there is now no doubt that he was a serious practitioner [Figure 9]. From 1669 (when he bought his first chemicals) until 1684 (when work on the *Principia* interrupted), Newton spent long hours in the "elaboratory." Newton again practiced alchemy from 1686 until 1696, but after he moved to London he never took it up again seriously [27, p. 121]. Newton did plan on adding alchemical references to the second edition of the *Principia* although he never did so. (For details on his alchemical work, see [5].) One benefit of this work was that he was able to cast the speculum for his first telescope.

In 1693, Newton suffered a nervous breakdown of uncertain duration and severity. There is no doubt that he frequently tasted his chemicals,

but whether it was caused by mercury poisoning is debatable [7, pp. 88–90].

When Newton wrote to Oldenburg in 1673 that he intended "to be no further solicitous about matters of Philosophy" [19, I, p. 294], and to Hooke in 1679 that "I had for some years past been endeavouring to bend my self from Philosophy to other studies in so much yt I have grutched the time spent in yt study" [19, II, p. 300], we must take him at his word. During most of the decade of the 1670s, Newton preferred theology and alchemy to physics and mathematics.

THE PRINCIPIA

In his old age, Newton liked to reminisce and he himself started the story of the falling apple. We have four independent accounts of the tale [7, p. 29–31]. Here is Conduitt's [28, p. 154]:

> In the year 1666 he retired again from Cambridge . . . to his mother in Lincolnshire & whilst he was musing in a garden it came into his thought that the power of gravity (wch brought an apple from the tree to the ground) was not limited to a certain distance from the earth but that this power must extend much farther than was usually thought. Why not as high as the moon.

So let us grant that a falling apple started Newton thinking about gravity during the plague years; even if he made up the story it is harmless. But his retelling of this event, in his 1718 letter to DesMaizeaux (which we quoted earlier), is not harmless. Newton attempted to push back the date of his discovery of the law of universal gravitation to the plague years. His papers tell quite a different story.

In late 1664, Newton learned Kepler's third law: the square of the time that it takes a planet to make one elliptical revolution around the sun is proportional to the cube of the mean distance from the sun, that is, $T^2 \sim R^3$. The following January, Newton discovered the Central Force Law (see the letter to DesMaizeaux), which Huygens independently discovered and first published without proof in his *Horologium Oscillatorium* (1673) (see [12]). The Central Force Law states that the centrifugal (center fleeing) force acting on a body traveling about a central point is proportional to the square of the speed and inversely proportional to the radius of the orbit: $F = S^2/R$. Strictly speaking, this "force" is an acceleration, but we shall follow Newton's usage.

Newton was able to discover that the gravitational force between a planet and the sun must be inversely proportional to the square of the distance between them. If a planet travels with uniform speed around a circular (not elliptical) orbit of radius R in time T, then its speed S is $2\pi R/T$. Thus,

$$F = \frac{S^2}{R} = \frac{(2\pi R/T)^2}{R} = 4\pi^2 \left(\frac{R^3}{T^2}\right)\left(\frac{1}{R^2}\right).$$

By Kepler's third law, R^3/T^2 is constant, and hence, $F \sim 1/R^2$. Newton left off at this point, devoting most of the next decade to alchemy and theology, though he was never completely divorced from mathematics (Figure 9).

Hooke, Halley, and Wren were able to make this same deduction by 1679, but the problem of explaining the elliptical orbits remained. On 24 November 1679, Hooke wrote to Newton suggesting a private "philosophical," that is scientific, correspondence on topics of mutual concern. In this letter, Hooke mentioned *his* hypothesis of "compounding the celestiall motions of the planetts [out] of a direct motion by the tangent & an attractive motion towards the centrall body." [19, II, p, 297]. This does not seem to be much of a hint for proving that if the inverse square law holds, then the planets must move in elliptical orbits. But it started Newton thinking about the question again. Hooke gave further encouragement on January 17, when he wrote "I doubt not but that by your excellent method you will easily find out what that Curve must be." [19, II, p. 313]. Newton did succeed in finding the answer, but he kept it to himself.

It was also in 1679 that Newton learned of Kepler's second law: the radius from the sun to a planet sweeps out equal areas in equal times. It seems strange that Newton would have learned of the third law as a student in 1664, but not about the second until years later. The explanation is that the third law was generally accepted in the scientific community because it could be empirically verified, whereas the second was much more of a conjecture.

In August 1684, Edmond Halley (1656–1743) visited the 41-year-old Lucasian Professor at Cambridge. He asked the question that had been consuming him and his friends Hooke and Wren at the Royal Society in London: What path would the planets describe if they were attracted to the sun with a force varying inversely as the square of the distance between them? Newton replied at once that the orbits would be ellipses. Since this was the expected answer, Halley asked Newton how he knew. Newton astonished him by answering that he had calculated it. Halley asked to see Newton's computation, but as Newton seemingly saved every scrap of paper he ever wrote on, he (not surprisingly) could not find it. Perhaps he did not want to find it; his desire to be left alone to pursue his own interests, his fear of controversy, and his reluctance to publish would all make Newton want to carefully check his proof over again before he showed it to anyone. In November of 1684, Newton did send the computation to Halley in London, who was so excited that he prompted Newton to expand his work. (Weinstock [25] has challenged the common view that this proof actually appears in the *Principia*.)

Newton put aside his alchemical and theological studies to work on what was to become the most significant scientific treatise ever written: *Philosophiae Naturalis Principia Mathematica*. It took Newton eighteen months of intense intellectual effort to compose his masterpiece, but the time was not just spent in writing up results that he had completed long ago. Many critical ideas in the *Principia* were not developed until the treatise

FIGURE 12

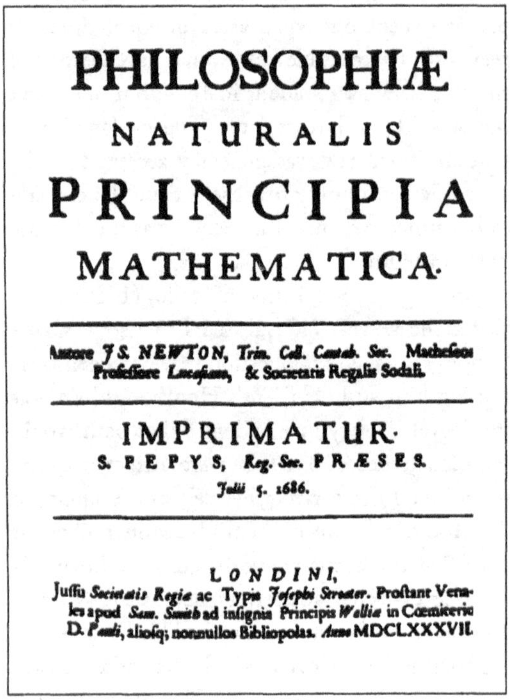

PHILOSOPHIÆ

NATURALIS

PRINCIPIA

MATHEMATICA.

Autore JS. NEWTON, Trin. Coll. Cantab. Soc. Mathefeos Profeffore Lucafiano, & Societatis Regalis Sodali.

IMPRIMATUR.

S. PEPYS, Reg. Soc. PRÆSES.

Julii 5. 1686.

LONDINI,

Juffu Societatis Regiae ac Typis Josephi Streater. Proftant Venales apud Sam. Smith ad infignia Principis Walliæ in Cœmiterio D. Pauli, aliosq; nonnullos Bibliopolas. Anno MDCLXXXVII.

was being written. In particular, Newton created the concept of universal gravity during this period.

Myth: *Though Newton used the notation of the calculus in arriving at his results, he was careful in the* Principia *to recast all the work in the form of classical Greek geometry understandable by other mathematicians and astronomers.*

Newton started this myth himself in the midst of the priority dispute with Leibniz. If he could argue that he had used his calculus in composing the *Principia,* then he could claim that he did not steal the calculus from Leibniz who published his first paper on the (differential) calculus in 1684.

The method of fluxions [Newton's calculus] is intrinsically algebraic rather than geometrical, and there is not the slightest reason—in the historical evidence or in logic—to suppose that the argument of the *Principia* was ever cast in an algebraic rather than the geometric mode in which it was published. [9, p. 28]

The geometrical format of the *Principia* is explained by the fact that around 1678, Newton became fascinated with classical geometry [Figure 9]. The *Principia* appears to be densely packed classical geometry, but that is only a façade. One need only read a bit to realize that it is packed with the informal geometrical *ideas* of the new analysis, the calculus. However, the formal machinery of the algebraic algorithms of the calculus is not to be found there. In order to make this point clear, it would help to look at the proof of a proposition from the *Principia.* Book I, Section II, Proposition I, Theorem I says:

The areas which revolving bodies describe by radii drawn to an immovable centre of force do lie in the same immovable planes, and are proportional to the times in which they are described. [17, p. 40]

That is, if the gravitational force (whatever it might be) always acts toward a fixed point *S,* then Kepler's equal area law holds.

Newton's proof begins with classical geometry [Figure 13]. Suppose we consider equal time intervals, and that the body moves from *A* to *B* in one of those intervals. In the next interval it would move, on the same straight line, from *B* to *c* if no external force acts on it. The triangles *SAB* and *SBc* that are swept out in these equal time intervals have equal areas since the bases *AB* and *Bc* are equal, and the triangles have the same altitude. However, if at *B* "a centripetal [center seeking] force acts at once with a great impulse," then the body moves to some other point *C* (in the same plane) in the next time interval. The Parallelogram Law of Forces determines the location of the point *C;* the lines *Cc* and *SB* are parallel. Triangles *SBc* and *SBC* also have the same area, since they have a common base *SB* and their altitudes are equal, namely, the distance between the parallel lines *SB* and *Cc.* By transitivity, the triangles *SAB* and *SBC* have equal areas. Similarly for other triangles in the diagram. So far the proof was easy geometry. Next, Newton used an idea from his calculus: "Now let the number of those triangles be augmented,

FIGURE 13   *Newton's* Principia, *p. 37*

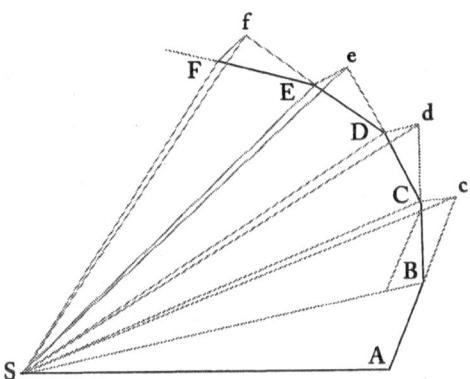

and their breadth diminished *in infinitum* . . . their ultimate perimeter *ADF* will be a curved line: and therefore the centripetal force, by which the body is continually drawn back from the tangent of this curve, will act continually" and the areas traced out in equal times will be equal [17, pp. 40–41].

So that was really quite easy. The geometric ideas of the calculus are used constantly in the *Principia,* but the algebraic notations are not.

Conclusion

On 5 February 1675/76, Newton wrote to Hooke [19, I, p. 416]:

> What Des-Cartes did was a good step. You have added much . . . If I have seen further it is by standing on ye shoulders of Giants.

While it is important to realize that Newton recognized the contributions of his predecessors, we must by now feel that Newton was the greatest giant of all. Just as Westfall, after twenty years of effort preparing *Never at Rest,* was more in awe of Newton when he finished than when he began, we too may realize that the closer we get to Newton, even when standing on the shoulders of Whiteside, the bigger the giant becomes.

Yes, Newton was a genius. That is undeniable. But he was not a Greek god. For all his faults, he displayed characteristics that we should tell our students about, for they are the keys to his, and their, success:

- He built on the best work of the past.

- He had brilliant insights.

- He worked "by thinking continually."

* He had stubborn perseverance.

- He steadily expanded his inquiries.

- He made mistakes—and learned from them.

Newton's success was a synergistic combination of innate genius and immense effort. This is the lesson of history.

ACKNOWLEDGMENTS

Portions of this survey paper were presented as lectures at an Ohio Section MAA Short Course on "The History of the Calculus" (Ashland College, July 1986), the University of Wisconsin/ Green Bay, Denison University, and Allegheny College. The many questions of those audiences and the detailed comments of friends and referees were most helpful.

Figures 2, 5, 12, and 13 were obtained courtesy of The Department of Rare Books and Special Collections, The University of Michigan Library; Figure 10 was obtained courtesy of the Center for Archival Collections, Bowling Green State University. [Figures 1 and 4 are reproductions of engravings from *A Portfolio of Eminent Mathematicians,* edited by David Eugene Smith (Chicago: Open Court Publishing Company, 1896).—*Frank Swetz*]

REFERENCES

1. H. J. M. Bos. "On the Representation of Curves in Descartes' *Géométrie.*" *Archive for History of Exact Sciences* 24 (1981): 295–338.

2. Florian Cajori. *William Oughtred, a Great Seventeenth-Century Teacher of Mathematics.* Chicago: Open Court: 1916.

3. René Descartes. *The Geometry of René Descartes.* Translated by D. E. Smith and Marcia L. Latham. Chicago: Open Court, 1925. Reprinted by Dover, 1954.

4. ———. *Discourse on Method, Optics, Geometry, and Meteorology.* Translated by Paul J. Olscamp. Bobbs-Merrill, Indianapolis: The Library of Liberal Arts, 1965.

5. Betty Jo Teeter Dobbs. *The Foundations of Newton's Alchemy or "The Hunting of the Green Lyon."* Cambridge University Press, 1975.

6. Charles Coulston Gillispie (editor). *Dictionary of Scientific Biography.* 16 vols. New York: Scribners, 1970–1980. If there are dates following an individual's name, then there is a signed article in this work concerning him.

7. Derek Gjersten. *The Newton Handbook.* London: Routledge and Kegan Paul, 1986. See abstract in the Media Highlights Column of this CMJ issue.

8. A. W. Grootendorst and J. A. van Maanen. "Van Heureat's Letter (1659) on the Rectification of Curves. Text, Translation (English, Dutch), Commentary." *Nieuw Archief voor wiskunde* 3:30 (1982): 95–113.

9. A. Rupert Hall. *Philosophers at War. The Quarrel Between Newton and Leibniz.* Cambridge University Press, 1980.

10. John Harrison. *The Library of Isaac Newton.* Cambridge University Press, 1978.

11. Joseph E. Hofmann. *Leibniz in Paris, 1672–1676. His Growth to Mathematical Maturity.* Cambridge University Press, 1974.

12. Christiaan Huygens. *The Pendulum Clock or Geometrical Demonstrations Concerning the Motion of Pendula as Applied to Clocks.* Translated by Richard J. Blackwell. Iowa State University Press, 1986.

13. Timothy W. Lenoir. *The Social and Intellectual Roots of Discovery in Seventeenth Century Mathematics.* Ph.D. dissertation. Indiana University, 573 pp., 1974. University Microfilms order number 75–1718.

14. Jan A. van Maanen. "Hendrick van Heureat (1634–1660?): His Life and Mathematical Work." *Centaurus* 27 (1984): 218–79.

15. Frank E. Manuel. *A Portrait of Isaac Newton.* Harvard University Press, 1968; and Washington, D.C.: New Republic Books (paperback), 1979.

16. J. E. McGuire and Martin Tamny. *Certain Philosophical Questions: Newton's Trinity Notebook.* Cambridge University Press, 1983.

17. Isaac Newton. *Sir Isaac Newton's Mathematical Principles of Natural Philosophy and his System of the World.* Revision of Andrew Motte's 1729 translation by Florian Cajori. University of California Press, 1934.

18. ———. *Isaac Newton's Papers and Letters on Natural Philosophy and Related Documents.* Edited by I. B. Cohen. Harvard University Press, 1958.

19. ———. *The Correspondence of Isaac Newton.* Edited by H. W. Turnbull et al. 7 vols. Cambridge University Press, 1959–1977.

20. ———. *The Mathematical Papers of Isaac Newton.* Edited by D. T. Whiteside. 8 vols. Cambridge University Press, 1967–1981. References are by volume and page number.

21. ———. *The Optical Papers of Isaac Newton.* Edited by Alan E. Shapiro. Cambridge University Press, 1984.

22. George Pólya. *Mathematical Discovery.* 2 vols. New York: John Wiley, 1962.

23. J. F. Scott. *The Scientific Work of René Descartes (1596–1650).* London, 1952.

24. John Wallis. *A Treatise on Algebra Both Historical and Practical . . . ,* London, 1685.

25. Robert Weinstock. "Dismantling a Centuries-Old Myth: Newton's *Principia* and Inverse-Square Orbits." *American Journal of Physics,* 50 (1982): 610–17). See abstract in the Media Highlights column of this CMJ issue.

26. Richard S. Westfall. "Award of the 1977 Sarton Medal to D. T. Whiteside." *Isis* 69 (1978): 86–87.

27. ———. "Newton's Marvelous Years of Discovery and Their Aftermath: Myth versus Manuscript." *Isis* 71 (1980): 109–121.

28. ———. *Never at Rest, A Biography of Isaac Newton.* Cambridge University Press, 1980.

29. Richard S. Westfall. "Newton's Theological Manuscripts." *Contemporary Newtonian Research.* Edited

by Z. Bechler. Dordrecht: D. Reidel, 1982, pp. 129–43.

30. Derek Thomas Whiteside. "Patterns of Mathematical Thought in the later Seventeenth Century." *Archive for History of Exact Sciences* 1 (1961): 179–388.

31. ———. "Newton's Discovery of the General Binomial Theorem." *Mathematical Gazette* 45 (1961): 175–80.

32. ———. "Sources and Strengths of Newton's Early Mathematical Thought." *The Annus Mirabilis of Sir Isaac Newton, 1666–1966.* Edited by Robert Palter. M.I.T. Press, 1970, pp. 69–85.

33. ———. "Newton the Mathematician." *Contemporary Newtonian Research.* Edited by Z. Bechler. Dordrecht: D. Reidel, 1982, pp. 109–27.

34. ———. "Newtonian Motion." *Isis* 73 (1982): 100–107. Essay review of [28].

Editor's Note: This article was written to celebrate the 300th anniversary of the publication of Newton's *Principia.* It won the Mathematical Association's 1987 George Pólya Award for exceptional expository writing. Another informative publication celebrating the anniversary of the *Principia* but published in 1989 is:
   Let Newton Be! A New Perspective on his Life and Works. John Fauval *et al.,* eds.

$\mathcal{N}$ewton's $\mathcal{M}$ethod of $\mathcal{F}$luxions

In perfecting his techniques of "differentiation," Isaac Newton borrowed heavily on the work of his predecessors, particularly John Wallis and Isaac Barrow. Combining their results with the insights of his own genius, Newton produced a technique of differentiation openly based on change. In his *Method of Fluxions* written in 1671, he considers a curve as the path of a moving point, thus the coordinates of the point are in a constant state of change. Newton termed a changing quantity a *fluent* and its rate of change the *fluxion* of the fluent. If a fluent is given by y, then its fluxion would be represented by $\dot{y}$ (in modern terms dy/dt). The fluxion of $\dot{y}$ would be represented by $\ddot{y}$ etc. Using the concept of fluent, Newton then devised the notion of the moment of a fluent, that is, the infinitely small amount of change experienced by a fluent in an infinitely small time interval O. He represented the moment of a fluent by $\dot{x}O$. In calculations, $\dot{x}O$ raised to powers 2 or higher would be zero. Consider how Newton's method of fluxions could be used to obtain the derivative of y with respect to x for $y = 3x^2 + 2x$.

Under the process of change y becomes $(y + \dot{y}O)$ and x becomes $(x + \dot{x}O)$ thus

$$(y + \dot{y}O) = 3(x + \dot{x}O)^2 + 2(x + \dot{x}O) \text{ expanding}$$

$$y + \dot{y}O = 3x^2 + 6x\,\dot{x}O + (\dot{x}O)^2 + 2x + 2\dot{x}O$$

now $(\dot{x}O)^2 = 0$ and $3x^2 + 2x = y$ so

$$\dot{y}O = 6x\,\dot{x}O + 2\dot{x}O \text{ or}$$

$$\dot{y}O = (6x + 2)\,\dot{x}O \text{ and}$$

$$\frac{\dot{y}O}{\dot{x}O} = 6x + 2 \quad \text{which in modern form}$$

is recognized as $\dfrac{dy}{dx} = 6x + 2$.

24

The Newton-Leibniz Controversy Concerning the Discovery of the Calculus

D O R O T H Y   V.   S C H R A D E R

The State of Analysis in the Seventeenth Century

The seventeenth century was one of activity and advancement in the world of mathematics. Analytic methods had become familiar tools to most of the mathematicians of the period; geometry was being employed to verify and demonstrate analytic conclusions; and special attention was focused on problems dealing with the infinite.

> The time was indeed ripe, in the second half of the seventeenth century, for someone to organize the views, methods, and discoveries involved in the infinitesimal analysis into a new subject characterized by a distinctive method of procedure.[1]

Unfortunately, not one but two men did just that. We say unfortunately, because the methods of the calculus developed by Sir Isaac Newton in England and Gottfried Wilhelm Leibniz on the continent were essentially the same, yet the dispute over the rights of the two discoverers developed into a controversy which has not yet been settled. Both these mathematicians and their followers stooped to tactics which were most unworthy of men of intelligence and honor; as a result, the development of mathematics in England was brought almost to a standstill for a full century.

Reprinted from *Mathematics Teacher* 55 (May, 1962): 385–96; with permission of the National Council of Teachers of Mathematics.

The judgment of history seems to be that credit belongs to both individuals equally.

> [I]t might be far better to speak of the evolution of the calculus. Nevertheless, inasmuch as Newton and Leibniz, apparently independently, invented algorithmic procedures which were universally applicable and which were essentially the same as those employed at the present time in the calculus, . . . there will be no inconsistency involved in thinking of these two men as the inventors of the subject.[2]

It must be remembered, however, that these two inventors are not responsible for the definitions and ideas which are considered basic to the calculus today. Only in the present generation, after more then two centuries of development, has there been laid the foundation of mathematical rigor on which the calculus now rests.

Newton's Method of Fluxions

Newton called his discovery the Method of Fluxions and described it in terms of geometry. The direct method of fluxions can be summarized in the solution of the mechanical problem: "The length of the Space described being continually given, to find the Velocity of the Motion at any time proposed,"[3] or "The relation of the flowing Quantities being given, to determine the relation of their Fluxions."[4] Elsewhere, Newton refers to these flowing quantities as Fluents. The inverse method of fluxions can be summarized in the inverse of the problem given above: "The Velocity of the Motion being continually given, to find the Length of the Space described at any Time proposed,"[5] or "An Equation being proposed including the Fluxions of Quantities, to find the Relation of those Quantities to one another."[6] These direct and inverse methods of fluxions are, of course, the familiar differential and integral calculus. Newton further described his fluents as quantities which are to be considered as gradually and indefinitely increasing; these he represented by the last letters of the alphabet: x, y, z. The velocities by which every fluent is increased by its generating motion, he called fluxions and designated by "pointed" or "prickt" letters, corresponding to the fluents involved: $\dot{x}$, $\dot{y}$, $\dot{z}$. The moment of a fluent, its velocity multiplied by an infinitely small quantity, o, he represented in the fluxional notation as $\dot{x}o$.[7]

Leibniz's Differential and Integral Calculus

Instead of the flowing quantities and velocities of Newton, Leibniz worked with infinitely small differences and sums. He used the now familiar $\frac{dy}{dx}$ instead of Newton's dotted letters for the derivative symbol; he used $\int$ for his integration symbol while Newton used either words or a rectangle enclosing the function. Newton himself asserted that his "prickt" letters were equivalent to Leibniz's $\frac{dy}{dx}$ and that by $\boxed{\dfrac{aa}{64x}}$ he meant the same thing that Leibniz meant by $\int \frac{aa}{64x}$.[8]

Leibniz was more interested in developing a notation for his new method than was Newton. The Englishman used the dot symbolism in his fluxionary calculus but did not employ it in his treatise on analysis nor in his famous *Principia*. In one report, printed anonymously but commonly believed to be written by Newton, we read, "Mr. Newton doth not place his Method in Forms of Symbols, nor confine himself to any particular

Sort of symbols for Fluents and Fluxions."[9] This independence of symbols, which Newton apparently thought praiseworthy, is today considered one of the major weaknesses of his method. Modern mathematics is almost wholly dependent upon the symbols by which it is expressed, so much so that someone has characterized mathematics as the science in which one operates with and on symbols, neither knowing nor caring what these symbols mean, if indeed they have any meaning. Felix Klein has lauded Leibniz for that very type of symbolism. He notes that $\int y$ and not $\int y \, dy$ was used in Leibniz's manuscripts, and credits him with being the founder of modern, formal mathematics for recognizing that it makes no difference what, if any, meaning is attached to the differentials, but that, if appropriate rules of operation are defined for them and the rules correctly applied, something reasonable and correct will result.[10]

Whether or not Klein errs in attributing too deep a perception to Leibniz is difficult to determine, but it is true that Leibniz was much concerned with finding the best possible notation for his calculus. He experimented with various symbols, explained them to different people, asked advice of a number of mathematicians, and used the dx and dy notation for a long time before he published it. He finally selected the particular form of notation because he saw the great need of being able to identify easily the variable and its differential.[11] Johann Bernoulli, who worked with him on integration, wanted to call the new branch of mathematics "calculus integralis" and use I as an integration symbol; Leibniz preferred the name "calculus summatorius" and the symbol $\int$. Today, we use Bernoulli's name and Leibniz's symbol.

The method of fluxions and the differential and integral calculus differ in more than their notation. While the two methods are essentially the same in that they can be reduced to a common method, they start from different principles. Newton made use of infinitely small quantities to find time derivatives, which he called fluxions; he stated specifically that his mathematical quantities were to be considered as described by continuous motion, not as existing in infinitely small parts. Leibniz, on the other hand, made the infinitely small quantities themselves the basic concepts in his differentials.[12] Newton dealt with a finite quantity which is the ratio of two infinitely small quantities, the ratio of velocities; Leibniz dealt with the finite sum of an infinite number of infinitely small quantities.[13]

The Quarrel

The famed Newton-Leibniz controversy concerning the discovery of the calculus involves more than merely the question of priority of time. Mutual accusations of plagiarism, secrecy which manifested itself in cryptograms, letters published anonymously, treatises withheld from publication, assertions of friends and supporters of the two men, national jealousies, and the efforts of would-be peacemakers, all serve to complicate the situation and make such a tangle of truth and falsehood, information and misinformation, that it can probably never be solved conclusively. We can at best review the major events and draw a few general conclusions.

In June, 1669, Newton, who had become interested in mathematical analysis during his days as an undergraduate at Cambridge, sent to Isaac Barrow, the geometer, a manuscript entitled "Analysis per Equations Numero Terminorum Infinitas," in which the underlying principles of the theory of fluxions were indicated. Barrow was impressed with the work, and, in a letter dated June 20, 1669, mentioned it to a mathematician friend, Collins. On July 31 of the same year, he sent the manuscript to Collins, who copied it and returned the original to Barrow. Collins was in correspondence with many of the leading mathematicians of England and the continent; in letters dated from 1669 to 1672, he communicated Newton's discoveries to Gregory, Bertet, Vernon, Slusius, Borelli, Strode, Townsend, and Oldenburg.[14]

At about the same time, Leibniz was working on problems suggested by the theories of Cavalieri; in 1671, he dedicated to the French Academy a paper, "Theoria Motus Abstracti," in which he showed that he was considering the use of infinitely small quantities in these problems.[15] This was not a well-developed theory of differential calculus, but it does indicate that Leibniz's mind was working along the lines of infinitesimal analysis, as was Newton's.

In 1672, Newton composed a treatise on fluxions, which, however, was not published until John Colson translated it from Latin into English and published it in London in 1736 under the title, *Method of Fluxions*. Why it was not published at the time it was written seems not to be known.

Early in 1673, Leibniz was in London where he visited Oldenburg, a fellow-countryman and the secretary of the Royal Society. As he attended meetings of the society and read mathematical papers before it, it is quite possible that he met Collins, who was a friend of Oldenburg. In 1890, notes of this London visit were discovered in the royal library at Hanover, where Leibniz had been librarian. These notes show extracts from Newton's work on optics, which was not published until 1704, but make no mention of mathematics. Hathaway[16] finds this omission very strange and somehow indicative that Leibniz saw Newton's paper, "De Analysis," during this visit.

By March of 1673, Leibniz was back in Paris, where he began a serious study of geometry with Huygens. In July of the same year, he wrote to Oldenburg, discussing his work on series. In reply, Oldenburg told him of some of Newton's and Gregory's discoveries on series and tangents. Leibniz, working with infinitely small sums and differences, defined the general problem of the area of the curve and from it developed the algorithm of the differential and the integral calculus, a logical outcome of the studies on which he had been engaged for several years. Manuscripts in the library at Hanover show that by the end of 1675 he had a clear idea of the principles of the calculus and had invented the notation.[17]

On June 13, 1671, Newton, answering a request from Leibniz, wrote a brief account of his method of quadrature by means of infinite series and discussed the binomial theorem. Leibniz replied, August 27, 1676, and asked for more details. In September of that year, Leibniz was again in London, where he spent a week with Collins, saw Newton's manuscript of the "De Analysis," and made copious notes from it. One author asserts that "since its pages were open freely to him at that time it is constructive proof that they were as freely open to him for the two months in 1673 that he was in London."[18] This reasoning seems obscure, but it may be that Hathaway's testimony is colored by anti-German feeling,[19] which is perhaps understandable, as he wrote shortly after World War I.

After his return to Paris, Leibniz received another letter from Newton, dated October 24 and sent through Oldenburg. This was fifteen closely written pages, discussing series and mentioning fluxions, but giving no detailed information. Newton himself said later that he had told Leibniz of his method of fluxions but disguised it in an anagram of transposed characters under the sentence, "Data aequatione quotcunque fluentes quantitates involvente, fluxiones invenire, et vice versa," which, transliterated, reads, "Given any equation whatsoever involving flowing quantities, to find the fluxions, and vice versa." This as an anagram would be: 6a 2c d ae 13e 2f 7i 3l 9n 4o 4q 2r 4s 9t 12v x. One wonders how much Leibniz could ever make out of that anagram, and having reconstructed Newton's sentence, could he deduce the method of fluxions from it? "Whoever can form a certain sentence properly out of 6 a's, 2 c's, a d and so on, will see as much as one sentence can show about Newton's mode of proceeding."[20]

The question immediately arises as to why Newton bothered to tell Leibniz anything at all if he was going to conceal it so thoroughly in his anagram. The device is not unprecedented. Galileo used to give his discoveries to his friends in the form of carefully dated cryptograms in order to establish priority. Sometimes academies and learned societies were the trustees for intellectual secrets. Such secrecy was considered necessary to protect the rights of the inventors. There was much jealousy among the learned who, in their desire to conceal their discoveries, were wont to publish theorems without proof or demonstration. Newton deposited his method, by anagram, in the hands of his rival.[21]

There is another question concerned with this famous anagram. If, as has been frequently stated, Newton had given a clear indication of his method of fluxions in his "De Analysis" of 1669, which was, with his knowledge, being circulated and discussed by Collins, why did he consider it necessary seven years later to conceal the method? Could it be that he had merely hinted at it in 1669 and did not have it so well-developed as has been supposed?

After receiving Newton's letter, Leibniz replied on June 21, 1677, through Oldenburg, that he too had a method of drawing tangents, not by fluxions but by differentials; he quite frankly and openly explained the differential calculus and its applications. However, he did not tell Newton that he had seen the 1669 manuscript in London but a few weeks before, that he knew of Newton's work, and that the anagram was useless. Was Leibniz being unfair to Newton or was the 1669 treatise less complete than it is reputed to have been? Newton did not answer and the correspondence ceased, perhaps due to the death of Oldenburg which occurred in August of 1678.

In 1683 Collins died; in 1684, in the *Acta Eruditorum* of Leipzig, Leibniz published his first paper on the calculus, an account similar to that which he had given Newton. If he had obtained his initial ideas from Newton via Collins, that might explain his desire to conceal the theft from Collins by not publishing the ideas as his own until after the death of Collins. Or is there another reason for the long delay in publication? Leibniz made a vague reference to Newton's having a method similar to his own but he made no claim to being the first or sole inventor. The first "Nova Methodus pro Maximis et Minimis" merely developed the rules for differentiation. Later works in the same publication gave an exposition of the principles from a formal viewpoint.[22]

Leibniz and the two Bernoullis were making rapid progress with the new and powerful analytic method when the first edition of Newton's *Principia* was published in 1687. In Book II, Lemma II, Newton explained the fundamental principle of the fluxionary calculus and in a scholium added:

> In letters which went between me and that most excellent geometer, G. W. Leibniz, ten years ago, when I signified that I was in the knowledge of a method of determining maxima and minima, of drawing tangents, and the like, and when I concealed it in transposed letters involving this sentence (Data aequatione, quotcunque, fluentes quantitates involunte, fluxiones invenire, et vice versa; that is, Having given any equation involving ever so many flowing quantities, to find the fluxions, and vice versa) that most distinguished man wrote back that he had also fallen on a method, which hardly differed from mine, except in his forms of words and symbols.[23]

Later, when the controversy was at its height, this scholium was quoted as evidence that Newton recognized Leibniz's rights as a second or simultaneous inventor. Such is not entirely the case; all Newton admitted was that Leibniz did have a method, however he learned it. Newton, after Leibniz's death, asserted that the scholium had been intended as a challenge to Leibniz to prove his priority if he could, not as an admission of his equality. In the second edition of the *Principia*, Newton added a phrase to the scholium, making it a bit more accurate. In that edition, the last words read, " . . . which hardly differed from mine except in the forms of words and symbols, and the concept of the generation of quantities."[24] In the third edition, the scholium was changed entirely and another subject inserted; neither Leibniz's name nor his work was mentioned.

In 1695, Dr. John Wallis chided Newton for being in possession of the method of fluxions for thirty years and never publishing it in its entirety; he had made reference to it in his own complete works, published in 1693, but felt that his treatment of the subject was most inadequate.

In the same year, John Bernoulli challenged Europe with two problems, to be solved in six months. To allow the mathematicians of the world time to work on these problems, Leibniz requested an additional year, which was granted. During this extension, Newton heard of the problems and solved them both in a single evening, sending his solutions to the president of the Royal Society the day after he had received the problems. Acknowledging the receipt of Newton's solutions, Leibniz, in the Leipzig *Acts*, managed to convey the impression that Newton was a pupil of his who, because he had mastered the calculus, was able to solve the problems.[25]

It was four years later, in 1699, that the hidden rivalry flared into open hostility. Fatio de Duillier, a Swiss mathematician from Geneva, who had been living in England for about ten years and was a close friend of Newton, published a memoir in which he claimed for himself independent invention of the calculus (which claim seems to have been completely ignored) and implicitly accused Leibniz of plagiarizing from Newton.

> I am bound to acknowledge that Newton was the first, and by many years the first, inventor of this calculus: from whom, whether Leibniz, the second inventor, borrowed anything, I prefer that the decision should lie, not with me, but with others who have had sight of the papers of Newton, and other additions to this same manuscript.[26]

Leibniz answered the charge, referred to Newton's scholium in the *Principia*, and, ignoring the question of priority, insisted upon his right to credit for the invention of the differential calculus. Newton ignored the entire situation and the Leipzig *Acts* refused to print De Duillier's reply to Leibniz.

Here the matter rested until Newton's *Opticks* came out in 1704. Published with the text on optics was a short treatise explaining the method of fluxions and commenting that, since theorems from the 1669 manuscript were appearing in various guises, it seemed best to the author to make the method public now. The next January, the *Acta Eruditorum* of Leipzig carried an anonymous review, later shown to have been written by Leibniz, stating:

> [t]he elements of this calculus have been given to the public by its inventor, Dr. Gottfried Wilhelm Leibniz in these *Acts*. . . . Instead of the Leibnizian differences, then, Dr. Newton employs and always has employed, fluxions, which are very much the same as the augments of fluents produced in the least intervals of time, and these fluxions he has used elegantly in his *Mathematical Principles of Natural Philosophy* and in other later publications, just as Honoratus Fabri, in his *Synopsis of Geometry* substituted progressive methods for the method of Cavalieri.[27]

This innocent-sounding review provoked a storm of opposition in England. Far from being a compliment to the acuteness of the Englishman, it was, by the comparison with Fabri, a scarcely veiled attack upon Newton's integrity. Fabri had been a notorious plagiarist, discredited and dishonored in the mathematical world for his theft of the ideas of another.[28] At the time, Leibniz denied authorship of the review, although later[29] he tacitly admitted it and gave his approval. While this attack may not have been entirely unprovoked, it most certainly was cowardly and unworthy of a man of Leibniz's standing.

John Keill, a friend of Newton, in a letter to Edmond Halley, on the laws of centripetal force, directly and openly accused Leibniz of plagiarism.

> All these laws follow from that very celebrated arithmetic of fluxions which, without any doubt, Dr. Newton invented first, as can readily be proved by anyone who reads the letters about it published by Wallis; yet the same arithmetic afterwards, under a changed name and method of notation, was published by Dr. Leibniz in *Acta Eruditorum*.[30]

When Leibniz received his copy of *Transactions*, in which the letter was published (Volume XXVI, # 317) he wrote to Sloane, the secretary of the Royal Society, demanding that Keill retract his accusation; he added that he felt sure that Keill had acted from rashness and not from improper motives, and that he would not consider the attack a matter of calumny. When the letter was read to the Royal Society, Newton, who was then the president, expressed displeasure at Keill's action. Keill justified himself by producing the review of Newton's *Opticks* in the Leipzig *Acts*, of which Newton had apparently been unaware until this time. Keill wrote to Leibniz on May 24, 1711, saying that he had not accused Leibniz of knowing the name or notation of Newton's method but that he had merely stated that Leibniz must have seen something from which he had been led to his own method. Instead of being mollified, Leibniz was outraged and declared that his honor was attacked even more openly than before; that Keill was an upstart and an unqualified judge; that he was acting without any authority from Newton; that it was the duty of the Royal Society to silence Keill; that he wanted Newton's own opinion directly expressed; and that in Leipzig *Acts* review, "no injustice had been done to any party, as everyone had received only his due."[31] By this last comment, Leibniz made that opinion his own and at once became the aggressor instead of the injured victim in the controversy.

An Attempt to Settle the Question

A committee of the Royal Society was appointed to investigate the situation, examine the documents which had been placed in the archives, and make a decision. The committee members, appointed on March 6, 20, 27, and April 17, 1712, were Halley, Jones, De Moivre, and Machin, friends of Newton and mathematicians; Brook Taylor, a friend of Keill; Robarts, Hill, Burnet, Ashton, Arbuthnot; and Bonet, the Prussian minister. With the exception of de Moivre and Bonet, they were all Englishmen, and almost all were friends of Newton; it was scarcely an unbiased group. There was no chance for Leibniz to give his side of the story or to produce any papers he might have to substantiate his claims. Burnet wrote to Bernouilli that the committee was busy about proving that Leibniz might have seen Newton's papers.[32] The business was handled quickly, and on April 24 of the same year, the report was read. On January 8, 1713, it was published under the title, *Commercium Epistolicum D. Johannis Collinsii et aliorum de analysi promota.* The report, anonymous but probably written by Newton,[33] contains the findings of the committee and copies of the letters and papers involved in the dispute. The actual report of the committee includes a chronology of Leibniz's contacts with Newtonian influences, an assertion that Newton was the prior inventor, and a statement that Keill had not injured Leibniz by his accusation. The plagiarism issue is not touched, except to clear Newton of any possibility by declaring his priority. The report does tell of one incident in Leibniz's career which, while it is not a direct accusation, implies that he is capable of stealing another's ideas.

> February 1672/3, meeting Dr. Pell at Mr. Boyle's, he pretended to the differential method of Mouton, on being shown that it was Mouton's, insisted that it was his because he hadn't known of Mouton's doing it and had much improved it.[34]

In retrospect, the *Commercium Epistolicum* seems to have been a grossly unfair way of handling the situation. It avoided the main issue and, in effect, told Leibniz that there had been no injustice done to him because Newton had had the method before the time when Leibniz was accused of having stolen it from him; this is a meaningless and utterly illogical conclusion.

Leibniz did not see a copy of the report, but Bernoulli wrote to him about it in a letter dated June 7, 1713, adding that Newton had admitted that he didn't think of his method of fluxions until he read of Leibniz's calculus, and that Newton, when he wrote the *Principia*, had no idea of how to find the fluxions of fluxions; Bernoulli also said that Leibniz might publish his letter if he chose, as

long as he did not disclose the identity of the author. Leibniz printed the letter, with comments, anonymously, without any hint as to where or by whom it was printed, under the title, *Charta Volans.* He circulated the little book among the continental mathematicians and scientists.[35] He seemed convinced that he had been injured and wrote to various mathematicians trying, unsuccessfully, to enlist them on his side; he even attacked, in a letter to the Princess of Wales, Newton's philosophy and religious orthodoxy. He wrote, but never published, *Historia et Origo Calculi Differentialis*[36] sometime between 1714 and his death in 1716, in which, writing in the third person, he gave his version of the discovery and the attempted theft of his method of the differential and integral calculus; this was apparently intended as an answer to the *Commercium Epistolicum.*

A Mr. Chamberlayne tried to make peace between the two warring mathematicians but received the comment from Newton that Leibniz had started the fuss in 1705 with his review of the *Opticks* and a similar statement from Leibniz that it was all Newton's fault due to the *Commercium Epistolicum.* Leibniz tried one last challenge in 1716 when, through Abbe Conti, a Venetian nobleman and priest, he sent a problem to test the ability of the English mathematicians. Newton solved the difficult problem in one evening.

The quarrel gradually subsided after Leibniz's death on November 14, 1716. Bernoulli made advances to Newton, vigorously declaring that he had never said anything against the Englishman and insisting that he had not given Leibniz permission to publish any of his letters. A reconciliation seems to have followed, but there is no record of any further correspondence.

The Judgment of History

It seems that the whole unsavory situation could have been avoided if both the men involved had been frank in their statements and prompt in publishing their findings. How can one account for

Newton's extreme secrecy? Did he hide his methods in order to perfect them before giving them to the public? A laudable but highly imprudent motive. Did he wish to have the methods for his own exclusive use? Inexcusable selfishness. Was he trying to avoid disputes and unpleasantness? Total failure. And how can one account for Leibniz's underhanded methods? Was he striving for the honor and glory of the fatherland? Hardly, since he spent much of his time in Paris and other cities outside of Germany. Was he determined to achieve fame at any cost? But he had won renown for his work. Was he utterly devoid of honor? This is scarcely possible, for he was respected and trusted by eminent friends.

It is obvious from the confused tangle of events, the accusations and counteraccusations, the doubtful statements and bold lies, that any definitive statement or decision is impossible. At best, one can say that Newton was probably the first inventor while Leibniz and Bernoulli were promoters and developers of the calculus. Newton seems to have invented fluxions at least ten years before Leibniz developed the calculus. There is no evidence that Newton borrowed from Leibniz; there is little evidence that Leibniz borrowed from Newton. In the hands of Newton and his followers, fluxions remained a relatively sterile theory, while Leibniz and his followers made of the calculus a powerful means of mathematical progress. To Leibniz and the Bernoullis belongs the credit for most of the vast superstructure which has been erected on the foundation laid by Newton.

> Both men owed a very great deal to their immediate predecessors in the development of the new analysis, and the resulting formulations of Newton and Leibniz were most probably the result of a common anterior, rather than a reciprocal coincident influence.[37]

Claims to the invention of the calculus have been made by or for several mathematicians. Fatio de Duillier's claim was apparently ignored by his contemporaries; indeed, he seemed never to have pressed the issue himself. It has been said that credit for the calculus should go to Barrow or even to Fermat, both of whom did admirable work in laying the foundation for the later discoveries of Newton and Leibniz. But the judgment of history still stands and Sir Isaac Newton and Gottfried Wilhelm Leibniz are generally considered as the two inventors of the differential and integral calculus.[38]

A Century of Isolation in England

Newton's influence on English mathematics was so great that, during the entire eighteenth century, especially at Cambridge, mathematics was confined to the study of optics, gravitation, geometry, and fluxional calculus. The English savants had regarded the struggle with Leibniz and his supporters as an attempt, even a plot, of the Germans to rob Newton of the credit for his invention. In loyalty to its famous son, Cambridge chose Newton's fluxional methods in preference to Leibniz's analytical ones. For some problems, either notation may be used, but for the calculus of variations and for most modern theoretical work, Newtonian notation is impossible. In fact, even Leibnizian notation is proving inadequate for some phases of modern calculus, and that on a fairly elementary level.[39] The relative merit of the two methods was completely obscured by the quarrel over the right to credit for the invention. Personal feelings and national jealousies made the decisions, and as a result, Cambridge withdrew into a sterile isolation.

A common language is essential in the development of a science. By accepting the isolation attendant on its adoption of the Newtonian notation and refusing to make an effort to keep up with the advances made on the continent, Cambridge rendered almost sterile the efforts of the group of truly brilliant followers who had gathered about Newton. The continental mathematicians kept up with whatever English advances there were, translating the Newtonian notation into Leibnizian and thus making the work available to all. However, the English, in the isolation of injured national pride, would not do likewise as the various continental developments were published. Cambridge was out of touch with the continental mathematicians for almost a century, although the journals in which the continental findings were published were circulated widely and gratuitously.[40] As Leibniz's work was interpreted by Bernoulli, Euler, D'Alembert, Lagrange, Laplace, and others, the knowledge of the calculus spread widely among those who would listen and learn. But England was passed by; the history of English calculus led nowhere.[41]

It is not true that the differential notation was entirely unknown in England. John Craig, a friend of Newton, used $dx, dy, dz,$ and $\int$ in articles printed in 1685, 1693, 1701, 1703, 1704, 1708, and 1713. Yet in 1718, he wrote *De Calculo Fluentium* using exclusively Newtonian notations.[42]

De Moivre, in 1702 and 1703, and John Keill, in 1713, used a mixed notation of the form $\int \dot{x}x$, which notation was still being used in some publications as late as 1815.[43] Joseph Fenn, an Irish writer with a fine disregard to national feelings, in a *History of Mathematics,* published in Dublin sometime after 1768, used the Newtonian terms, fluxion and fluent, and the Leibnizian notation.[44]

Maclaurin in Scotland and Clairaut in France seem to be the only non-English mathematicians who used the Newtonian notation. The works of Newton and other Englishmen appearing on the continent were published as they were written, in the fluxional notation, at least until they had been "translated." It is interesting to note that there was one Dutch journal, *Maandelykse Mathematische Liefhebbery* ("Monthly Mathematical Recreations"), published in Amsterdam from 1754 to 1769, which used the Newtonian notation exclusively.[45]

The quality of the work of the English mathematicians declined rapidly after the break with the continent. Isolation accounted for part of that decline, but there was also another reason. When Newton began his work on fluxions, he was aware

that he was dealing with new concepts which would be accepted only if the proofs were unimpeachable. In order to avoid having his ideas rejected because of questionable proofs, he shunned the new (1637) analytic geometry of Descartes and used only the methods of classical geometry.[46] At times he used other methods to discover theorems and derive proofs, but always he confined his final demonstrations to geometry and elementary algebra. Geometric proofs are, in themselves, adequate, but they are often labored and unnecessarily complex. Moreover, separate demonstration is required for each kind of problem; the processes are not general as they are in analysis.[47] However, long after the principles of analytic geometry and analysis were commonly accepted and freely used by the continental mathematicians, the English analysts remained true to the traditions of the master. Thus the situation stood for almost a century.

Some records of the Senate House examinations at Cambridge have come down to us, showing the general tenor of the calculus work being done at the university which was Newton's Alma Mater, a stronghold of Newtonian mathematics and physics. The 1772 examinations included:

> the doctrine of fluxions, and its application to the solution of questions de maximis et minimis, to the finding of areas, . . . as unfolded and exemplified, in the fluxional treatises of Lyons, Saunderson, Simpson, Emerson, Maclaurin, and Newton . . .[48]

In 1785, the examinees were required to

> find the fluent of $\dot{x}\sqrt{a^2 - x^2}$

and to find, by the method of fluxions,

> the number from which, if you take its square, there shall remain the greatest difference possible.[49]

The 1786 examination contained problems which were a little more difficult:

> To find the fluxion of $x^2(y^n + z^n)^{1/q}$.
>
> To find the fluxion of the mth power of the Logarithm of x.

To find the fluent of $\dfrac{ax}{a+x}$.[50]

By 1802, the students were subjected to the following types of problems:

Find the fluents of the quantities

$$\frac{d\dot{x}}{x(a^2 - x^2)} \quad \text{and} \quad \frac{h\dot{y}}{y(a+y)^{3/2}} \ .$$

Given the fluent: $(a + cz^n)^m \times z^{pn+n-1} \dot{z}$, find the fluent: $(a + cz^n)^{m+1} \times z^{pn-1} \dot{z}$.

Required also the fluent of

$$\frac{\dot{x} \sqrt{a^2 + x^2}}{x^3} \quad \text{and of} \quad \frac{z^\theta \dot{z}}{1 + mz} \ ,$$

θ being a whole positive number.[51]

The Return to Analysis

Towards the end of the eighteenth century, the more thoughtful mathematicians at Cambridge began to suspect the evils that were consequent upon their separation from news of continental mathematics. The logical thing to do was to adopt the Leibnizian notation and methods, but there was a sentimental objection to such an action; would not such a move be an act of disloyalty to the memory of the great Newton? Finally, in 1803, Robert Woodhouse, then a tutor and later a professor at Cambridge, wrote *Principles of Analytic Calculation,* a work which explained the differential notation and advocated its adoption. Woodhouse criticized the continental methods in some points, especially in the use of principles which were neither obvious nor enunciated. By exposing some of the errors of the system, he gave the impression that he was as much against as he was for the analytic system, thus gaining a hearing among those who would have opposed the system on the basis of those very errors. His writings seem to have been ignored by most of the professors but some of the more serious students read and wondered. As soon as they were aware of the great amount of mathematics which had been closed to them, they obtained the continental books and

began to read and study. Unusual answers began to appear on some of the examinations.[52]

"A man like Woodhouse, of scrupulous honor, universally respected, a trained logician and with a caustic wit, was well fitted to introduce the new system."[53] Nevertheless, the movement might have died with him if it had not been for George Peacock, who, with Herschel and Babbage, formed an Analytical Society in 1812. As undergraduates, they habitually breakfasted together on Sunday mornings, and out of these meetings and their common interest in mathematics, the society grew.

George Peacock (1791–1858) received his B.A. from Trinity College in 1813 as second wrangler.[54] He received a fellowship in 1814 and later became a tutor. Well loved by his students, he was a brilliant lecturer and a kindly and practical tutor, indeed a rare combination of qualities. The establishment of the University Observatory was largely due to his efforts.[55]

Sir John Frederick William Herschel (1792–1871), the son of an astronomer, entered St. John's College at Cambridge in 1809 and graduated as senior wrangler in 1813. While still an undergraduate, he wrote a paper on Cotes's theorem; he published several other papers later. He left the University about 1816 and became an astronomer and chemist.[56]

Charles Babbage (b. 1792), who entered Trinity in 1810, had had a good mathematical education before his arrival at Cambridge, having studied the works of Ditton, Maclaurin, and Simpson on fluxions, Agnesi's *Analysis* in fluxional notation, Woodhouse's *Principles of Analytic Calculation,* and Lagrange's *Théorie des Fonctions.* In 1813, Babbage transferred to Peterhouse because he wanted a chance to be first in the examinations and knew that Peacock and Herschel would surpass him if he tried to compete against them. A many-sided personality, he held a professorship, invented a machine for arithmetical processes, and wrote several scientific papers.[57]

These three eager young students, together with Maule, Ryan, Robinson, and D'Arblay, formed the original membership of the Analytical Society, whose aim was, according to Babbage, to advocate "the principles of pure d-ism as opposed to the dot-age of the university."[58] The Society published, in 1819, a translation of Lacroix's *Elementary Differential Calculus*. Peacock, who, as moderator of the Senate House examinations, was in a position to advance the cause, introduced the differential notation in the examination of 1817. In the same year, a colleague, John White, used the fluxional notation. Peacock was criticized but went on his way, feeling that the younger generation of students was ready to accept the change and that the time was right to

> reduce the many-headed monster of prejudice and make the university answer her character as the loving mother of good learning and science.[59]

The differential notation was again used by Peacock in the 1819 examinations, by Whewell in 1820, and by Peacock again in 1821. Whewell published a work on mechanics in 1819, using the differential notation; Peacock's volume on differential and integral calculus was published by the Analytical Society in 1820; Herschel's work on the calculus of finite differences, illustrative of the new method, came out in the same year; Airy, a pupil of Peacock, published *Tracts* in 1826, a work in which the new method was successfully applied to mechanics. By this time, the exclusive use of fluxions had disappeared among all but a few of the older professors.[60]

From Cambridge, the use of analytical methods spread rapidly through the rest of England. By 1830, the fluxional methods and geometric proofs had very largely disappeared, for, while the geometric demonstration is a useful auxiliary to analysis, it is almost useless as a research device.

The results of the Newton-Leibniz controversy in terms of the personal pain and mental disturbance suffered by the two principal protagonists cannot, of course, be adequately judged. The effect of the controversy on the mathematical world seems to be twofold. As far as credit for the discoveries is concerned, today the two men are honored equally as two independent inventors. Concerning the development of analysis, England seems to have been the loser. The world of mathematics progressed. German and French mathematicians established reputations for themselves and their countries, while England remained insular and isolated. What was the cost to English mathematics? Perhaps no one will ever know. British mathematicians and scientists have done much in the last hundred years for the honor of England and the advancement of knowledge. Nevertheless, one cannot but regret the irreparable loss occasioned by that "dark age" in intellectual history.

NOTES

1. Carl B. Boyer, *The Concepts of the Calculus* (Wakefield, Mass.: 1949), p. 187.

2. *Ibid.*, pp. 187–88.

3. Sir Isaac Newton, *The Method of Fluxions* (London: 1736), p. 19.

4. *Ibid.*, p. 21.

5. *Ibid.*, p. 19.

6. *Ibid.*, p. 25.

7. *Ibid.*, p. 20.

8. J. Edleston (ed.), *Correspondence of Sir Isaac Newton and Professor Cotes* (London: 1850), p. 169.

9. "Commercium Epistolicum," *Philosophical Transactions*, No. 342, January, February 1714/15 (London: 1717), p. 204.

10. Felix Klein, *Elementary Mathematics from an Advanced Standpoint*, Part I (New York: 1932), p. 215.

11. Florian Cajori, *A History of Mathematical Notation*, Vol. II (Chicago: Open Court Publishing Co., 1929), p. 181.

12. Sir Isaac Newton, *Mathematical Principles of Natural Philosophy* (Berkeley, Calif.: University of California, 1947), pp. 655–56.

13. John Theodore Merz, *Leibniz* (New York: 1948), p. 60.

14. David Brewster, *The Life of Sir Isaac Newton* (New York: 1831), pp. 175–76.

15. Augustus De Morgan, *Essays on the Life and Work of Newton* (Chicago: Open Court Publishing Co., 1914), pp. 95–96.

16. Arthur S. Hathaway, "The Discovery of the Calculus," *Science*, New Series, Vol. L, No. 1280 (July–December, 1919), pp. 41–43.

17. Merz, *op. cit.*, pp. 54–55.

18. Hathaway, *op. cit.*, p. 42.

19. Hathaway says that Leibniz's methods here described are like the methods of German propaganda in world War I and that Leibniz deserves no honor at all even if he were an independent discoverer because "he does not come into court with clean hands." *Op. cit.*, p. 43.

20. De Morgan, *op. cit.*, p. 25 (note).

21. Merz, *op. cit.*, pp. 58–59.

22. Klein, *loc, cit.*

23. Newton, *Mathematical Principles, loc. cit.*

24. *Ibid.*

25. J. W. N. Sullivan, *Isaac Newton* (New York: 1938), p. 229.

26. Gottfried Wilhelm Leibniz, *The Early Mathematical Manuscripts of Leibniz*, J. M. Child (ed.) (Chicago: Open Court Publishing Co., 1920), p. 8.

27. Sullivan, *op. cit.*, p. 232.

28. *Ibid.*

29. See below.

30. Sullivan, *op. cit.*, p. 234.

31. Brewster, *op. cit.*, p. 189.

32. De Morgan, *op. cit.*, pp. 27–28 (note).

33. Augustus De Morgan, "On the Authorship of the Account of the *Commercium Epistolicum* in the *Philosophical Transactions*," in *Philosophical Magazine*, 4th series, Vol. 3 (1852), pp. 440–43.

34. "Commercium Epistolicum," *Philosophical Transactions*, p. 183.

35. Brewster, *op. cit.*, pp. 192–93.

36. Leibniz, *op. cit.*, pp. 22–57.

37. Boyer, *op. cit.*, p. 188.

38. For a discussion of the validity of other claims, see Florian Cajori, "Who Was the First Inventor of Calculus?" *American Mathematical Monthly*, XXVI (1919), 15–20; and J. M. Child, "Barrow, Newton, and Leibniz, in Their Relation to the Discovery of the Calculus," *Science Progress*, XXV (1930–31), 295–307.

39. R. Creighton Buck, *Advanced Calculus* (New York: McGraw-Hill Book Co., Inc., 1956), pp. 58–59.

40. W. W. Rouse Ball, *A History of the Study of Mathematics at Cambridge* (Cambridge: 1889), p. 98.

41. *Ibid.*

42. Florian Cajori, "The Spread of Newtonian and Leibnizian Notations of the Calculus," *Bulletin of the American Mathematical Society*, XXVII (June–July, 1921), 453.

43. *Ibid.*, p. 454.

44. *Ibid.*

45. *Ibid.*, p. 455.

46. Ball, *op. cit.*, p. 69.

47. *Ibid.*, p. 98.

48. *Ibid.*, p. 192.

49. *Ibid.*, pp. 195–96.

50. *Ibid.*

51. *Ibid.*, pp. 200–09.

52. *Ibid.*, p. 119.

53. *Ibid.*

54. Wrangler—an honors student, in the first class in the mathematical Tripos. First ranking man is designated as senior wrangler, next as second, third, fourth, etc., wranglers.

55. Ball, *op. cit.*, p. 124.

56. *Ibid.*, pp. 126–27.

57. *Ibid.*, pp. 125–26.

58. *Ibid.*, p. 126.

59. *Ibid.*, p. 121.

60. *Ibid.*, p. 122.

Mengoli's Proof for the Divergence of the Harmonic Series

A nemesis of every first year calculus student is the harmonic series, an innocent looking sum of the form:

$$1 + \frac{1}{2} + \frac{1}{3} + \frac{1}{4} + \frac{1}{5} + \ldots$$

which proceeds forever in the same pattern. Its individual terms become very small approaching zero but what of the sum itself? Can it be found?—that is, is there a finite number that this sum approaches in its total? If such a finite number exists then the series is said to *converge*; if no such number exists, then the series is said to approach infinity or to *diverge*. Intuition would lead many people to believe that this series approaches some definite sum and therefore converges; however, this is not the case. Since at least the fourteenth century, mathematicians have attempted to ascertain the behavior of the harmonic series. An early proof for the divergence of the harmonic series was given by the French cleric and natural philosopher Nicole Oresme (ca. 1323–1382) and later in the eighteenth century one was devised by the Swiss mathematician Jakob Bernoulli (1645–1705). Over the years, these have remained the most popular two proofs used to demonstrate the divergence of the harmonic series; however, other proofs exist. In 1647, the Italian mathematician Pietro Mengoli (1625–1686) devised a noteworthy proof. It goes as follows:

Given a number n, where $n > 1$, then $\dfrac{1}{n-1} + \dfrac{1}{n} + \dfrac{1}{n+1} > \dfrac{3}{n}$

The truth of this statement can be shown by simple algebra. Let us call this mathematical fact proposition A. Now consider the sum of the harmonic series as symbolized by S where

$$S = 1 + \frac{1}{2} + \frac{1}{3} + \frac{1}{4} + \frac{1}{5} + \frac{1}{6} + \frac{1}{7} + \frac{1}{8} + \frac{1}{9} + \frac{1}{10} + \ldots$$

Using brackets, the sum, S, can be divided into a series of numerical groupings.

$$S = 1 + \left(\frac{1}{2} + \frac{1}{3} + \frac{1}{4}\right) + \left(\frac{1}{5} + \frac{1}{6} + \frac{1}{7}\right) + \left(\frac{1}{8} + \frac{1}{9} + \frac{1}{10}\right) + \ldots$$

(continued)

Applying proposition A to each of these groupings, we find:

$$\left(\frac{1}{2} + \frac{1}{3} + \frac{1}{4} \right) > \frac{3}{3} = 1$$

$$\left(\frac{1}{5} + \frac{1}{6} + \frac{1}{7} \right) > \frac{3}{6} = \frac{1}{2}$$

$$\left(\frac{1}{8} + \frac{1}{9} + \frac{1}{10} \right) > \frac{3}{9} = \frac{1}{3}$$

therefore:

$$S > 1 + 1 + \frac{1}{2} + \frac{1}{3} + \dots$$

This process can be repeated over and over again, theoretically forever showing that S is larger than any finite number. Conclusion, the harmonic series diverges.

25

The Bernoulli Family[*]

HOWARD EVES

*T*HERE is a general rule to the effect that any given family possesses at most one outstanding mathematician and that, in fact, most families possess none. Thus a search through the ancestors, descendants, and relatives of Isaac Newton fails to turn up any other great mathematician. There are exceptions to this general rule. For example we have, here in the United States, the two Lehmers (father and son) and the two Birkoffs (father and son). One also recalls the two Cassinis (father and son) of the late seventeenth and early eighteenth centuries, and perhaps one can build a case for the two Clairaut children of the eighteenth century. And of course there were Theon and Hypatia (father and daughter), who lived during the closing years of ancient Greek mathematics. But such cases are relatively rare. All the more striking, then, is the Bernoulli family of Switzerland, which in three successive generations produced no less than eight noted mathematicians.

The Bernoulli family record starts with the two brothers Jakob Bernoulli (1654–1705) and Johann Bernoulli[†] (1667–1748). These two men gave up earlier vocational interests and became mathematicians when Leibniz's papers began to appear in the *Acta eruditorum*. They were among the first mathematicians to realize the surprising power of the calculus and to apply the tool to a great diversity of problems. From 1687 until his death, Jakob occupied the mathematical chair at Basel University. Johann, in 1697 , became a professor at Groningen University, and then, on Jakob's death in 1705, succeeded his brother in the chair at Basel University, to remain there for the rest of his active life. The two brothers, often bitter rivals, maintained an almost constant exchange of ideas with Leibniz and with each other.

Among Jakob Bernoulli's contributions to mathematics are the early use of polar coordinates, the derivation in both rectangular and polar coordinates of the formula for the radius of curvature of a plane curve, the study of the catenary curve with extensions to strings of variable density and strings under the action of a central force, the study of a number of other higher plane curves, the discovery of the so-called *isochrone*—or curve along which a body will fall with uniform vertical velocity (it turned out to be a semicubical parabola with a vertical cusptangent), the determination of the form taken by an elastic rod fixed at one end and carrying a weight at the other, the form assumed by a flexible rectangular sheet having two opposite edges held horizontally fixed at the same height and loaded with a heavy liquid, and the shape of a rectangular sail filled with wind. He also proposed and discussed the problem of isoperimetric figures (planar closed paths of given species with fixed perimeter which include a maximum area), and was thus one of the first mathematicians to work in the calculus of variations. He was also one of the

[*] Largely adapted from the author's *An Introduction to the History of Mathematics* (revised edition), Holt, Rinehart and Winston, Inc., 1964.

[†] Referred to by various authors as Johann, Jean, and John.—F.S.

Reprinted from *Mathematics Teacher* 59 (Mar., 1966): 276–78; with permission of the National Council of Teachers of Mathematics.

FIGURE 1 *Jakob Bernoulli*
(Source: *A Portfolio of Eminent Mathematicians*, edited by
David Eugene Smith (Chicago: Open Court Publishing
Company, 1908)

FIGURE 2 *Johann Bernoulli*
(Source: *A Portfolio of Eminent Mathematicians*, edited by
David Eugene Smith (Chicago: Open Court Publishing
Company, 1908)

early students of mathematical probability; his book in this field, the *Ars conjectandi*, was posthumously published in 1713.

There are several things in mathematics which now bear Jakob Bernoulli's name. Among these are the *Bernoulli distribution* and *Bernoulli theorem* of statistics and probability theory, the *Bernoulli equation* met by every student of a first course in differential equations, the *Bernoulli numbers* and *Bernoulli polynomials* of number-theory interest, and the *lemniscate of Bernoulli* encountered in any first course in the calculus. In Jakob Bernoulli's solution to the problem of the isochrone curve, which was published in the *Acta eruditorum* in 1690, we meet for the first time the word "integral" in a calculus sense. Leibniz had called the integral calculus *calculus summatorius;* in 1696

Leibniz and Johann Bernoulli agreed to call it *calculus integralis.*

Jakob Bernoulli was struck by the way the equiangular spiral reproduces itself under a variety of transformations and asked, in imitation of Archimedes, that such a spiral be engraved on his tombstone, along with the inscription *Eadem mutata resurgo* ("I shall arise the same, though changed").

Johann Bernoulli was an even more prolific contributor to mathematics than was his brother Jakob. Though he was a jealous and cantankerous man, he was one of the most successful teachers of his time. He greatly enriched the calculus and was very influential in making the power of the new subject appreciated in continental Europe. It was his material that the Marquis de l'Hospital (1661–1704), under a curious financial agreement with

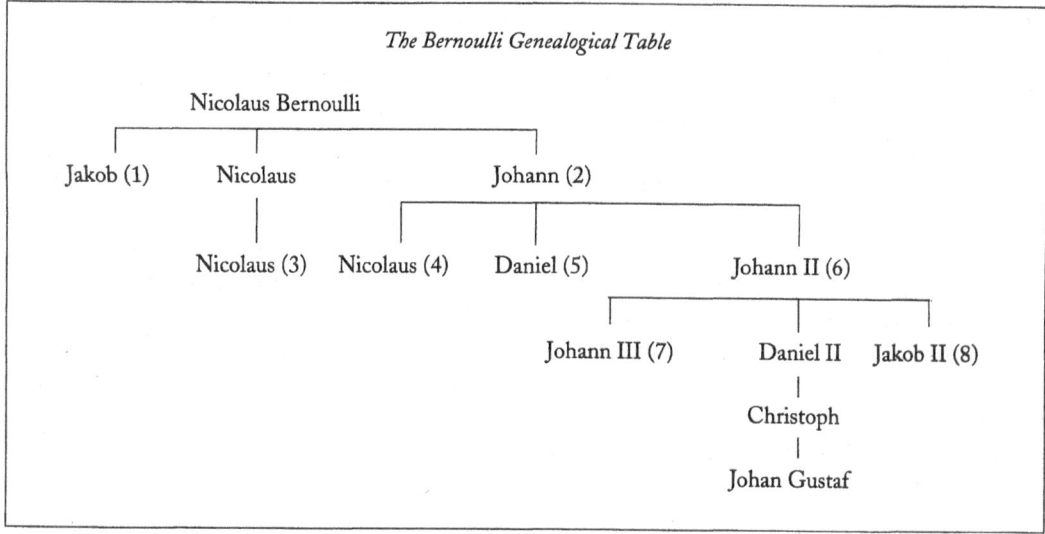

The Bernoulli Genealogical Table

Johann, assembled in 1696 into the first calculus textbook. It was in this way that the familiar method of evaluating the indeterminate form 0/0 became incorrectly known in later calculus texts, as *l'Hospital's rule*.

Johann Bernoulli wrote on a wide variety of topics, including optical phenomena connected with reflection and refraction, the determination of the orthogonal trajectories of families of curves, rectification of curves and quadrature of areas by series, analytical trigonometry, the exponential calculus, and other subjects. One of his more noted pieces of work is his contribution to the problem of the *brachystochrone*—the determination of the curve of quickest descent of a weighted particle moving between two given points in a gravitational field; the curve turned out to be an arc of an appropriate cycloid curve. This problem was also discussed by Jakob Bernoulli. The cycloid curve is also the solution to the problem of the *tautochrone*—the determination of the curve along which a weighted particle will arrive at a given point of the curve in the same time interval no matter from what initial point of the curve it starts. This latter problem, which was more generally discussed by Johann Bernoulli, Euler, and Lagrange, had earlier been solved by Huygens (1673) and Newton

(1687), and applied by Huygens in the construction of pendulum clocks.

Johann Bernoulli had three sons, Nicolaus (1695–1726), Daniel (1700–1782), and Johann (II) (1710–1790), all of whom won renown as eighteenth-century mathematicians and scientists. Nicolaus, who showed great promise in the field of mathematics, was called to the St. Petersburg Academy, where he unfortunately died—by drowning—only eight months later. He wrote on curves, differential equations, and probability. A problem in probability, which he proposed from St. Petersburg, later became known as the *Petersburg paradox*. The problem is: If A receives a penny should heads appear on the first toss of a coin, 2 pennies if heads does not appear until the second toss, 4 pennies if heads does not appear until the third toss, and so on, what is A's expectation? Mathematical theory shows that A's expectation is infinite, which seems a paradoxical result.

The Petersburg paradox problem was investigated by Nicolaus's brother Daniel, who succeeded Nicolaus at St. Petersburg. Daniel returned to Basel seven years later. He was the most famous of Johann's three sons, and devoted most of his energies to probability, astronomy, physics, and hydrodynamics. In probability he devised the concept of *moral expectation,* and in his *Hydrodynamica,* of 1738, appears the principle of hydrodynamics that bears his name in all present-day elementary physics texts. He wrote on tides, established the kinetic theory of gases, studied the vibrating string, and pioneered in partial differential equations. He was awarded the prize of the French Academy no less than ten times.

Johann (II), the youngest of Johann Bernoulli's three sons, studied law but spent his later years as a professor of mathematics at the University of Basel, succeeding his father in that position in 1743. He was particularly interested in the mathematical theory of heat and light. He received the prize of the French Academy on three occasions.

There was another eighteenth-century Nicolaus Bernoulli (1687–1759), a nephew of Jakob and Johann, who achieved some fame in mathematics. This Nicolaus held, for a time, the chair of mathematics at Padua once filled by Galileo. He wrote extensively on geometry, differential equations, infinite series, and probability. Later in life he taught logic and law.

Johann Bernoulli (II) had three sons, Johann (III) (1744–1807), Daniel (II) (1751–1834), and Jakob (II) (1759–1789). Johann (III), like his father, studied law but then turned to mathematics. When barely 19 years old, he was called as a professor of mathematics to the Berlin Academy. He wrote on astronomy, the doctrine of chance, recurring decimals, and indeterminate equations. Jakob (II) also first studied law, but then became a professor of mathematics at the St. Petersburg Academy. His works are related to those of his uncle and teacher, Daniel Bernoulli.

Lesser Bernoulli descendants are Daniel (II), Christoph (1782–1863), a son of Daniel (II), and Johann Gustav (1811–1863), a son of Christoph.

26

The Bernoullis and the Harmonic Series

WILLIAM DUNHAM

$\mathcal{A}$NY INTRODUCTION to the topic of infinite series soon must address that first great counter-example of a divergent series whose general term goes to zero—the harmonic series $\sum_{k=1}^{\infty} 1/k$. Modern texts employ a standard argument, traceable back to the great fourteenth-century Frenchman Nicole Oresme (see [3], p. 92), which establishes divergence by grouping the partial sums:

$$1 + \frac{1}{2} > \frac{1}{2} + \frac{1}{2} = \frac{2}{2}$$

$$1 + \frac{1}{2} + \left(\frac{1}{3} + \frac{1}{4} \right) > \frac{2}{2} + \left(\frac{1}{4} + \frac{1}{4} \right) = \frac{3}{2}$$

$$1 + \frac{1}{2} + \frac{1}{3} + \frac{1}{4} + \left(\frac{1}{5} + \frac{1}{6} + \frac{1}{7} + \frac{1}{8} \right) > \frac{3}{2} +$$
$$\left(\frac{1}{8} + \frac{1}{8} + \frac{1}{8} + \frac{1}{8} \right) = \frac{4}{2},$$

and in general

$$1 + \frac{1}{2} + \frac{1}{3} + \ldots + \frac{1}{2^n} > \frac{n+1}{2},$$

from which it follows that the partial sums grow arbitrarily large as n goes to infinity.

Reprinted from *College Mathematics Journal* 60 (Jan., 1987): 18–23; with permission.

It is possible that seasoned mathematicians tend to forget how surprising this phenomenon appears to the uninitiated student—that, by adding ever more negligible terms, we nonetheless reach a sum greater than any preassigned quantity. Historian of mathematics Morris Kline ([5], p. 443) reminds us that this feature of the harmonic series seemed troubling, if not pathological, when first discovered.

JACOBI BERNOULLI,
Profeſſ Baſil. & utriuſque Societ. Reg. Scientiar.
Gall. & Pruſſ. Sodal.
MATHEMATICI CELEBERRIMI,

ARS CONJECTANDI,
OPUS POSTHUMUM.

Accedit

TRACTATUS
DE SERIEBUS INFINITIS,

Et EPISTOLA Gallicè ſcripta

DE LUDO PILÆ
RETICULARIS.

BASILEÆ,
Impenſis THURNISIORUM, Fratrum.
cIɔ Iɔcc xIII.

Courtesy of the Lilly Library, Indiana University, Bloomington, IN

So unusual a series could not help but attract the interest of the preeminent mathematical family of the seventeenth century, the Bernoullis. Indeed, in his 1689 treatise "Tractatus de Seriebus Infinitis," Jakob Bernoulli provided an entirely different, yet equally ingenious proof of the divergence of the harmonic series. In "Tractatus," which is now most readily found as an appendix to his posthumous 1713 masterpiece *Ars Conjectandi*, Jakob generously attributed the proof to his brother ("Id primus deprehendit Frater"), the reference being to his full-time sibling and part-time rival Johann. While this "Bernoullian" argument is sketched in such mathematics history texts as Kline ([5], p. 444) and Struik ([6], p. 321), it is little enough known to warrant a quick reexamination.

The proof rested, quite unexpectedly, upon the *convergent* series

$$\frac{1}{1} + \frac{1}{6} + \frac{1}{12} + \frac{1}{20} + \ldots = \sum_{k=1}^{\infty} \frac{1}{k(k+1)}.$$

The modern reader can easily establish, via mathematical induction, that

$$\sum_{k=1}^{n} \frac{1}{k(k+1)} = \frac{n}{n+1},$$

and then let n go to infinity to conclude that

$$\sum_{k=1}^{\infty} \frac{1}{k(k+1)} = 1.$$

Jakob Bernoulli, however, approached the problem quite differently. In Section XV of *Tractatus*, he considered the infinite series

$$N = \frac{a}{c} + \frac{a}{2c} + \frac{a}{3c} + \frac{a}{4c} + \ldots,$$

then introduced

$$P = N - \frac{a}{c} = \frac{a}{2c} + \frac{a}{3c} + \frac{a}{4c} + \frac{a}{5c} + \ldots.$$

and subtracted termwise to get

$$\frac{a}{c} = N - P = (\frac{a}{c} - \frac{a}{2c}) + (\frac{a}{2c} - \frac{a}{3c}) + (\frac{a}{3c} - \frac{a}{4c}) + \ldots$$
$$= \frac{a}{2c} + \frac{a}{6c} + \frac{a}{12c} + \frac{a}{20c} + \ldots \tag{1}$$

Thus, for $a = c$, he concluded that

$$\frac{1}{2} + \frac{1}{6} + \frac{1}{12} + \frac{1}{20} + \ldots = \frac{1}{1} = 1. \tag{2}$$

Unfortunately, Bernoulli's "proof" required the subtraction of two divergent series, N and P. To his credit, Bernoulli recognized the inherent dangers in his argument, and he advised that this procedure must not be used without caution ("non sine cautela"). To illustrate his point, he applied the previous reasoning to the series

$$S = \frac{2a}{c} + \frac{3a}{2c} + \frac{4a}{3c} + \ldots,$$

and

$$T = S - \frac{2a}{c} + \frac{3a}{2c} + \frac{4a}{3c} + \frac{5a}{4c} + \ldots.$$

Upon subtracting termwise, he got

$$\frac{2a}{c} = S - T = \frac{a}{2c} + \frac{a}{6c} + \frac{a}{12c} + \frac{a}{20c} + \ldots, \tag{3}$$

which provided a clear contradiction to (1).

Bernoulli analyzed and resolved this contradiction as follows: the derivation of (1) was valid since the "last" term of series N is zero (that is, $\lim_{k\to\infty} a/(kc) = 0$), whereas the parallel derivation of (3) was invalid since the "last" term of series S is non-zero (because $\lim_{k\to\infty}(k+1)a/(kc) = a/c \neq 0$). In modern terms, he had correctly recognized that, regardless of the convergence or divergence of the series $\sum_{k=1}^{\infty} x_k$, the new series $\sum_{k=1}^{\infty} (x_k - x_{k+1})$ converges to x_1 *provided* $\lim_{k\to\infty} x_k = 0$. Thus, he not only explained the need for "caution" in his earlier discussion but also exhibited a fairly penetrating insight, by the standards of his day, into the general convergence/divergence issue.

Having thus established (2) to his satisfaction, Jakob addressed the harmonic series itself. Using his brother's analysis of the harmonic series, he proclaimed in Section XVI of *Tractatus*:

> **XVI.** *Summa seriei infinita harmonicè progressionalium,*
> $\frac{1}{1} + \frac{1}{2} + \frac{1}{3} + \frac{1}{4} + \frac{1}{5}$ *&c. est infinita.*

He began the argument that "the sum of the infinite harmonic series

$$\frac{1}{1} + \frac{1}{2} + \frac{1}{3} + \frac{1}{4} + \frac{1}{5} \text{ etc.}$$

is infinite" by introducing

$$A = +\frac{1}{2} + \frac{1}{3} + \frac{1}{4} + \frac{1}{5} + \frac{1}{6} + \frac{1}{7}\ldots,$$

which "transformed into fractions whose numerators are 1, 2, 3, 4 etc" becomes

$$\frac{1}{2} + \frac{2}{6} + \frac{3}{12} + \frac{4}{20} + \frac{5}{30} + \frac{6}{42} + \ldots.$$

Using (2), Jakob next evaluated:

$$C = \frac{1}{2} + \frac{1}{6} + \frac{1}{12} + \frac{1}{20} + \ldots = 1$$

$$D = \frac{1}{6} + \frac{1}{12} + \frac{1}{20} + \ldots = C - \frac{1}{2} = 1 - \frac{1}{2} = \frac{1}{2}$$

$$E = \frac{1}{12} + \frac{1}{20} + \ldots = D - \frac{1}{6} = \frac{1}{2} - \frac{1}{6} = \frac{1}{3}$$

$$F = \frac{1}{20} + \ldots = E - \frac{1}{12} = \frac{1}{3} - \frac{1}{12} = \frac{1}{4}$$

$$\vdots \qquad \vdots \quad \vdots \qquad \qquad \vdots$$

By adding this array columnwise, and again implicitly assuming that termwise addition of infinite series is permissible, he arrived at

$$C + D + E + F + \ldots$$

$$= \frac{1}{2} + \left(\frac{1}{6} + \frac{1}{6}\right) + \left(\frac{1}{12} + \frac{1}{12} + \frac{1}{12}\right) + \ldots$$

$$= \frac{1}{2} + \frac{2}{6} + \frac{3}{12} + \frac{4}{20} + \ldots$$

$$= A.$$

On the other hand, upon separately summing the terms forming the extreme left and the extreme right of the arrayed equations above, he got

$$C + D + E + F + \ldots$$

$$= 1 + \frac{1}{2} + \frac{1}{3} + \frac{1}{4} + \ldots = 1 + A.$$

Hence, $A = 1 + A$. In Jakob's words, "The whole" equals "the part"—that is, the harmonic series $1 + A$ equals its part A—which is impossible for a finite quantity. From this, he concluded that $1 + A$ is infinite.

Jakob Bernoulli was certainly convinced of the importance of his brother's deduction and emphasized its salient point when he wrote: "The sum of an infinite series whose final term vanishes perhaps is finite, perhaps infinite."

Obviously, this proof features a naive treatment both of series manipulation and of the nature of "infinity." In addition, it attacks infinite series "holistically" as single entities, without recourse to the modern idea of partial sums. Before getting overly critical of its distinctly seventeenth-century flavor, however, we must acknowledge that Bernoulli devised this proof a century and a half before the appearance of a truly rigorous theory of series. Further, we cannot deny the simplicity and cleverness of his reasoning nor the fact that, if bolstered by the necessary supports of modern analysis, it can serve as a suitable alternative to the standard proof.

Indeed, this argument provides us with an example of the history of mathematics at its best—paying homage to the past yet adding a note of freshness and ingenuity to the modern classroom.

XVI. *Summa seriei infinitæ harmonicè progressionalium,* $\frac{1}{1} + \frac{1}{2} + \frac{1}{3} + \frac{1}{4} + \frac{1}{5}$ *&c. est infinita.*

Id primus deprehendit Frater: inventa namque per præced. summa seriei $\frac{1}{2} + \frac{1}{6} + \frac{1}{12} + \frac{1}{20} + \frac{1}{30}$, &c. visurus porrò, quid emergeret ex ista serie, $\frac{1}{2} + \frac{2}{6} + \frac{1}{12} + \frac{4}{20} + \frac{5}{30}$; &c. si resolveretur methodo Prop. XIV. collegit propositionis veritatem ex absurditate manifesta, quæ sequeretur, si summa seriei harmonicæ finita statueretur. Animadvertit enim,

Seriem A, $\frac{1}{2} + \frac{1}{3} + \frac{1}{4} + \frac{1}{5} + \frac{1}{6} + \frac{1}{7}$, &c. ∞ (fractionibus singulis in alias, quarum numeratores sunt 1, 2, 3, 4, &c. transmutatis)

seriei B, $\frac{1}{2} + \frac{2}{6} + \frac{3}{12} + \frac{4}{20} + \frac{5}{30} + \frac{6}{42}$, &c. ∞ C + D + E + F, &c.

C. $\frac{1}{2} + \frac{1}{6} + \frac{1}{12} + \frac{1}{20} + \frac{1}{30} + \frac{1}{42}$, &c. ∞ per præc. $\frac{1}{1}$

D. $+ \frac{1}{6} + \frac{1}{12} + \frac{1}{20} + \frac{1}{30} + \frac{1}{42}$, &c. ∞ C $- \frac{1}{2}$ ∞ $\frac{1}{2}$

E. $+ \frac{1}{12} + \frac{1}{20} + \frac{1}{30} + \frac{1}{42}$, &c. ∞ D $- \frac{1}{6}$ ∞ $\frac{1}{3}$

F. $+ \frac{1}{20} + \frac{1}{30} + \frac{1}{42}$, &c. ∞ E $- \frac{1}{12}$ ∞ $\frac{1}{4}$

&c. ∞ &c.

∞ G; unde sequitur, se-

seriem G ∞ A, totum parti, si summa finita esset.

Ego

Courtesy of the Lilly Library, Indiana University, Bloomington, IN

Perhaps, in contemplating this work, some of today's students might even come to share a bit of the enthusiasm and wonder that moved Jakob Bernoulli to close his *Tractatus* with the verse [7]

So the soul of immensity dwells in minutia.
And in narrowest limits no limits inhere.
What joy to discern the minute in infinity!
The vast to perceive in the small, what divinity!

Remark. Jakob Bernoulli, eager to examine other infinite series, soon turned his attention in section XVII of *Tractatus* to

$$1 + \frac{1}{4} + \frac{1}{9} + \frac{1}{16} + \ldots = \sum_{k=1}^{\infty} \frac{1}{k^2}, \qquad (4)$$

the evaluation of which "is more difficult than one would expect" ("difficilior est quam quis expectaverit"), an observation that turned out to be quite an understatement. He correctly established the convergence of (4) by comparing it termwise with the greater, yet convergent series

$$1 + \frac{1}{3} + \frac{1}{6} + \frac{1}{10} + \ldots$$

$$= 2 \left(\frac{1}{2} + \frac{1}{6} + \frac{1}{12} + \frac{1}{20} + \ldots \right) = 2(1) = 2.$$

But evaluating the sum in (4) was too much for Jakob, who noted rather plaintively

If anyone finds and communicates to us that which up to now has eluded our efforts, great will be our gratitude.

The evaluation of (4), of course, resisted the attempts of another generation of mathematicians until 1734, when the incomparable Leonhard Euler devised an enormously clever argument to show that it summed to $\pi^2/6$. This result, which Jakob Bernoulli unfortunately did not live to see, surely ranks among the most unexpected and peculiar in all of mathematics. For the original proof, see ([4], pp. 83–85). A modern outline of Euler's reasoning can be found in ([2], pp. 486–487).

REFERENCES

1. Jakob Bernoulli. *Ars Conjectandi*. Basel, 1713.

2. Carl B. Boyer. *A History of Mathematics*. Princeton University Press, 1985.

3. C. H. Edwards. *The Historical Development of the Calculus*. Springer-Verlag, New York, 1979.

4. Leonhard Euler. *Opera Omnia* (1), Vol. 14 (C. Boehm and G. Faber, editors), Leipzig, 1925.

5. Morris Kline. *Mathematical Thought from Ancient to Modern Times*. Oxford University Press, New York, 1972.

6. D. J. Struik (editor). *A Source Book in Mathematics (1200–1800)*. Harvard University Press, 1969.

7. Translated from the Latin by Helen M. Walker, as noted in David E. Smith's *A Source Book in Mathematics*. New York: Dover, 1959, p. 271.

27

The First Calculus Textbooks

CARL B. BOYER

THE YEAR 1946 marks just a quarter of a millennium since the appearance of the *Analyse des infiniment petits* of L'Hospital.[1] This book, the first published text on the calculus, could boast truly that in its day it filled with distinction that ubiquitous lure of textbook writers—the long-felt need. Moreover, its influence and popularity dominated the whole of the eighteenth century, the period during which the new analysis developed until it completely overshadowed other branches of mathematics.

Historically integration antedates differentiation by some two thousand years. The ancient Greek method of exhaustion and the infinitesimal mensurations of Archimedes represent early examples of limits of integral sums; but it was not until the seventeenth century that Fermat found tangents and critical points through methods equivalent to evaluating limits of difference quotients. It was the discovery of the inverse nature of these two processes, together with the consequent exploitation of the anti-derivative in determining limits of sums, which leads one to call Newton and Leibniz the inventors of the calculus. Differentiation, both direct and inverse, became the basic algorithm in a new and powerful part of mathematics. Integration was nothing but "the memory of differentiation," and it was not until a century and a half later that some attention again was directed to the summation concept in the calculus.

Reprinted from *Mathematics Teacher* 39 (Apr., 1946): 159–67; with permission of the National Council of Teachers of Mathematics.

The earliest expositions of the new analysis were but the barest outlines of the rules of differentiation. The calculus first appeared in print in a six-page memoir by Leibniz in the *Acta Eruditorum* of 1684. This contained a definition of the differential and gave brief rules for determining the differentials of sums, products, quotients, power, and roots. It included also a few applications to problems of tangents and critical points. This account was so short and contained so many misprints as to be largely enigmatic to the best mathematicians of the age.[2] The basis of Newton's method of fluxions was first published unobtrusively in the form of lemmas in the *Principia* of 1687, more than a score of years after his discovery of the calculus. Here one finds some properties of limits, as well as cryptically brief directions for finding infinitely small "moments" of products, powers, and roots.[3]

The early years of the calculus resemble the infancy of analytic geometry in that little was done to popularize the subject. It is said that when Huygens about 1690 wished to master the new methods, there were not half a dozen men qualified to expound this analysis. Newton between 1669 and 1676 had composed several treatises on the elements of fluxions, but these did not begin to appear in print until 1704. John Craig in 1685 and 1693 published two works based in part on the method of Leibniz, but these were not intended as introductions to the subject and were moreover difficult to read because of the geometrical form in which they were cast. On the Continent the Bernoulli brothers with some effort achieved a thor-

ANALYSE

DES

INFINIMENT PETITS,

Pour l'intelligence des lignes courbes.

A PARIS,

DE L'IMPRIMERIE ROYALE.

M. DC. XCVL.

*Title-page of the first edition of
L'Hospital's textbook*

able largely to the fact that it was the first book to supply what the many had long awaited—an elementary introduction to a new and fascinating branch of mathematics.

Guillaume-François-Antoine de L'Hospital, Marquis de St. Mesme,[6] was born in 1661. His interest in geometry was aroused at an early age, and by the time he was fifteen years old he solved difficult problems on the cycloid which Pascal had proposed. He became a captain of cavalry, but later gave up army life to devote more time to the study of mathematics. When Jean Bernoulli was in Paris in 1692, L'Hospital for four months studied the new infinitesimal geometry under him. From that time on he joined the ranks of the mathematical elite and became the chief exponent of the calculus in France. Recognizing the lack of elementary expository works, he wrote to Bernoulli in 1695 that he was on the point of publishing a work on conic sections and that he proposed to add a little treatise on the differential calculus. L'Hospital indicated that he would render his master the justice he deserved, and he added modestly that the projected work would serve merely as an introduction to a more elaborate treatise, *De scientia infiniti*, which Leibniz planned to write.[7]

The promised book on conics was delayed by the author's illness, and appeared posthumously in 1707; but the *Analyse des infiniment petits* was published promptly in 1696. In the preface L'Hospital points out that he owed much to Leibniz and the Bernoullis, especially "the young professor at Groningen"; and he says that they can take whatever credit they wish, leaving to him what they will.[8] After Jean Bernoulli had received a copy of the book, he wrote and thanked L'Hospital for mentioning him in the volume, and he promised to return the compliment when he in turn should have composed something. He wrote that the work was admirably done, and praised it for the sound arrangement of propositions and for the intelligibility of the exposition. To L'Hospital's suggestion that he should write an integral calculus as a sequel to the *Analyse*, in view of the

ough understanding of the subject, and it was Jean Bernoulli who then instructed L'Hospital, while the latter in turn passed his knowledge on to Huygens. That there was an atmosphere of enthusiasm surrounding the calculus toward the close of the seventeenth century is to be ascribed in large part to the research articles published in the learned periodicals of their day by Leibniz, L'Hospital, and the Bernoullis. The large mathematical public had, before 1696, no easy introduction to the subject. Jean Bernoulli about 1691–1692 had composed two little textbooks, but their publication was long delayed: that on the integral calculus finally appeared in 1742,[4] and that on the differential calculus almost two hundred years later still.[5] That the text of L'Hospital achieved such prompt success is ascrib-

protracted delay on the part of Leibniz, Bernoulli replied that he was so distraught by domestic embarrassments that he could not carry out the project.[9]

Seldom has a textbook in mathematics been received by savants with such eagerness as was that of L'Hospital. It opened the door to a powerful tool which previously had been accessible only to the very gifted, and thus made it possible for the average reader to solve problems which had defied the best efforts of the geometers of Greece. But the reception was not one of unmixed acclaim. Echoing the contemporary literary controversy on "ancients vs. moderns," some admirers of the classic synthetic geometry cast doubt on the soundness of the new analytic approach. The Académie des Sciences in 1701 was sharply divided over the *Infiniment petits*. The Abbé Gallois led a polemic attack on the methods of the book, ably supported by Rolle.[10] L'Hospital, although he was a member of the Académie, did not reply; but his cause was warmly and effectively sustained by Varignon, Saurin, and others. The controversy reached such a pitch that the Académie finally felt obliged to name a commission to judge the question. Interest in the attack now began to wane, but L'Hospital did not witness the ultimate triumph of the infinitely small, for he died in 1704.

Jean Bernoulli meanwhile had observed with apparent jealousy the growing success of his protégé's book. During the years immediately following the author's death, he wrote letters attacking the work and practically accusing L'Hospital of plagiarism. L'Hospital's general prefatory acknowledgment, Bernoulli argued, did not do him justice inasmuch as he was in effect the true author of almost all the substance of the book. Historians of mathematics have been divided in judgment on this matter, but a thorough analysis of the question made by Eneström reaches the conclusion that the broad claims of Bernoulli with respect to the authorship of the material are not substantiated.[11]

The *Analyse des infiniment petits* represents the first systematic treatment of the calculus, and as such it presents a good picture of the level of the subject at that time. The preface includes a brief historical account in which the author, while admitting that "Newton also had a kind of calculus," prefers and emphasizes the contributions of Leibniz. The treatise itself is divided into ten chapters or sections, of which the first (pp. 1–10) is devoted to the fundamental definitions, assumptions, and rules of procedure. A variable quantity is defined loosely as one which continually increases or diminishes; and the infinitely small portion by which a variable quantity changes is called the "difference" or differential. Two postulates are laid down: that one can take as equal two quantities which differ only by an infinitely small quantity; and that a curved line can be considered as an infinite assemblage of infinitely small straight lines which determine, by the angles which they make with each other, the curvature of the curved line. To L'Hospital these premises appeared to be "so self-evident as not to leave the least scruple about their truth and certainty on the mind of an attentive reader," but he adds that if space permitted he could prove them in the manner of the ancients. The crudeness of these fundamental principles is an indication of the infancy of the subject rather than of carelessness on the part of the author. The language of L'Hospital is eminently restrained in comparison with that of contemporaries who wrote such things as

$$\frac{d^3y}{d^2x} = d^3y \int^2 x \text{ or } \infty \cdot \infty^{\infty-1} = \infty^\infty.$$

The basic differential formulas for algebraic functions—sums, products, quotients, powers, and roots—are derived by L'Hospital in the customary Leibnizian manner, infinitesimals of higher order being neglected. Independent and dependent variables were not always clearly distinguished at the time, and here L'Hospital took a step in advance by stating formally the essential implicit function formula in the case of powers of expressions involving one or more variables. A modern reader will be struck by the absence of rules for the differentiation of transcendental functions. These were known at the time, but they had not been popularly formalized. L'Hospital had suggested to Bernoulli that he might wish to append to the *Analyse* his rules for the calculus of logarithms, but this was not done. However, the equivalent of these was implied by the well-known results on the area between the hyperbola and its asymptotes. The differentials of the trigonometric functions were at the time likewise geometrically inferred from the application of the characteristic triangle to the circle. The inverses offered no difficulty in view of the known formula for implicit functions.

The second section of the *Infiniment petits* (pp. 11–40) is devoted to problems of tangency. Inasmuch as the tangent is defined as the prolongation of an infinitely small side of the polygon which makes up the curved line, one sees immediately from the similarity of triangles that the subtangent is given by $y(dx/dy)$. Through this relationship, the author finds the tangents to the curves of his day: the higher hyperbolas and parabolas, the spirals of various kinds, the quadratrix of Dinostratus, the conchoid of Nicomedes, the cissoid of Diocles, the cycloid, the tractrix, and the curve of logarithms. Some of these curves are defined analytically, while others are described kinematically or in terms of geometrical properties. L'Hospital follows the Cartesian classification of curves into geometrical and mechanical, but inasmuch as no special notations are used for transcendental functions, only in the former category is "the nature of a curve" given by an ordinary equation.

The third chapter (pp. 41–54), on maxima and minima, opens with a definition of the basic terms: If the ordinate (appliquée) increases, as the abscissa (coupée) increases to a given value E, and thereafter decreases, then the ordinate at E is called the greatest; and similarly, *mutatis mutandis,* for the least ordinate. The treatment here differs from that in modern textbooks not only in being less rigorous, but also in that cusp maxima and minima

are treated as extensively as extrema having horizontal tangents. L'Hospital points out that the differential of an increasing quantity is positive, that of a decreasing quantity is negative. Moreover, no quantity which varies continuously can go from positive to negative without going through zero or infinity. Hence to find maxima and minima of an algebraic expression, L'Hospital equates to zero both the numerator and the denominator of the differential, discarding results which lead to a contradiction. A few cases of applications in geometry and science are included. He does not explicitly state the various tests distinguishing maxima from minima, but these were well-known to the author and his times.

Points of inflection and of *rebroussement* (cusp points for which the tangents are horizontal) are treated in section four (pp. 55–70), and here the thorny question of second order differences arose. Leibniz had been unable to give a satisfactory definition of differentials of order greater than one, and the first attack on the calculus, launched by Nieuwentijdt in 1695–1696 while L'Hospital was preparing the *Analyse,* was centered on this inadequacy. In his preface (p. eiij) L'Hospital claimed that Leibniz had satisfactorily answered Nieuwentijdt's doubts about the existence of higher differentials, but this is belied by the author's own definition: "The infinitely small portion by which the difference of a variable quantity increases or decreases continually is called the difference of the difference of this quantity, or its second difference." That he obtained correct results in this connection is due less to his definition than to the pragmatic rule which follows it: "One takes as constant whichever difference one wishes, and treats the others as variable quantities." For example, the differential of the function xy being $xdy + ydx$, the second differential is either $dd(xy) = xddy + 2dxdy$ or $dd(xy) = yddx + 2dxdy$. Here one sees the need to distinguish between the differentials of independent and dependent variables. This distinction was not at the time sufficiently emphasized, and even today numerous textbooks founder on the differential through neglect in this connection. L'Hospital then gives the geometrical description of points of inflection and *rebroussement* in terms of convexity and concavity, and he points out that these points are determined by making the second differential either zero or infinitely great. A more rigorous treatment is not, of course, to be expected of that period.

The concept of radius of curvature as such is not presented in the modern manner but is given in the context in which it arose historically. Huygens had been led by his pendulum clock to study the subject of evolutes and involutes (*developpées* and *developpants*), and this material is presented by L'Hospital in section five (pp. 71–103). Toward the close of this chapter, the longest of all, the author points out in a casual sort of way that $dx^2+dy^{2\frac{3}{2}}/-dxddy$ is a general expression for "the radius of an evolute." This is, of course, the equivalent of the modern formula for radius of curvature, where y is a function of x.

Sections six and seven (pp. 104–119, 120–130) are on caustics by reflection and refraction. These are not now traditional topics in elementary calculus, but they were popular new topics in the late seventeenth century. Moreover, L'Hospital and his master, Bernoulli, had shared with Tschirnhaus in their development. Section eight (pp. 131–144) is on envelopes of families of straight lines, and includes the well-known method of Leibniz of differentiating with respect to the parameter.

The portion of the book which in its day roused most discussion was section nine (pp. 145–163). This chapter carries the unenlightening heading, "Solution of some problems depending on the preceding methods," but it involves what would now be called indeterminate forms. The presentation is largely geometrical, but the basic result is the so-called "rule of L'Hospital." To find "the value" of a rational expression in x which for the abscissa in question takes the form 0/0, he determines the ratio of the "differences" of the numerator and denominator for this abscissa. This rule appears not to have been familiar at the time, and

so the author here might well have indicated the source of his material. The author never claimed the rule as his own, but when his friend Saurin implied that it was due to Leibniz, protest was raised by Jean Bernoulli that here, as elsewhere, L'Hospital had not given him due credit. Bernoulli's claim to the rule seems to be substantiated, and his name, rather than L'Hospital's, should be applied to it. But it should be pointed out that the traditional nomenclature is due to the fickleness of fortune rather than to any ill intent on the part of L'Hospital. The *Analyse des infiniment petits* occupies a place in the calculus somewhat analogous to that of Lavoisier's *Traité élémentaire* in chemistry almost a century later. Both books would have been more attractive had they included a substantial historical background, but in neither case should the omission of this be interpreted as an argument *e silencio* that the author claims credit for more than the general arrangement and exposition of the subject matter. One does not demand of writers of textbooks the same scrupulous regard for the citation of sources that one expects in the publication of the results of research work.

The tenth and last section of L'Hospital's treatise (pp. 164–181) is somewhat out of harmony with the remainder. It is a clever bit of salesmanship in which the author contrasts the elegant methods of the calculus with the awkward anticipatory procedures of Descartes and Hudde for determining maxima and minima. On the basis of this comparison L'Hospital emphasizes that the method of Leibniz "gives general solutions, whereas the other furnishes only particular ones, and it extends to transcendental curves and it is not at all necessary to avoid incommensurables, which often is impracticable." This closing sentence of the first textbook on the calculus reminds one of the title of Leibniz's first paper on the new analysis a dozen years before: "A new method for maxima and minima, as well as tangents, which is not impeded by irrational quantities."

The *Analyse* of L'Hospital was composed two hundred and fifty years ago, and hence the mate-

Marquis de L'Hospital

dynamics afforded a golden opportunity for apt and timely illustrations of the new calculus, but we are told in the preface (p. eij) that illness prevented the author from fulfilling his intention to add a section on applications to physics. The language and notation of the *Analyse* differ slightly from that now in use; there is an anachronistic emphasis on the ideas of ratio and proportion; and the material is divided into propositions, corollaries, and scholia. Nevertheless, a modern reader should have no difficulty in following the author's exposition.

The *Analyse des infiniment petits* appeared in a second edition at Paris in 1715. This was in reality a reissue of the original, but it contained numerous typographical errors. The work was reprinted the following year, and again in 1720. By this time the popularity of the calculus was such that a textbook on a lower level was desired. Publishers sought to capitalize on this desire and on the popularity of L'Hospital's work by providing commentaries. In 1721 Crousaz issued at Paris his *Commentaire sur l'analyse des infiniment petits*, consisting of an elementary introduction and a series of supplementary notes. This book was damned by faint praise on the part of Jean Bernoulli and met with severe criticism from Saurin. Four years later there appeared a posthumous and successful supplementary volume by Varignon, *Eclaircissemens sur l'analyse des infiniment petits*. The author appears to have had in mind the preparation of a new edition of L'Hospital's work, together with added notes. The publisher, however, felt that inasmuch as L'Hospital's *Analyse* already "had appeared several times in France and in foreign countries," and since "most men in the profession already had copies," he could save the public some expense by issuing Varignon's *Eclaircissemens* as a separate volume rather than as part of a new edition. This volume is on a much higher level than that of Crousaz, and although it follows L'Hospital section by section, it adds to this fresh problems and methods resulting from later research.

rial it contains does not in all cases parallel that which appears in modern textbooks. One fails to find, for example, the notions of function and limit. The first of these was in the air in L'Hospital's day, but it was popularized half a century later in Euler's *Introductio ad analysin infinitorum*; the latter was likewise known in a general sense through the calculus of Newton, but it was a hundred years before it became basic in the treatises of Cauchy. One misses also the now traditional work on Taylor's series. This, too, was known to James Gregory and Jean Bernoulli, but it was formalized after L'Hospital's time through Brook Taylor's work on finite and infinitesimal differences. Curve tracing is a further topic which is lacking in the *Analyse*. The seventeenth century here invented the analytical tools, but it was the eighteenth century which first applied them to the systematic study of a wide variety of curves. It is strange to relate also that L'Hospital's text is deficient in applications to science. Seventeenth-century

The rift between British and Continental mathematicians during the eighteenth century was not sufficiently wide to prevent Stone in 1730 from publishing an English translation of L'Hospital's *Infiniment petits*, ten years before Buffon reciprocally prepared a French translation of Newton's *Method of Fluxions*. The full title of Stone's book is, *The method of fluxions both direct and inverse. The former being a translation from the celebrated Marquis de L'Hospital's Analyse des infiniments petits: and the latter supply'd by the translator*. The translator opens with praise of the author's work, "the Character whereof is so well establish'd," but adds that out of regard to Sir Isaac Newton he has altered some of the language and notation of the original to conform to the Newtonian. Increments or differentials, for example, are taken as fluxions, inasmuch as they are proportional to the latter; and these are denoted by $\dot{x}$, $\ddot{x}$, $\dddot{x}$, etc., instead of by dx, ddx, $dddx$, etc. A few years later Bishop Berkeley took full advantage, in the famous *Analyst* controversy of the confusion then existing in Britain between fluxions and infinitely small quantities.

Stone's translation, in part I, does not depart much from the original, and the diagrams are faithful reproductions of those of L'Hospital. Some new material is added, however, on logarithms, exponentials, and trigonometric functions. The second part or appendix (separately paginated) is on the inverse method of fluxions, and this constitutes a new work. L'Hospital apparently had hoped that his treatise would sometime be completed by a second part on the integral calculus. The pages of the *Analyse* (even in later editions) carry the heading *I. Part.*, but the author never completed another *partie*. Stone was but one of several who undertook to compose a suitable second half. His attempt was so successful that it was translated into French and published in 1735, "servant de suite aux infiniment petits de M. le Marquis de l'Hôpital."

The *Analyse des infiniment petits* continued throughout the eighteenth century to be the standard elementary manual on higher mathematics.[12] During the latter half of the century it found a

worthy rival in the *Instituzioni analitiche* of Maria Agnesi (Milan, 1748), but frequency of publication indicates that the vogue for L'Hospital's work continued. It appeared in new French editions, revised and enlarged, in 1758 (Paris), 1768 (Avignon), and 1781 (Paris), and in Latin translations in 1764 and 1790 (both at Vienne).[13]

Toward the turn of the century, however, Lacroix[14] began the publication of his renowned series of textbooks, and one of these, the *Traité élémentaire de calcul différentiel et de calcul intégral*, (Paris, 1802) took the place during the nineteenth century which L'Hospital had held a hundred years earlier. The concepts of function and limit now took over the field which the infinitesimal "ghosts of departed quantities" had so long dominated. But the work of the first textbook on the calculus had been well done. The enthusiasm for the subject which it kindled and nourished has never abated, and modern textbooks are but new monuments to the influence of the *Analyse des infiniment petits*.

NOTES

1. G.F.A. de L'Hospital, *Analyse des infiniment petits, pour l'intelligence des lignes caurbes*, Paris, 1696. Page references indicated in this paper refer to the original edition.

2. See Gustav Eneström, "Über die erste Aufnahme der Leibnizschen Differentialrechnung," *Bibliotheca Mathematica* (3), IX (1908–1909), 309–320. An English translation of Leibniz's paper of 1684 is found in D. E. Smith, *A Source Book in Mathematics* (New York, 1929), pp. 619–626.

3. Sir Isaac Newton, *Opera quae exstant omnia* (ed. by Samuel Horsley, 5 vols., Londini, 1779–1785), I, 251, II, 277–280.

4. Jean Bernoulli, *Opera omnia* (4 vols., Lausanne and Geneva, 1742), vol. III; also *Die erste Integralrechnung. Eine Auswahl aus Johann Bernoullis mathematischen Vorlesungen über die Methode der Integrale und anderes aufgeschrieben zum Gebrauch des Herrn Marquis de l'Hospital in den Jahren 1691 und 1692. Translated from the Latin*, Leipzig and Berlin, 1914.

5. Jean Bernoulli, *Die Differentialrechnung aus dem Jahre 1691/92,* Ostwald's klassiker, No. 211, Leipzig, 1924.

6. A satisfactory biographical account is found in J. F. Michaud, *Biographie Universelle* (2nd ed., Paris, 1880), XXIII, 448–451. In later years the family name came to be spelled L'Hôpital instead of L'Hospital.

7. See L'Hospital, *op. cit.*, preface.

8. *Ibid.*

9. Eneström, "Sur le part de Jean Bernoulli dans la publication de l'Analyse des infiniment petits," *Bibliotheca Mathematica* (new series), VIII (1894), 65–72. Cf. O. J. Rebel, *Der Briefwechsel zwischen Johann* (I). *Bernoulli und dem Marquis de l'Hospital,* Bottrop i.w., 1934.

10. Académie des Sciences, *Histoire et mémoires,* 1701 (Amsterdam, 1735), Histoire, pp. 114 ff. This controversy was the second of three important attacks on the calculus. The first was that of Nieuwentijdt against the work of Leibniz in 1695–1696. The third was the famous *Analyst* controversy of 1734–1735 concerning the calculus of Newton.

11. Eneström, *op. cit.*

12. The Abbé Sauri in his *Cours complèt de mathéma-tiques* (5 vols., Paris, 1774), III, xij, refers to L'Hospital's *Analyse* as "connue de tout le mond."

13. This paper does not pretend to give a complete listing of editions. I have examined those of 1696, 1715, 1730, 1768, and 1781, and the commentaries by Crousaz and Varignon, all of which are available at the New York Public Library or at Columbia University. Editions other than these have been listed on the basis of references found in standard sources, such as Poggendorff and the catalogues of the British Museum and the Bibliothèque Nationale. It is of interest to note that L'Hospital's *Traité analytique des sections coniques* likewise enjoyed a considerable popularity during the eighteenth century, appearing in editions of 1707, 1720, 1723 (English translation by Stone), 1770 (Venice), and 1776.

14. S. F. Lacroix was perhaps the most prolific textbook writer of modern times, if allowance is made for multiple editions. In 1848 there appeared at Paris the 20th edition of his *Traité élémentaire d' arithmétique* and the 16th edition of his *Élémens de géométrie.* The 20th edition of his *Élémens d' algèbre* was published at Paris in 1858, and the ninth edition of the *Traité élémentaire de calcul* in 1881. And the above figures do not include the large number of editions in other languages.

28

The Origin of L'Hôpital's Rule

D. J. S T R U I K

$\mathcal{T}$HE SO-CALLED RULE of L'Hôpital, which states that

$$\lim_{x \to a} \frac{f(x)}{g(x)} = \frac{f'(a)}{g'(a)}$$

when $f(a) = g(a) = 0$, $g'(a) \neq 0$, was published for the first time by the French mathematician G. F. A. de l'Hôpital (or De Lhospital) in his *Analyse des infiniment petits* (Paris, 1696).[1] The Marquis de l'Hôpital was an amateur mathematician who had become deeply interested in the new calculus presented to the learned world by Leibniz in two short papers, one of 1684 and the other of 1686. Not quite convinced that he could master the new and exciting branch of mathematics all by himself, L'Hôpital engaged, during some months of 1691–92, the services of the brilliant young Swiss physician and mathematician, Johann Bernoulli, first at his Paris home and later at his château in the country.[2] When Bernoulli left for his home town Basel, the Marquis kept up correspondence with his tutor, at the same time publishing some original contributions of his own findings. When, in 1696, L'Hôpital's book appeared, he acknowledged his indebtedness to Leibniz and Bernoulli, but only in general terms: "I have made free with their discoveries (*je me suis servi sans façon de leur découvertes*), so that whatever they please to claim as their own I frankly return to them."

Reprinted from *Mathematics Teacher* 56 (Apr., 1963): 257–60; with permission of the National Council of Teachers of Mathematics.

The question of the actual dependence of L'Hôpital on Bernoulli remained unanswered, and acquired in the course of the years somewhat the character of a mystery. Bernoulli, after L'Hôpital had sent him a copy of the book, thanked him courteously and praised it. But subsequently, in some private letters written during the lifetime of the Marquis, he claimed that much of the content of the *Analyse des infiniment petitis* was really his own property. In 1704, after L'Hôpital's death, he made a public claim to that section, No. 163, which contains the rule for 0/0.[3] Mathematicians interested in such kind of priority puzzles have been philosophizing about this supposed dependence of L'Hôpital on Johann Bernoulli ever since, weighing Bernoulli's acknowledged reputation for nastiness.[4] A generally acceptable conclusion was not reached until recent times.

Considerable clarification came in 1922, when Johann Bernoulli's manuscript on the differential calculus, dating from 1691–92, was at last published (the corresponding manuscript on the integral calculus was known from Johann Bernoulli's *Opera*, published in 1742 during the lifetime of the author).[5] Comparison of these notes by Bernoulli and the text of L'Hôpital's book revealed that there was a considerable overlapping, so that it seemed that Bernoulli had fathered much of the nobleman's intellectual offspring. But the true situation came to light only in 1955, when Bernoulli's early correspondence was published.[6] It then appeared that in 1694 a deal was actually made between the Marquis and his former tutor, by which L'Hôpital offered him a yearly allowance of 300 livres (and more later) provided that Bernoulli agreed to three conditions:

1 To work on all mathematical problems sent to him by the Marquis,
2 To make all the discoveries known to him, and
3 To abstain from passing on to others a copy of the notes sent to L'Hôpital.

This settled the priority question. Here is a translation of the section of the letter which contains the unusual proposition, sent by L'Hôpital in Paris to Johann Bernoulli in Basel, March 17, 1694:

> I shall give you with pleasure a pension of three hundred livres, which will begin on the first of January of the present year, and I shall send two hundred livres for the first half of the year because of the journals that you have sent, and it will be one hundred and fifty livres for the other half of the year, and so in the future. I promise to increase this pension soon, since I know it to be very moderate, and I shall do this as soon as my affairs are a little less confused. . . . I am not so unreasonable as to ask for this all your time, but I shall ask you to give me occasionally some hours of your time to work on what I shall ask you—and also to communicate to me your discoveries, with the request not to mention them to others. I also ask you to send neither to M. Varignon[7] nor to others copies of the notes that you let me have, for it would not please me if they were made public. Send me your answer to all this and believe me, *Monsieur tout à vous.*

le M. de Lhopital

Bernoulli's answer has not been found, but from a letter of July 22, 1694, we know that he had accepted the proposal. It must have been a little windfall for the impecunious young scientist, just married and still looking for a position (which he obtained the following year at the University of Groningen in the Netherlands). How long this interesting relationship lasted we do not know, but Bernoulli's finances improved and those of L'Hôpital did not become any better. By 1695 it may have come to an end.

Several letters from Bernoulli to his patron with answers to questions have now been published, and the one dated July 22, 1694, contains the rule for 0/0. The formulation is very much like the one we find in the *Analyse des infiniment petits*, and is based on a geometrical consideration. In our words, if

$$y = \frac{f(x)}{g(x)}$$

and both curves $y = f(x)$ and $y = g(x)$ pass through the same point P on the x axis, $OP = a$, so that $f(a) = g(a) = 0$, and if we take an ordinate $x = a + h$, then the figure shows immediately that

$$\frac{f(a + h)}{g(a + h)}$$

is almost equal to the quotient of $hf'(a + h)$ and $hg'(a + h)$ when h is small. In the limit we find, now in Bernoulli's words:

"In order to find the value of the ordinate (*appliquée*) of the given curve

$$\left[y = \frac{f(x)}{g(x)} \right]$$

in this case it is necessary to divide the differential (*la différentielle*) of the numerator of the general fraction by the differential of the denominator."

Bernoulli's examples are almost the same that L'Hôpital uses:

1) $y = \dfrac{\sqrt{2a^3x - x^4} - a\sqrt[3]{a^2x}}{a - \sqrt[4]{ax^3}}$ for $x = a$. Then $x = a$

$$y = \left(\frac{16}{9} \right) a.$$

This example is used by both Bernoulli and L'Hôpital.

2) $y = \dfrac{a\sqrt{ax - xx}}{a - \sqrt{ax}}$ for $x = a$. Then $y = 3a$.

This example of Bernoulli is changed by L'Hôpital into

$$y = \frac{aa - ax}{a - \sqrt{ax}} \text{ for } x = a. \text{ Then } y = 2a.$$

The situation has thus been clarified. When L'Hôpital's book appeared, Bernoulli was bound by his promise not to reveal which sections of the book belonged to him. He could only express himself privately. Then, after the death of the Marquis, he felt that he need not be so silent any more, and claimed as his own the most striking result of the book—the rule for 0/0. But he could not prove his assertion. At present he stands vindicated.

From this discovery of the origin of L'Hôpital's rule we should not conclude that from now on it should be called after Bernoulli. First of all, there are already plenty of rules and theorems called after Bernoulli (due to at least three members of the Bernoulli family—Jakob, Johann, and Daniel). But there is a more weighty consideration. When we begin to change the names of rules and theorems in accordance with the strict laws of priority, we soon come to the dreary conclusion that our science will lose many of its most familiar expressions. Pythagoras's theorem was known to the Babylonians more than a millennium before the sage of Crotona lived. The Cauchy-Riemann equations were known to D'Alembert and Euler. Taylor's theorem would be Gregory's until another claimant pops up—as a matter of fact, he already exists in the person of the Indian Nilakantha (ca. 1500). The Indians also played with Pell's equation long before John Pell studied it—or better, did not study it, since Pell's connection with the equation is rather remote. Fourier's series were used by Euler and Daniel Bernoulli. Pascal's triangle was known to the Chinese mathematician Yang Hui (thirteenth century), and probably is even older, and his contemporary Chhin Chin-Shao worked with Horner's method in the theory of algebraic equations as with an ancient tool. And so on.

The names attached to mathematical discoveries are often the names of persons who made these results better known or understood through their own outstanding work. Any new historical discovery may disturb the delicate balance of nomenclature again. Let the good Marquis keep his elegant rule; he paid for it and made it public property. After all, he deserves some fame; his book on the new calculus was not only the first to be published, and contained contributions of his own, but it was good enough to hold its prominent position for half a century and longer. Even after other and better textbooks appeared, it continued to be used as a good first introduction into the calculus; we know of an edition as late as 1790. It appeared in an English and a Latin translation, and there also exist commentaries, like that of L'Hôpital's friend Varignon. We should have some respect for it.

NOTES

1. Details are given in C. B. Boyer, "The First Calculus Textbooks," *The Mathematics Teacher*, XXXIX (April, 1946), 159–67.

2. Johann Bernoulli (1667–1748) and his older brother Jakob (1655–1705) were the earliest and most prominent students of Leibniz's mathematical discoveries. Jakob was a professor at the University of Basel until his death, when he was succeeded by his brother, who, from 1695, had been teaching at the University of Groningen. Between ca. 1685 and ca. 1700, Leibniz and the Bernoulli brothers developed most of our elementary differential calculus, advancing also into the integral calculus, differential geometry, and even the calculus of variations.

3. *Acta eruditorum*, 1704; more details in the *Acta eruditorum*, 1721, pp. 223–27.

4. Eneström, G. *Bibliotheca mathematica*, VIII (1894), 65–72.

5. Bernoulli, Jakob, "Die Differential-rechnung (aus dem Jahre 1691/92)," Leipzig, *Ostwald's Klassiker 211* (1922), 56 pp., editor: J. Schafheitlin.

6. *Der Briefwechsel von Johann Bernoulli*, Band I, Basel 1955, 531 pp., editor: O. Spiess. Section B, pp. 121–383, contains the correspondence of L'Hôpital with Bernoulli; it is introduced by a detailed account of the relationship between the two men.

7. Pierre Varignon (1654–1722), French mathematician, was, like his friend L'Hôpital, an early student of the calculus. In 1725, his commentary on L'Hôpital's book was published.

29

Euler, the Master Calculator

JERRY D. TAYLOR

ON THE BICENTENNIAL of Leonard Euler's (pronounced oiler) death, we should review some of his contributions, both to get an idea of the debt we owe to Euler and to get a feeling for the methods of this eighteenth-century mathematician. I shall present some problems and then suggest how Euler might have solved them.

Euler could have worked all the problems in table 1 without pen and paper and without the use of any mathematical tables. It has been said that he could calculate as easily as others breathe. His feats of memory are legendary—he once learned by heart a complete book of Virgil, and in his youth he prepared himself to teach in a medical school in only a few months (Bell 1937, p. 149). He memorized all the logarithmic and trigonometric tables as aids in mathematical calculations (Kline 1972, p. 401).

A good memory was especially helpful to Euler because he was blind for more than one-fourth of his adult life. Being blind did not stop him from becoming the most published mathematician in history. Although estimates vary a little from source to source, his collected works number about a hundred volumes. They include textbooks in algebra, analytic geometry (plane and solid), calculus, and differential equations. He also published about five hundred articles on research in such areas as complex variables, topology, algebra, probability, number theory, mechanics, and optics. Euler's out-

Leonard Euler
(15 April 1707–18 September 1783)

put was amazing, especially when one realizes that during the midyears he wrote while tending to thirteen children and that during his later years he did the mathematics mentally and dictated the results.

In one respect Euler was fortunate to have lived in the eighteenth century. During his lifetime it was fashionable for governments to support royal academies, where some of the best scholars

Reprinted from *Mathematics Teacher* 76 (Sept., 1983): 424–28; with permission of the National Council of Teachers of Mathematics.

TABLE I

Some Easy Problems for Euler

1. Simplify.

$$3 + \cfrac{1}{1 + \cfrac{1}{1 + \cfrac{1}{1 + \cfrac{1}{1 + \cfrac{1}{6}}}}}$$

2. Find the sum.

$$1 - \frac{1}{4} + \frac{1}{16} - \frac{1}{64} + \frac{1}{256} - \ldots$$

3. Approximate to five decimal places
$$(1.01)^{10}$$

4. Find the sum.

$$1 + \frac{1}{2} + \frac{1}{3} + \frac{1}{4} + \frac{1}{5} + \frac{1}{6} + \frac{1}{7} + \frac{1}{8} + \frac{1}{9} + \frac{1}{10}$$

5. A dodecahedron contains how many vertices?

6. Simplify.
$$\ln \ln \ln e^{1/e}$$

7. Simplify.
$$e^{i\pi}$$

8. Find the last two digits of 17^{42}

in Europe were brought together. Euler was a member of the St. Petersburg Academy in Russia from 1727 to 1741, the Berlin Academy from 1741 to 1766, and the St. Petersburg Academy again from 1766 to 1783. For this entire fifty-six-year period he received at least one stipend from an academy. While in Berlin, he drew salaries from both academies.

At the Berlin Academy, Euler was friendly with fellow mathematicians like D'Alambert and Lambert. However, he was not sophisticated enough for the philosopher Voltaire or Frederick the Great, who referred to Euler as the *mathematical cyclops* (Boyer 1949, p. 483). In contrast, Catherine the Great treated Euler like royalty on his return to St. Petersburg, where he remained until his death in 1783 (Bell 1937, pp. 148–49). With this condensed biography of Euler in mind, let us return to the problems listed in the table to see how he might have solved them.

Solutions

1. Simplify.

$$3 + \cfrac{1}{1 + \cfrac{1}{1 + \cfrac{1}{1 + \cfrac{1}{1 + \cfrac{1}{6}}}}}$$

This problem might be found in an algebra text today and called a *complex fraction*. Euler would have recognized this problem as a *continued fraction*. Because he was so familiar with continued fractions, Euler would have instantly recognized that the answer was 119/33. Moreover, he could have started with the answer and found the problem. Note that

$$\frac{119}{33} = 3 + \frac{20}{33}$$

$$= 3 + \cfrac{1}{\cfrac{33}{20}}$$

$$= 3 + \cfrac{1}{1 + \cfrac{13}{20}}$$

$$= 3 + \cfrac{1}{1 + \cfrac{1}{\cfrac{20}{13}}}$$

$$= 3 + \cfrac{1}{1 + \cfrac{1}{1 + \cfrac{7}{13}}}$$

$$= 3 + \cfrac{1}{1 + \cfrac{1}{1 + \cfrac{1}{\cfrac{13}{7}}}}$$

$$= 3 + \cfrac{1}{1 + \cfrac{1}{1 + \cfrac{1}{1 + \cfrac{6}{7}}}}$$

$$= 3 + \cfrac{1}{1 + \cfrac{1}{1 + \cfrac{1}{1 + \cfrac{1}{\cfrac{7}{6}}}}}$$

$$= 3 + \cfrac{1}{1 + \cfrac{1}{1 + \cfrac{1}{1 + \cfrac{1}{1 + \cfrac{1}{6}}}}}$$

We sometimes write the final form of this continued fraction as 3;11116 for brevity. If the fraction were extended to the repeating continued fraction 3;111161111611116 ... , the value would be $\sqrt{13}$ (NCTM 1969, pp. 269–70). The finite continued fraction 3;11116 = 119/33 = 3.6060 ... is a reasonably good approximation for $\sqrt{13} \doteq 3.6055....$

Euler was very proficient with continued fractions and could easily generate them. Continued fractions were helpful in obtaining approximations for $\sqrt{13}$, e, π, or other irrational numbers. Euler showed that rational numbers can always be represented as finite or terminating continued fractions, and irrational numbers always have infinite continued fractions. Euler was the first to prove, from their expansions, that numbers like e and e^2 are irrational.

2. Find the sum.

$$1 - \frac{1}{4} + \frac{1}{16} - \frac{1}{64} + \frac{1}{256} - \cdots$$

Most students today would think of this problem as a geometric series with ratio − 1/4. Although Euler was familiar with this method, he preferred to divide 1 by $(1 + x)$, using the common division algorithm to obtain the results

$$\frac{1}{1 + x} = 1 - x + x^2 - x^3 + \cdots$$

If x is replaced by 1/4 in this expression, the original problem is solved quickly. This formula appeared to be general and was certainly obtained by legitimate means. Hence, Euler and many of his contemporaries were puzzled by the fact that replacing x by 1 gives

$$\frac{1}{2} = 1 - 1 + 1 - 1 + \cdots$$

Many eighteenth-century mathematicians were convinced that the sum had to be zero, whereas others thought that the answer was one, a solution arrived at by regrouping as follows:

$$1 - (1 - 1) - (1 - 1) - \ldots = 1.$$

We know that none of these answers is correct, because the series does not converge, but presenting the possibilities to a class can be interesting.

3. Approximate to five decimal places.

$$(1.01)^{10}$$

Problems of this nature have been popular for a long time, since they arise in compound interest and related problems. The problem can be done mentally by thinking in terms of $(1 + .01)^{10}$ and using the binomial theorem:

$$(a + b)^n = a^n + na^{n-1}b$$

$$+ \frac{n(n-1)}{2} a^{n-2}b^2$$

$$+ \frac{n(n-1)(n-2)}{2 \times 3} a^{n-3}b^3 + \ldots + b^n.$$

When $a = 1$, $b = 0.01$, and $n = 10$, the expansion becomes

$$(1 + 0.01)^{10} = 1 + (10)(0.01) + (45)(0.01)^2$$
$$+ (120)(0.01)^3 + \ldots$$
$$= 1 + 0.1 + 0.0045 + 0.00012 + \ldots$$
$$\doteq 1.10462.$$

Being a formalist, Euler liked to explore such expressions. He knew the generalized form, of course, but he was concerned about the result

$$(1 - 2)^{-1} = 1 + 2 + 4 + 8 + \ldots$$

For some mathematicians of Euler's time, such expressions simply confirmed their belief that negative numbers often led to absurdities and should be avoided whenever possible. Some even wanted to avoid multiplying two negative quantities, since no physical interpretation of this phenomenon was known. Euler was not troubled by the lack of physical models for negative or even imaginary numbers, and he used both freely. He found expressions like

$$(1 - 2)^{-1} = 1 + 2 + 4 + \ldots$$

to be intriguing, and our students might also. They might be interested to discover that sometimes even well-known formulas have restrictions. In this expression, if 2 is replaced by a number between 1 and $^-1$, then the formula works.

4. Find the sum.

$$1 + \frac{1}{2} + \frac{1}{3} + \frac{1}{4} + \frac{1}{5} + \frac{1}{6} + \frac{1}{7} + \frac{1}{8} + \frac{1}{9} + \frac{1}{10}$$

Euler would have worked this problem just like the rest of us, albeit faster. If you solve the problem by getting a common denominator, you will better understand why Euler tried but failed to find a formula for the sum of the harmonic progression

$$1 + \frac{1}{2} + \frac{1}{3} + \ldots + \frac{1}{n} .$$

No one else has found a formula either. Euler did show that

$$\lim_{n \to +\infty} \left(1 + \frac{1}{2} + \frac{1}{3} + \ldots + \frac{1}{n} \right) - \ln n = \gamma,$$

called *Euler's constant*. Euler found the approximate value of γ to be 0.577218, but it is not yet known whether the actual constant is rational or irrational. This observation is interesting in view of the fact that he did find the exact sum for such related problems as

$$1 + \frac{1}{2^2} + \frac{1}{3^2} + \ldots = \frac{\pi^2}{6}$$

and

$$1 + \frac{1}{2^4} + \frac{1}{3^4} + \ldots = \frac{\pi^4}{90}.$$

5. A dodecahedron contains how many vertices? The problem can be simplified and solved without counting the vertices on a model, if the Euler-Descartes formula

$$V + F = E + 2$$

is used. A dodecahedron has twelve pentagonal faces and, hence, five edges per face. Since each edge is on exactly two faces,

$$E = \frac{5 \times 12}{2} = 30.$$

Thus, $V + 12 = 30 + 2$, or $V = 20$. Euler's proof of the formula $V + F = E + 2$ and his work on the now-famous Königsberg bridge problem mark him as a pioneer in topology.

6. Simplify.

$$\ln \ln \ln e^{1/e}$$

When entered on a calculator, this problem causes the calculator to display an error message. The problem goes nicely at first; that is, $\ln \ln \ln e^{1/e} = \ln \ln (1/e) \ln e = \ln \ln 1/e = \ln \ln e^{-1} =$

$$\ln (^-1) \ln e = \ln (-1) = ?$$

In the eighteenth century quite a discussion ensued among Euler and other prominent mathematicians as to the resolution of $\ln (^-1)$. Consider the following argument of John Bernoulli (Kline 1972, pp. 409–10):

$$
\begin{array}{rcl}
(^-1)^2 & = & 1^2 \\
\ln (^-1)^2 & = & \ln (1)^2 \\
2 \ln (^-1) & = & 2 \ln 1 \\
\ln (^-1) & = & \ln 1 = 0
\end{array}
$$

In general, many mathematicians wanted to define the logarithm of a negative number to be the same as the logarithm of its additive inverse. Euler showed that, to define a logarithm of a negative number, one must employ complex numbers.

The flaw in Bernoulli's argument is that $2(\ln x) = \ln x^2$ only when the logarithms exist in the field of real numbers. To overcome a possible flaw in the thinking of some students and to acknowledge the work of Euler, it would be good to point out that logarithms of negative and even complex numbers do exist in the field of complex numbers, but the rules of operation are different.

7. Simplify.

$$e^{i\pi}$$

If you know Euler's relation

$$e^{ix} = \cos x + i \sin x,$$

then it follows that $e^{i\pi} = {}^-1$. But even today, many students are jolted by the fact that a real number raised to an imaginary power could be real. Once Euler discovered how to find logarithms of complex numbers, he was able to evaluate even more novel expressions, like $1i$, $(^-1)i$, and i^i.

8. Find the last two digits of 17^{42}.

Although Euler might have been able to evaluate 17^{42} mentally, using only arithmetic, he would surely have used what is now called Euler's ϕ-function, which gives the number of positive integers less than m that are relatively prime to m. For example $\phi(10) = 4$, $\phi(11) = 10$, and $\phi(12) = 4$. Euler showed that

$$a^{\phi(m)} \equiv 1 (\bmod\ m),$$

or that $a^{\phi(m)} - 1$ is divisible by m whenever a and m are relatively prime. To find the last two digits of 17^{42}, it is helpful to know that $\phi(100) = 40$ and $17^{40} \equiv 1 (\bmod\ 100)$. Thus 17^{42} has the same last two digits as 17^2, or 8 and 9. Euler developed formulas for calculating $\phi(m)$, such as

$$\phi(P) = P - 1$$

and

$$\phi(P^n) = P^n - P^{n-1},$$

whenever P is a prime. The value of $\phi(m)$ can also be calculated by simply counting the positive integers less than m that are relatively prime to m. Problems involving number theory were Euler's favorites, and he wrote extensively in this area.

Closing Remarks

Euler was neither too proud to discuss the simple problems in his textbooks nor too content to tackle the unsolved problems of his day. A gifted linguist, Euler published his results in Latin, French, German, or Russian as circumstances dictated. Many of his terms and symbols have become standard. These include e, π, i, and $f(x)$. He defined function, logarithms, and trigonometric and other transcendental functions much as we do today. We owe Euler a great debt for his contributions in mathematics, and we should stop at least every century or so to acknowledge this debt.

BIBLIOGRAPHY

Bell, Eric Temple. *Men of Mathematics.* New York: Simon & Schuster, 1937.

Boyer, Carl B. *The History of the Calculus and Its Conceptual Development.* New York: Dover Publishing Co., 1959. (Reprint of *The Concepts of the Calculus,* 1949).

———. *A History of Mathematics.* New York: John Wiley & Sons, 1968.

Eves, Howard. *An Introduction to the History of Mathematics,* rev. ed. New York: Holt, Rinehart & Winston, 1964.

Kline, Morris. *Mathematical Thought from Ancient to Modern Times.* Oxford University Press, 1972.

Struik, Dirk J. *A Concise History of Mathematics,* 3d rev. ed. New York: Dover Publications, 1967.

National Council of Teachers of Mathematics. *Historical Topics for the Mathematics Classroom.* Thirty-first Yearbook. Washington, D. C.: The Council, 1969.

Editor's Note: In honor of Euler's 300th birthday (2007),the Mathematical Association of America (MAA) published a five volume series on the life and work of this great mathematician:

    Volume I: *The Early Mathematics of Leonhard Euler.* Edward Sandifer.
    Volume II: *The Genius of Euler: Reflections on His Life and Work.* William Dunham (ed.).
    Volume III: *How Euler Did It.* Edward Sandifer.
    Volume IV: *Euler and Modern Science.* N. N. Bogolyubov, et al (eds.).
    Volume V: *Euler at 300: An Appreciation.* Robert E. Bradley, et al (eds.).

Gaspard Monge and Descriptive Geometry

LEO GAFNEY

$\mathcal{M}$EANDERING about a second-hand book-store some time ago, I came upon a musty volume entitled *Géométrie Descriptive* by G. Monge, Paris, 1827 (see Fig. 1). Out of curiosity, I bought the old tome, but gave neither subject nor author another thought until the following summer when I found the name "Monge" sprinkled through a course in differential geometry. He was credited, for example, with having given the first analytical expression of a tangent and with having done pioneering work in the field of partial differential equations. With renewed interest, I dusted off my French dictionary and returned to *Géométrie Descriptive*.

Gaspard Monge

Gaspard Monge was born in Beane, France, in 1746, and died in Paris in 1818. While still in his teens, he constructed a "blueprint" of his home town which won him an appointment to the college of engineers at Mézières. Since he was the son of a peddler, it was a somewhat limited appointment, admitting him not to the school for officer training but to the annex where surveying and drawing were taught. Fortunately for the history of geometry, he was at length permitted to con-

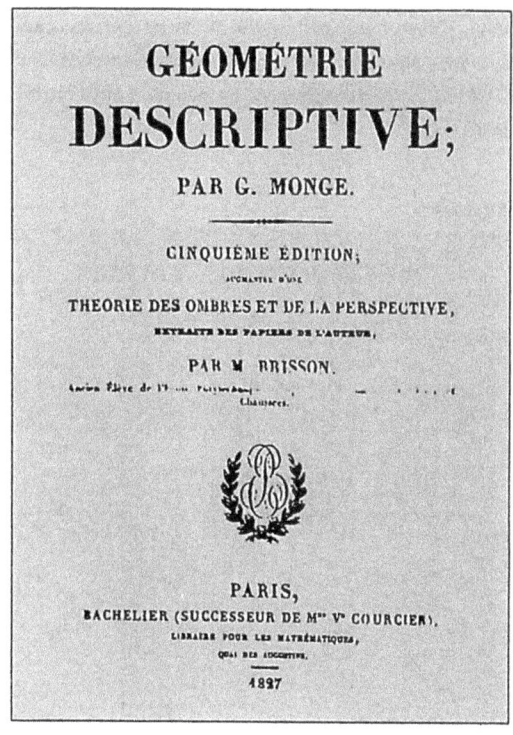

FIGURE I

struct the plan of a fortress from observed data. Rejecting the cumbersome arithmetic calculations then demanded for such a project, he produced a solution using only ruler and compass. With geometrical methods, remarkable in their elegance and simplicity, Monge at once contributed to the solution of practical military problems of his day

Reprinted from *Mathematics Teacher* 58 (Apr., 1965): 338–44; with permission of the National Council of Teachers of Mathematics

215

and at the same time became the father of a new theoretical study—descriptive geometry.

Due both to the rivalry between French military schools and to the agitated times surrounding the French Revolution, Monge was not permitted to publish or teach his discoveries from 1768, when he was made professor at Mézières, until 1794, when he was elected professor at the short-lived normal school of the Republic. In the intervening years, Monge was not only a teacher but also an active French revolutionary. In 1792 he was minister of naval affairs and director for the manufacture of cannon and powder. These offices, however, did not protect him from the terrorists, and he was forced to flee the guillotine. Returning in 1794, he helped organize the normal school mentioned above.

FIGURE 2

AVERTISSEMENT DE L'ÉDITEUR.

Ce Traité de *Géométrie descriptive* a été composé pour l'usage des Élèves de la première École normale, créée par une loi du 9 brumaire an 3 (30 octobre 1794).

Cette École, qui n'a subsisté que pendant les quatre premiers mois de l'année 1795, et qui était destinée à régénérer l'instruction publique, anéantie en France sous le règne de la terreur, comptait quinze cents Élèves pris dans tous les départemens, parmi les instituteurs et les hommes cultivant les Sciences et les Lettres; elle avait pour professeurs

MM. Lagrange, Laplace .... pour les Mathématiques,
Monge .... pour la Géometrie descriptive,
Haüy .... pour la Physique,
Berthollet .... pour la Chimie,
Daubenton .... pour l'Histoire naturelle,
Thouin .... pour l'Agriculture,
Buache, Mentelle, Volney .... pour la Géographie et l'Histoire,
Laharpe, Garat, Bernardin-de-St.-Pierre, Sicard .... pour la Grammaire, la Littérature et la Morale.

MM. Lacroix et Hachette étaient Professeurs-Adjoints pour la *Géométrie descriptive*. Les leçons orales sur cette science se donnaient, ainsi que pour les autres parties de l'enseignement, à l'amphithéâtre du Jardin des Plantes. On avait disposé, dans les bâtimens de la Sorbonne, de grandes salles de dessin, où les Élèves de l'École normale étaient exercés aux constructions graphiques.

Cette nouvelle édition des leçons de Géométrie descriptive

a

The book which initiated this article begins with an interesting reference to the normal school (see Fig. 2): "This treatise on descriptive geometry was written for the use of students in the first normal school, established by the law of the 9th of Brumaire in the year 3 (30th of October, 1794). This school, which existed only during the first four months of the year 1795, and which was intended to revive public education, annihilated in France under the Reign of Terror, . . . had as professors. . . ." There are two things of note on this page: first, the strange month and year—French Revolution time—and second, the extraordinary mathematics faculty.

Monge continued to support the Republic until its collapse, when he became a zealous partisan of Napoleon and even accompanied Napoleon on his Egyptian campaign, although he did not follow his emperor on the Russian march. With the fall of Napoleon, Monge could not redeem himself once more, and, deprived of his previous honors by Louis XVIII, he died in a relatively degraded state.

Descriptive Geometry

The objectives of descriptive geometry," Monge begins his work, "are two: first, to give an understanding of the methods of representing on a two-dimensional surface objects which in nature have three dimensions. . . . The second objective is to teach the way to determine the forms of objects and to deduce all of the properties resulting from their respective positions."

Monge achieved his first goal with virtually unrevisable success; in striving for the second he opened many doors for future mathematicians.

In a certain sense, Monge begins where the artist Albrecht Dürer (1471–1528) let off. A careful examination of Figures 3 and 4 will reveal the similarity. In each case, a point is projected onto several planes, and these are rotated so as to form a right angle (wall and floor) about the ground line. The method of Monge, therefore, consists chiefly

FIGURE 3

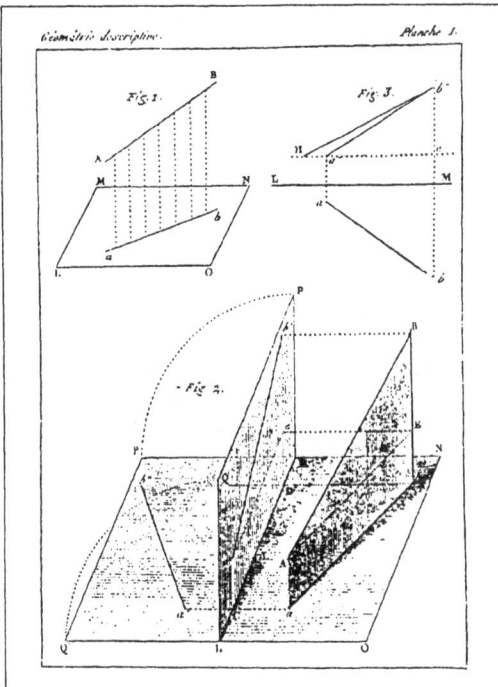

FIGURE 4*

in making orthographic projections of an object on two (or possibly more) planes and establishing a definite relation between the different projections. The two planes are ordinarily the horizontal and a vertical, and the line of intersection is the ground line.

Some of the projections and constructions discussed by Monge, which are accompanied by elegant diagrams, are as follows:

1. Given a point and a line (or plane), to find a line (plane) through the given point parallel (perpendicular) to the given line (plane).

2. Given a point and a surface, to find tangents to the surface from the given point.

3. Given a cylinder or conic, to find the tangent at a particular point.

4. To describe the intersections of surfaces—cylinders, conics, etc.

5. To investigate the curve which is described on a three-dimensional surface when the surface is "sliced" at a point by a plane. (See Figures 46 and 47 in Figure 5. This is actually a projection, but, by considering it on the surface, it is possible to examine tangents, as in Figures 46 and 47, and to discuss the shape of the intersection of two surfaces, as in Figure 49 in Figure 5.)

The surfaces considered by Monge are of two general types. The first is that which can be conceived of as being formed by bending a plane surface without, however, stretching, crumpling, or tearing it. This is called a developable surface. The central problem under this topic is to determine the shape of the plane figure formed by the boundary of a piece of developable surface as the latter is rolled out into a plane. A circular cylinder slit along an element, for example, will become a rectangular figure, and a circular cone slit along an element will become a sector of a circle.

The second type of surface is that which cannot be formed by bending a piece of a plane. A

---

*D. J. Struik, *Lectures on Analytic and Projective Geometry*, p. 236. Reading, Mass.: Addison-Wesley Publishing Company, 1953. The page is reproduced with permission of the publisher.

FIGURE 5

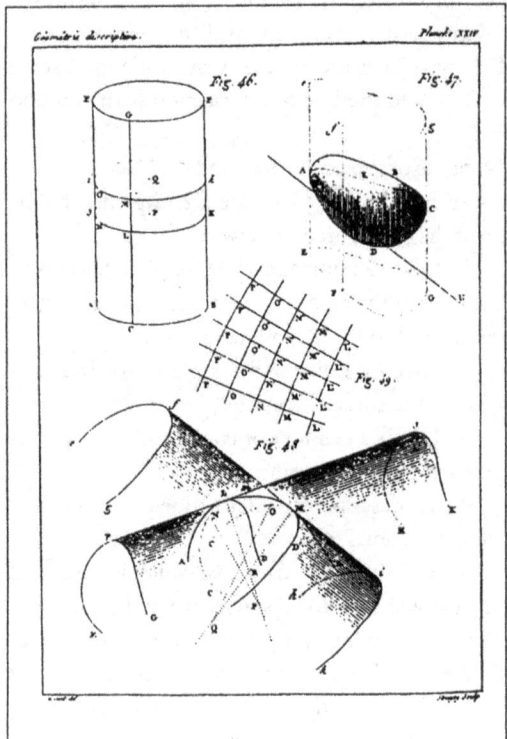

FIGURE 6

Gaspard Monge

(Source: *A Portfolio of Eminent Mathematicians*, edited by David Eugene Smith (Chicago: Open Court Publishing Company, 1908)

sphere and an ellipsoid are two such figures. A practical problem which will throw light on our discussion is that of making a world map. As we know, there are various types of projections, but none of them perfectly represents on a plane the continental figures we find on the sphere of our Earth. The reason is that a sphere cannot be "rolled out smoothly" so as to form a plane surface; it is not a developable surface.

The work of Monge contained in the book we are considering, and which we have outlined above, virtually completes the subject of three-dimensional representation in the plane from the point of view of double projection. Teachers of mechanical or engineering drawing even today will recognize the principles of Monge as forming the backbone of their subject. The actual completion of the subject in the form of a few finishing touches was added by two students of Monge—S. F.

Lacroix and J. N. P. Hachette. Their only really new contribution falls under the topic of shadows.

Two other outstanding pupils of Monge were Charles Dupin, who contributed to the field of differential geometry and after whom the Dupin indicatrix was named, and Victor Poncelet, who, as a prisoner of war in Napoleon's Russian campaign, undertook an investigation of the properties of plane configurations left invariant under projections.

Monge the Teacher

It would be unfair to complete even this brief study of Monge without some discussion of him as a teacher. His biographers acclaim him as one of those very special mathematicians who not only possess the fire of genius but can also ignite it in others. About two-thirds of the way through the book we have been discussing (p. 111), there is evidence of this when Monge takes time out to talk about students and teachers:

If schools have been established ... where young people can exercise themselves in mathematical constructions and become familiar with the more important phenomena of nature and thus gain most necessary knowledge, a knowledge which will develop their intelligence, and give them the habit and feeling for precision so that these students then will most assuredly contribute to the progress of the national industry, and be guaranteed forever safety from false teachers, and if I were here proposing the basic text for instruction in such schools, the abstract treatment would end right here and we would pass immediately to applications which are more useful and to those which are more frequently used. But we ought not write only for students of the schools; we ought to write also for their professors.

It is possible to conduct ordinary instruction on the level of simple cases which are of use in daily life; but if a student comes just once in his life upon a difficulty which was not treated in class, to whom shall he turn if not to the professor? And what shall the professor answer if he has not familiarized himself with considerations of a more general nature than those which comprise the ordinary content of the subject?

BIBLIOGRAPHY

1. Aleksandrov, O. D., "Curves and Surfaces," *Mathematics, Its Content, Methods and Meaning* (Vol. 1, part 3). Providence, R. I.: American Mathematical Society, 1963.

2. Arago, D. F. J., "Biographie de Gaspard Monge," *Mémoires de l'Acad. d. Sci.*, Paris, XXIV (1853).

3. ———, *Notices Biographiques*, Paris, Vol. 2, 1853, 427–592.

4. Archibald, R. C., "Centers of Similitude and Certain Theorems Attributed to Monge. Were They Known to the Greeks?" *American Mathematical Monthly*, XXII (1915), 6–12.

5. ———, "Historical Note on Centers of Similitude of Circles," *American Mathematical Monthly*, XXIII (1916), 159–61.

6. Aubry, P. V., *Monge, le Savant Ami de Napoleon Bonaparte*. Paris, 1954.

7. Ball, W. W. R., *A Short Account of the History of Mathematics*. New York: Dover Publications, Inc., 1960 (reprinted from 1908 edition).

8. Beumer, M. G., "Gaspard Monge as a Chemist," *Scripta Mathematica*, XIII (1944), 122–23.

9. Bell, E. T., *Men of Mathematics* (Chapter 12). New York: Simon and Schuster, Inc., 1937.

10. Cajori, F., *A History of Mathematics*. New York: The Macmillan Company, 1919.

11. Coolidge, J. L., *A History of Geometrical Methods*. Oxford: The Clarendon Press, 1940.

12. De Launay, L., *Monge, Fondateur de l'École Polytechnique*. Paris, 1933.

13. De Morgan, A., "Monge, Gaspard," *Engl. Cycl.- Biog.*, Vol. 4, 1857.

14. Dupin, C., *Essai Historique sur les Services et les Travaux Scientifiques de Gaspard Monge*. Paris, 1819.

15. Kreyszig, E., *Differential Geometry*. Toronto: University of Toronto Press, 1959.

16. Monge, G., *Géométrie Descriptive*, 5th ed. Paris, 1827.

17. Nielson, N., *Géométrie Français sous la Révolution*, "Monge," 182–90. Copenhagen, 1929.

18. Sanford, V., "Gaspard Monge," *The Mathematics Teacher*, XXVIII (1935), 238–40.

19. Simon, L. G., "The Influence of French Mathematicians at the End of the Eighteenth Century upon the Teaching of Mathematics in American Colleges," *Isis*, XV (1931), 115, 119.

20. Smith, D. E., "Gaspard Monge, Politician," *Scripta Mathematica*, I (1932), 111–22.

21. ———, "Among My Autographs: Monge and the American Colonies" and "Monge the Lesser," *American Mathematical Monthly*, XXVIII (1921), 166, 208–9.

22. Struik, D. J., *Lectures on Analytic and Projective Geometry*. Reading, Mass.: Addison-Wesley Publishing Company, 1953.

23. ———, *Lectures on Classical Differential Geometry*. Reading, Mass.: Addison-Wesley Publishing Company, 1950.

24. ———, "Outline of a History of Differential Geometry," *Isis*, XIX (1933), 92–120; XX (1934), 161–91. Part 4, "Monge and the École Polytechnique," XIX, 113–20.

25. Taton, R., "Gaspard Monge," *Elemente der Mathematik, Beihefte*, IX (1950).

26. ———, *L'Oeuvre Scientifique de Gaspard Monge*. Paris, 1951.

Mathematicians of the
French Revolution

CARL B. BOYER

THE EIGHTEENTH century has had the misfortune to come after the seventeenth and before the nineteenth. How could any period which followed the "Century of Genius" and which preceded the "Golden Age" of mathematics be looked upon as anything but an inconsequential interlude? Analytic geometry and the calculus were invented in the seventeenth century; the rise of mathematical rigor and the flowering of geometry are associated with the nineteenth. It is not easy to find comparable developments in the eighteenth century. H. G. Zeuthen wrote a *Geschichte der Mathematik im XVI. und XVII. Jahrhundert* (German ed., Leipzig, 1903), and Felix Klein published *Volesungen über die Entwiklung der Mathematik im 19. Jahrhundert* (2 vols., Berlin, 1926–1927)—both unusually scholarly works—but who ever heard of a history of mathematics in the eighteenth century? Not that the period is a complete void, for the veriest tyro knows that in the very middle of the century which began with the closing years of Newton and ended with the early years of Gauss there lived the prolific Euler—a man for whom calculation was as easy and natural as breathing is for us. But one does not look to the latter part of the eighteenth century for significant trends or discoveries in mathematics. This is in marked contrast to what is true in other fields. For Americans the date 1776 was

Reprinted from *Scripta Mathematica* 25 (Spring, 1960): 11–31; with permission of *Scripta Mathematica* and Yeshiva University Press.

decisive; in France the year 1789 was crucial. Nor was the Age of Revolutions confined to the sphere of politics. The Industrial Revolution has changed the whole fabric of Western society, and the thermotic revolution during the same years laid the foundations of modern chemistry. Can it be possible that mathematics during these stirring events was enjoying a Rip Van Winkle nap, uninfluenced by the world about it? It is the purpose of this paper to show that this was far from the case—that the mathematicians of France at the time of the Revolution not only contributed handsomely to the fund of knowledge, but that they were in large measure responsible for the chief lines of development in the explosive proliferation of mathematics during the succeeding century. We shall even be tempted to add to the already impressive list of revolutions of the time two more, a "geometrical revolution" and an "analytical revolution." In developing this argument considerable use has been made of Niels Nielsen's *Géomètres français sous la révolution* (Copenhague, 1929); but whereas Nielsen has catalogued alphabetically the contributions of some four score French mathematicians, it is the object of this paper to furnish a synthesis of the mathematical milieu of the time with emphasis upon but a handful of individuals.

Every age is inclined to think of itself as one of revolution—a period of tremendous change. But almost every age of rapid change has been preceded by a long period in which preparations for the revolution are made, sometimes consciously, more often unconsciously; and in this respect the French political and mathematical revolutions were far from exceptional. Among the heralds of the French Revolution were Voltaire, Rousseau, D'Alembert, and Diderot—not one of whom lived to see the fall of the Bastille (Voltaire and Rousseau died in 1778; D'Alembert died in 1783; and Diderot a year later)—and their associate Condorcet, who fell a victim in the holocaust which he helped to father. In mathematics six figures who were to show the way—Monge, Lagrange, Laplace, Legendre, Carnot, and Condorcet—were to be in the midst of the turmoil, and it is primarily about these men that we shall build our account.

The half dozen mathematicians who will be the center of our attention were almost of an age: Lagrange, the oldest, was born in 1736; Condorcet was born in 1743; Monge in 1746; Laplace in 1749; Legendre in 1752; Carnot, the youngest, was born in 1753. With the exception of Condorcet, who died a suicide in prison, these mathematicians all lived to be septuagenarians, and one, Legendre, an octogenarian. And it is a comfort to us to note, in these days of surveys proving that it is mostly the young men who contribute significantly to mathematics, that the chief influence of our five heroes came during their forties, fifties, and sixties. But before we tell about their role as mature leaders, let us look at their early training.

Today we take it as a matter of course that anyone who has unusual ability and interest in mathematics can follow his inclinations and find a place for himself in the scholarly world. In France of the eighteenth century, however, this was far from the case. Universities were not the mathematical foci that they are today; and one is hard put to it to name even one eighteenth-century mathematician at, say, the University of Paris. During the fourteenth century Paris had been one of the mathematical centers of the world, the other being at Oxford; but it had long since lost this position. It was always behind the times: when Europe turned to Cartesianism, Paris clung to Peripatetic Scholasticism; and when most of the

scientific world had turned to Newtonionism, Paris fought a rear-guard action for Cartesianism. Most of the French mathematicians of the time were associated not with the universities, but with either the church or the military; others found royal patronage or became private teachers. Only families of some means or position were able to provide their sons with mathematical training, generally through a church or military school. Lagrange, the only one of our group who was not strictly a Frenchman, was born of a once prosperous family who provided for him an education at the college in Turin. As a young man he became professor of mathematics in the military academy of Turin, but later he found royal patrons in Frederick the Great and Louis XVI. Condorcet's family included influential members in the cavalry and the church, and hence his education presented no problem. At Jesuit schools and later at the Collège de Navarre he made an enviable reputation in mathematics; but instead of becoming a captain of cavalry, as his family had hoped, he lived the life of a scholar in much the same sense as Voltaire, Diderot, and D'Alembert. The third of our sextet, Gaspard Monge, was the son of a poor tradesman. However, through the influence of a lieutenant colonel who had been struck by the boy's ability, Monge was permitted to attend some courses at the École Militaire de Mézières; and he so impressed those in authority that he soon became a member of the teaching staff—the only one of our group of six who was primarily a teacher, and as we shall presently see, perhaps the most influential mathematics teacher since the days of Plato. Laplace, too, was born without wealth; but, like Monge, he found influential friends who saw that he got an education—again in a military academy. And for a while Laplace, like Monge, taught mathematics in the school in which he had been educated. Legendre was another member of the sextet who experienced no difficulty in securing an education; but even he was not a university teacher in the strict sense, although for five years he taught in the École Militaire at Paris. The youngest of our group,

Lazare Carnot (who is not to be confused with his son, Sadi Carnot), was sufficiently above bourgeois standing to be permitted to enter the École Militaire at Mézières, where Monge was one of his teachers. Upon graduation Carnot entered the army, although, lacking a title, he could not, under renewed rule of the *ancien régime*, aspire to a rank above that of captain. This must have rankled in his mind—as it did in the case of so many others that the proverb arose at examination time that "the competent were not noble and the noble were not competent." The economic wastefulness of the government may have been the immediate cause of the French Revolution, but it was far from the only one. The enormous waste of human resources was also an important factor, and symptomatic of this was the failure at first of the men of our group to win positions commensurate with their ability. Is it any wonder that not one of the six expressed regret later when the old order passed away? But before we turn to the events which brought about the new order of things, let us glance at the mathematics education our sextet inherited.

In the early eighteenth century France had taken the lead in the exploitation of two capital inventions of the preceding century—analytic geometry and the calculus. Neither of these subjects, however, resembled closely those now given in our textbooks. Representative of the forms in which they appeared in France throughout the eighteenth century are the two works of L'Hospital—*Traité des sections coniques* (Paris, 1707) and *Analyses des infiniment petits* (Paris, 1696). The author of these two books was not a teacher; but no other textbooks could rival his during the eighteenth century. This is only one instance of the observation we shall frequently make that the good teacher and the good textbook writer are not generally the same person. L'Hospital was not responsible in any significant measure for the material in his books; he cribbed from others—in true textbook tradition—and not even the rule on indeterminate forms which bears his name was his own discovery. But he did give the subjects a much-needed

conventional arrangement which enabled the noviate to digest the new subjects more easily. Nevertheless, neither of L'Hospital's texts would satisfy an Anglo-Saxon instructor of today, for two deficiencies in particular would be immediately apparent. First, there were no lists of exercises to be worked out by the student; more importantly, very little attention was paid to fundamental principles—and this at a time when the new disciplines were badly in need of a sound foundation. Even apart from such criticisms of the L'Hospital textbooks, they would scarcely fit into modern courses. The *Conics* in particular would be a poor substitute for our analytic geometries, for there are no generalities on coordinates, no analytic proofs of elementary theorems, no formulas for slope, distance, angle, no consideration of the straight line, no sketching of higher plane curves, and nothing on problems in three dimensions. L'Hospital's calculus comes somewhat closer to modern material, but here too one finds substantial differences: there are no rules for the differentiation of transcendental functions, and L'Hospital never got around to writing a complementary book on the integral calculus.

L'Hospital's works were not the greatest textbooks of the century—they were simply the most popular. The book which, in didactic quality, stands out above all others of the eighteenth century was Euler's *Introductio in analysin infinitorum* (2 vols., Lausanne, 1748). This added to analytic geometry the sketching of higher plane curves, both algebraic and transcendental; and, together with Euler's treatises on the calculus, it made the logarithmic and trigonometric functions a traditional part of college courses in analysis. Even on the lower level Euler ultimately had a deep influence, for trigonometry became less and less the geometry of half-chords and more and more the study of the now familiar goniometric functions, with the full recognition of their periodicity. The Euler function-concept approach, however, took hold but slowly. The textbooks of L'Hospital reappeared in just as many editions after Euler's *Introductio* as before, whereas the *Introductio* itself did not appear in another edition until almost half a century later. The medium through which the influence of Euler was most felt was that of the multi-volume compendium—the *Cours complet de mathématiques*—which made its appearance at about the time that our sextet was studying its mathematics.

Of the mathematical encyclopedias of the later eighteenth century the most successful, judging from repeated editions, was that by Étienne Bézout, instructor in the school at Mézières which both Monge and Carnot attended. The *Cours de mathématique* of Bézout appeared in six volumes at Paris from 1764–1769, and a second edition began to appear only a year after the completion of the first. During the first third of the nineteenth century it still was a very influential work, especially in America where parts of it appeared in English translation at West Point and other academies. Bézout's *Cours* was intended to cover virtually the entire spectrum of mathematical instruction from arithmetic through the applications of the calculus. His preface claimed that he presupposed of his readers nothing beyond the names of the numbers. Even the techniques involved in the arithmetic operations on integers are explained at length in the first volume, together with some short-cuts in calculation and the proof by nines. The whole of the first volume is devoted to arithmetic techniques. The second volume contains the elements of geometry and trigonometry. The geometry is a thoroughly emasculated structure, with little in the way of proof. Following the definition of parallel lines as those which never meet no matter how far one imagines them prolonged, the theorem on the equality of corresponding angles formed by a transversal is "proved" by the simple observation that the parallel lines, not having any inclination with respect to each other, must necessarily have the same inclination with respect to the transversal. There are no formal proofs of theorems, but angle-measure looms large—as one should expect in a volume intended for future navigators—as does also solid mensuration. The definition of trigo-

nometry is interesting for its emphasis: the subject is defined as "a part of geometry which teaches how to determine or calculate three of the six parts of a triangle [*triangle rectiligne*] from the knowledge of three other parts, when this is possible." There is nothing here of Euler's ratios or the periodic function approach. Anticipating the needs of students of navigation, Bézout hurried on to spherical trigonometry.

Part III of Bézout's *Cours* covers algebra and its application to arithmetic and geometry: and here too the author holds to the belief that the study of mathematics "is less the accumulation of a great number of propositions than the acquisition of the spirit of research and discovery." Admitting that the chief object of his volume on algebra is to prepare students for the study of mechanics, Bézout apologizes for spending some time on applications to arithmetic and geometry, justifying the digression by the need for developing facility in algebra. One sees here, almost a century and a half since the appearance of Descartes's *Géométrie*, that analytic geometry was indeed a step-child in mathematics. One had to apologize for teaching it; and perhaps one reason for this situation was that the purpose of the subject was not what we know it today. The object of "the application of algebra to geometry," as inherited from Descartes, was the geometrical construction of algebraic quantities—such as the roots of the equation $4u^3 - 3r^2u - cr^2 = 0$. Here, too, the influence of Euler was slow in making itself felt.

The fourth part of Bézout's *Cours* is the *raison d'être* of the program—the principles of mechanics—but the author necessarily had to introduce this with a substantial development of the differential and integral calculus, including differential equations. This is one place where the ubiquitous influence of Euler is really felt, especially in the analysis of transcendental functions. The emphasis given to mechanics and to the closing section on navigation is in keeping with the use of the *Cours de mathématiques* as a text in a military academy. The mathematical preeminence of France (and, indeed, of continental Europe as a whole) in the eighteenth century is, in fact, based in large measure on the application of analysis to mechanics as taught in technical schools, and it was under this influence that the mathematicians of the French Revolution had been brought up. This is in marked contrast to the situation in England, which remained a stronghold of synthetic geometry. One should naturally expect the contrast in mathematical spirit to become sharper during the Revolution, for France had greater need for technical training and England became more thoroughly isolated from the continent. But let us look more closely at the mathematical situation in France.

Everyone of the six men we have named as the mathematical leaders during the Revolution had produced abundantly before 1789. Lagrange had published his *Méchanique analytique* (1788), as well as frequent papers on algebra, analysis, and geometry. Condorcet, perhaps the most interesting of the six because of the breadth of his interests, had published *De calcul intégral* as early as 1765, and *Essai sur l'application de l'analyse à la probabilité des décisions rendues à la pluralité des voix* in 1785. A firm believer in the perfectibility of man, a basic tenet of the Philosophes, Condorcet was the only one of our six who can be said to have played an anticipatory role in the events leading to 1789. (It is ironic to note that of our mathematical sextet the one who did most to bring about the Revolution was the only one to lose his life through it, although two others, Carnot and Monge, were not always safe from the guillotine.) Monge had contributed numerous mathematical articles to the *Mémoires* of the *Académie des Sciences*. Inasmuch as he succeeded Bézout as examiner for the School of the Marine, Monge was urged by those in authority to do what Bézout had done—write a *Cours de mathématiques* for the use of candidates. Monge however, was interested in teaching and research rather in writing textbooks, and he completed only one volume of the project—*Traité élémentaire de statique* (Paris, 1788). He was attracted not only to both pure and applied mathematics but also to

physics and chemistry. In particular, he participated with Lavoisier in experiments, including those on the composition of water, which led to the chemical revolution of 1789. Through his numerous activities Monge had become, at the time of the revolution, one of the best known of French scientists. In fact, his reputation as a physicist and chemist was perhaps greater than that as a mathematician, for his geometry had not been properly appreciated. His chief work, the *Géométrie descriptive*, had not been published because his superiors felt that it was in the interests of national defense to keep it confidential. (Classified material is not, you see, a monopoly of the mid-twentieth century.) Laplace and Legendre were regular contributors to learned periodicals, and Carnot by 1786 had published a second edition of his *Essai sur les machines en général* as well as some verses and a work on fortifications.

In looking closely at the achievements of the six men with whom we are concerned, one is struck by a lack of utilitarian motive in their work. Carnot's *Essai* would appear, from the title, to be most technically oriented, but a glance at the book will show that it deals with broad principles, not with technology. The *Mécanique* of Lagrange likewise is concerned with a postulational treatment of the subject, far removed from criteria of practicability. The beauty of Lagrange's work is apparent not to the engineer but to the pure mathematician; even in the more elementary portions of his work there is an aesthetic quality. It is primarily to him that we owe such compact results as

$$\frac{1}{2!} \begin{vmatrix} x_1 & y_1 & 1 \\ x_2 & y_2 & 1 \\ x_3 & y_3 & 1 \end{vmatrix}$$

for the area of a triangle and

$$\frac{1}{3!} \begin{vmatrix} x_1 & y_1 & z_1 & 1 \\ x_2 & y_2 & z_2 & 1 \\ x_3 & y_3 & z_3 & 1 \\ x_4 & y_4 & z_4 & 1 \end{vmatrix}$$

for the volume of a Tetrahedron.

The work of Lagrange on the volume of a tetrahedron looks pretty, but inconsequential; yet it contained an idea which was to become, through the educational reforms of the Revolution, of considerable importance. As Lagrange expressed it, "I flatter myself that the solutions which I am going to give will be of interest to geometers as much for the methods as for the results. These solutions are purely analytic and can even be understood without figures." And true to his promise, there is not a single diagram throughout the work. Monge, too, seems somehow to have come to the conclusion, whether independently or not one cannot tell, that one should avoid the use of diagrams. Even Carnot seems vaguely to have felt somewhat the same way, and his *Essai,* antedating the *Mécanique* of Lagrange, contains not a single diagram.

Laplace, of all the members of our sextet, came closest to being an applied mathematician; but even in his case one must interpret the phrase in a very broad sense. After all, how "practical" in those days was the theory of probability or celestial mechanics? One can safely conclude that, in spite of their education in predominantly technical schools, the great figures in mathematics just before the Revolution had shown remarkable "purity" of interest. Was this to be lost during the national emergency?

The fall of the Bastille in 1789 found our six men divided into two categories: the three L's (Lagrange, Laplace, and Legendre) took no active part in shaping the political events which were to follow; the other three (Carnot, Condorcet, and Monge) welcomed the changed outlook and played definite roles in revolutionary activities. Men from both groups, however, played roles in at least one mathematical project during the Revolution.

The reform of the system of weights and measures is an especially appropriate example of the way in which mathematicians patiently persisted in their efforts in spite of confusion and political difficulties. As early in the Revolution as 1790 Talleyrand proposed the reform of weights and measures. The problem was referred to the

Académie des Sciences, in which a committee, of which Lagrange and Condorcet were two of the members, was established to draw up a proposal. Legendre should have been a member, for he had achieved quite a reputation for his triangulation of France; but revolutionary politics seem to have been responsible for his being overlooked. The Committee agreed on a decimal system, although there appear to have been some earnest supporters of a duodecimal scheme. Lagrange firmly supported the decimalists against the duodecimalists, for he was not greatly impressed by the argument about divisibility. (He is reported to have almost regretted not adopting as a base for the system some *prime* number, such as eleven; but it has been suggested that he may have done this simply to obstruct the duodecimalists.) As is well known, the Committee considered two alternatives for the basis of the new system. One was the length of the pendulum which should beat seconds. The equation for the pendulum being $T = 2\pi\sqrt{l/g}$, this would make the standard length g/π^2. But the Committee was so impressed by the accuracy with which Legendre and others had measured the length of a terrestrial meridian that in the end the meter was defined to be the ten-millionth part of the distance between the equator and the pole. The resulting metric system was ready in most respects in 1791, but there was confusion and delay in establishing it. The National Convention in 1793 suppressed the Académie des Sciences, while the Jardin des Plantes was greatly expanded. This inconsistency seems to have been the result of political forces. The Académie was led by older and more conservative men, the Jardin by younger scientists who were eager in their support of the new government. There was, moreover, quite a cult of Robespierre which represented a back-to-nature attitude derived in part from Rousseau. Evidently there was in France an attitude toward physical science something like Goethe's belligerency toward Newtonian physics. The Jardin des Plantes represented "safe" science, that of the Académie was suspect. The closing of the Académie was a blow to mathematics; but the Convention continued the Committee on Weights and Measures, although it purged the Committee of some members, such as Lavoisier, and enlarged it by adding others, including Monge. At one point Lagrange was very nearly lost to the Committee, for the provincially minded Convention had banned foreigners from France; but Lagrange was specifically exempted from the decree, and he remained to serve as head of the Committee. Still later the Committee was placed under the Institut National which had replaced the Académie des Sciences; and Lagrange, Laplace, Legendre, and Monge all served on the Committee at this stage. By 1779 the work of the Committee had been completed, and the metric system as we have it today became a reality. It will be noted that five of our group of six revolutionary mathematicians took active part in this project, only Carnot being unconnected with it; but we shall find that Carnot was engaged in many other essential activities, both political and mathematical. The metric system is, of course, one of the more tangible mathematical results of the Revolution; but in terms of the development of our subject it cannot be compared in significance with other results we shall point to.

Condorcet was, of all the members of our group, the only one who can be said to have worked to bring about the Revolution. A physiocrat, a philosophe, and an encyclopedist, he belonged to the circle of Voltaire and D'Alembert. He was a capable mathematician who had published books on probability and the integral calculus; but he was also a restless visionary and idealist who was interested in anything related to the welfare of mankind. He, like Voltaire, had a passionate hatred of injustice; and although he held the title of marquis, he saw so many inequities in the *ancien régime* that he wrote and worked toward reform. With implicit faith in the perfectibility of mankind and belief that education would eliminate vice, he argued for free public education, something which we in America now take for granted. He was, in short, what would today be known as a liberal.

Like our own Thomas Jefferson, he believed that error, like truth, had a right to freedom. The extent of his inspiration can well be gauged by the fact that an impressive tome of 891 pages was published by Franck Alengry in 1904 with the title *Condorcet: Guide de la révolution française*. This book, however, is on his social and legal philosophy, rather than on his mathematics. Condorcet is perhaps best remembered mathematically as a pioneer in social mathematics, especially through the application of probability and statistics to social problems. When, for example, conservative elements (including the Faculty of Medicine, as well as the Faculty of Theology) attacked those who advocated innoculation against small pox, Condorcet (along with Voltaire and Daniel Bernoulli) came to the defence of variolation. He argued the case through statistics, just as we do today in justifying vaccination against polio; but his approach did not meet with general approval. Even the mathematician D'Alembert was not convinced, schooled as he had been in an absolutistic attitude toward mathematics.

With the opening of the Revolution, Condorcet's thoughts turned almost exclusively to administrative and political problems. The educational system had collapsed under the effervescence of the Revolution, and Condorcet saw that that was the time to try to introduce the reforms he had in mind. He presented his plan to the Legislative Assembly, of which he became President; but agitation over other matters precluded serious consideration of it. Condorcet published his scheme in 1792, but the provision for free education became a target of attack. Instead of adopting constructive measures, the extremists added to the problems by abolishing the schools of law, theology, medicine, and arts. Not until years after his death did France achieve Condorcet's ideal of free public education. Condorcet boldly denounced the Septembrists, and was ordered arrested for his pains. He sought hiding, and during the long months of concealment he composed the celebrated *Sketch for a historical picture of the progress of the human mind,* indicating nine steps in the rise of mankind from a tribal stage to the founding of the French Republic, with a prediction of the bright tenth stage which he believed the Revolution was about to usher in. Shortly after completing this work (in 1794), believing that his presence endangered the lives of his hosts, he left his hiding place. Promptly recognized as an aristocrat, he was arrested; and the following morning he was found dead on the floor of his prison, presumably a suicide.

Condorcet had been sympathetic to the moderate Gironde wing of the Revolution. (As is well known, extremists often are more bitter against the middle of the political spectrum than they are against the opposite end.) Monge was plebian and an important member of the more radical Jacobin Club; but he, too, was to have some trouble. Monge was an enthusiastic partisan and joined patriotic organizations—as did also Vandermonde and Meusnier. He was assigned a role in the reform of weights and measures ordered by the Constituent Assembly in 1790, but his post as examiner for the navy kept him from Paris for a couple of years. On his return to the city in 1792 he was named Ministre de la Marine, apparently on the suggestion of Condorcet. (It had been Condorcet who earlier had secured for him the chair of hydraulics created by Turgot and had aided in his election to the Académie des Sciences. Condorcet had been offered the post of Minister of the Navy, but he declined the responsibility, recognizing the handicap of his aristocratic title.) It was in his capacity as minister that to Monge fell the task of signing the official record of the trial and execution of the king. The French fleet, however, was so poorly organized and so ineffectual that Monge was unable to accomplish anything significant, and within a year he demanded that he be replaced. He nevertheless remained active in politics and governmental operations. He was a prominent member of the Jacobin Club, serving as vice-president; and he devoted an enormous amount of energy to meeting the needs for gunpowder of the revolutionary

arsenal. At the instance of the Committee of Public Safety he published also a *Description de l'art de fabriquer les canons.* Throughout the Revolution Monge sometimes found himself in a precarious position, for he was too liberal for the conservatives and too conservative for the extremists. At one point, after 9 Thermidor, Monge felt that he had to leave Paris to escape the guillotine; but he was soon back again, zealous for public service.

More important for the future of mathematics were the efforts of Monge, after the crisis of foreign invasion had subsided, to establish a school for the preparation of engineers. As Condorcet had been the guiding spirit in the Committee on Instruction, so Monge was the leading advocate of institutions of higher learning. The result was the formation in 1794 of a Commission of Public Works, of which Monge was an active member, charged with the establishment of an appropriate institution. The school was the famous École Polytechnique which took form so rapidly that students were admitted in the following year. At all stages of its creation the role of Monge was essential, both as administrator and as teacher. It is gratifying to note that the two functions are not incompatible, for Monge was eminently successful in both. He was even able to overcome his reluctance to write textbooks, for in the reform of the mathematics curriculum the need for suitable books was acute. Monge found himself lecturing on two subjects both essentially new to a university curriculum. The first of these was known as stereotomy, now more commonly called descriptive geometry. Monge gave a concentrated course in the subject to four hundred students, and a manuscript outline of the syllabus survives. This shows that the course was of wider scope, both on the pure and the applied side, than is now usual. Besides the study of shadow, perspective, and topography, attention was paid to the properties of surfaces, including normal lines and tangent planes to these, and to the theory of machines. Among the problems set by Monge, for example, was that of determining the curve of intersection of two gauche surfaces each of which is generated by a line which moves so as to intersect three skew lines in space. Another was the determination of a point in space equidistant from four lines. Such problems point up a change in mathematical education which was sponsored primarily by the French Revolution. As long ago as the Golden Age of Greece it had been pointed out by Plato that the state of solid geometry was deplorable. Euclid, schooled by the pupils of Plato, had done something about this, climaxing his *Elements* with the proof of the famous theorem that there are five and only five regular solids. Archimedes followed with his mensuration of the volumes of the sphere, cone, and paraboloid of revolution, and Pappus added his theorems relating centers of gravity to areas and volumes of solids. But the decline in mathematics had hit solid geometry much harder than it had plane geometry. One who could not cross the *pons asinorum* could scarcely be expected to reach the study of three dimensions. The inventors of analytic geometry, Descartes and Fermat, had been well aware of the fundamental principle of solid analytic geometry—that every equation in three unknowns represented a surface, and conversely—but they had not taken steps to develop it. One can say that whereas the seventeenth century was the century of curves—the cycloid, the limaçon, the catenary, the lemniscate, the equiangular spiral, the hyperbolas, parabolas, and spirals of Fermat, the pearls of Sluse, and many others—the eighteenth was the century which really began the study of surfaces. Coolidge, in an article in the *American Mathematical Monthly* a few years before he died, called attention to the slow progress made in solid analytic geometry before the days of Euclid. It was Euler who called attention to the quadric surfaces as a family analogous to the conics. Whereas the conics had been exhaustively treated by Apollonius, before the middle of the eighteenth century the general hyperboloid of one sheet and the hyperbolic parabolid had not been identified. Euler's *Introductio* in a sense established the subject of solid analytic geometry (although one must perforce mention Clairaut as a precursor); but Euler was not a proselytizer and hence his subject found no place in the school curriculum. One reason may have been that, like Descartes, he did not begin with the simplest rectilinear cases. We have seen that Lagrange, influenced perhaps by his calculus of variations, manifested interest in problems in three dimensions and emphasized their analytic solution. He was first, for example, to give the formula

$$D = \frac{ap + bq + cr - d}{\sqrt{a^2 + b^2 + c^2}}$$

for the distance D from a point (p, q, r) to the plane $ax + by + cz = d$. But Lagrange did not have a geometer's heart, nor did he have enthusiastic disciples. Monge, by contrast, was a specialist in geometry—almost the first since Apollonius—as well as a superior teacher and a curriculum builder. (Parenthetically it may be mentioned that Monge had two brothers who also were Professors of Mathematics, thus putting the name of Monge in a class with that of the Bernoullis and the Cassinis as designating a family of mathematicians). The rise of solid geometry consequently was due in large part to the mathematical and revolutionary activities of Gaspard Monge. Had he not been politically active, the École Polytechnique might never have come into being; and had he not been an inspiring teacher, the revival of geometry of three dimensions might not have taken place.

It should be pointed out that the École Polytechnique was not the only school created at the time. The École Normale had been hastily opened to some fourteen or fifteen hundred students less carefully selected than at the École Polytechnique, and it boasted a mathematical faculty of high calibre, Monge, Lagrange, Legendre, and Laplace being among the instructors. It was the lectures of Monge at the École Normale that finally were published in 1799 as his *Géométrie descriptive;* but administrative difficulties made the school short-lived.

Descriptive geometry was not the only contribution of Monge to three-dimensional mathematics, for at the École Polytechnique he taught also the course in "application of analysis to geometry." Just as the abbreviated title "analytic geometry" had not yet come into general use, so also there was no "differential geometry"; but the course given by Monge was essentially an introduction to the latter field. Here, too, no textbook was available, and so Monge found himself compelled to write up and print his *Feuilles d'analyse* (1795) for the use of students. Here the analytic geometry of three dimensions really came into its own; and it was this course, required of all students at the École Polytechnique, which formed the prototype of the present program in solid analytic geometry. Students, however, evidently found the course difficult, for the lectures skimmed very rapidly over the elementary forms of the line and plane, the bulk of the material being on the applications of the calculus to the study of curves and surfaces in three dimensions. Monge ever was reluctant to write textbooks on the elementary level, or to organize material which was not primarily his own. However, he found collaborators ready to edit material which he included in his course; and so in 1802 there appeared in the *Journal de l'École Polytechnique* an extensive mémoire by Monge and Hachette on *Application d'algèbre à la géométrie.* The first theorem in this is typical of a more elementary approach to the subject. It is the well-known eighteenth-century generalization of the Pythagorean theorem: The sum of the squares of the projections of a plane figure upon three mutually perpendicular planes is equal to the square of the area of the figure. By Monge and Hachette the theorem is proved just as in modern courses; and, in fact, the whole volume could serve without difficulty as a text in the twentieth century. Equations for transformations of axes, the usual treatment of lines and planes, the determination of the principal planes of a quadric are treated fully. The one thing that might be missed is the use of determinants, for this, despite the anticipation

by Lagrange, was the work of the nineteenth century.

Among the results first given by Monge are two theorems which bear his name: (1) The planes drawn through the midpoints of a tetrahedron perpendicular to the opposite edges meet at a point M (which has since been called the Monge point of the tetrahedron). M turns out to be the midpoint of the segment joining the centroid and the circumcenter. (2) The locus of the vertices of the trirectangular angle whose faces are tangent to a given quadric surface is a sphere (known as the Monge sphere or director sphere of the quadric).

As we have already indicated, Monge possessed a quite unusual combination of talents, for he was at once a capable administrator, an imaginative research mathematician, and an inspiring teacher. The one trait of a pedagogue he might have had, but lacked, was that of a textbook compiler. But if Monge here showed a deficiency, it was more than made up for by his young and eager students. One can say with no fear of contradiction, that the pupils of Monge let loose a spate of elementary textbooks on analytic geometry such as has never been equalled—not even in our own day, deluged as we are with new books. If one is to judge from the sudden appearance of so many analytic geometries beginning with 1798, a revolution had taken place in mathematical instruction. Analytic geometry, which for a century and more had been overshadowed by the calculus, suddenly achieved a recognized place in the schools; and this "analytical revolution" can be credited primarily to Monge. Between the years 1798 and 1802 four elementary analytic geometries appeared from the pens of Lacroix, Puissant, Lefrançois, and Biot, all directly inspired by the lectures at the École Polytechnique. And Polytechnicians were responsible for about again as many in the next decade. Most of these were eminently successful, appearing in many editions. The volume by Biot achieved a fifth edition in less than a dozen years; that by Lacroix, student and colleague of Monge, appeared in twenty-five editions within ninety-nine years! Perhaps we should speak instead of the "textbook revolution" for Lacroix's other textbooks were almost as spectacularly successful, his *Arithmetic* and his *Geometry* appearing in 1848 in the 20th and 16th editions, respectively. The 20th edition of his *Algebra* was published in 1859, and the 9th edition of his calculus in 1881. And these figures do not include translations into other languages.

Monge is known to most readers as a founder of modern pure geometry. Through Poncelet and other *anciens élèves* of the École Polytechnique pure geometry did undergo a glorious renaissance, largely through the inspiration of Monge; but it may be well for us to dwell for a moment on an aspect of Monge's work which is less well known. Virtually without exception, the textbook writers ascribe the inspiration for their work to Monge, although Lagrange occasionally is mentioned also. Lacroix most clearly expressed the point of view as follows.

> In carefully avoiding all geometric constructions, I would have the reader realize that there exists a way of looking at geometry which one might call *analytic geometry,* and which consists in deducing the properties of extension from the smallest possible number of principles by purely analytic methods, as Lagrange has done in his mechanics with regard to the properties of equilibrium and movement.

Lacroix held that algebra and geometry "should be treated separately, as far apart as they can be; and that the results in each should serve for mutual clarification, corresponding, so to speak, to the text of a book and its translation." Lacroix pointed to Lagrange's work on the tetrahedron as an instance of this point of view, but he believed that Monge "was the first one to think of presenting in this form the application of algebra to geometry." (Delambre likewise ascribed to Monge the "resurrection of the alliance of algebra and geometry.") His section on solid analytic geometry Lacroix admitted to be almost entirely the work of Monge. Perhaps teachers today can take satisfaction in the

thought that analytic geometry as presented by Fermat and Descartes, a lawyer and a philosopher, respectively, remained ineffectual, and only when it was given a new form by genuine pedagogues— Monge and those of his students who in turn became teachers at the École Polytechnique—did it show vitality.

It is interesting to note that Lacroix declined to use the name "analytic geometry" as a title for this textbook, and edition after edition carried the ponderous title *Traité élémentaire de trigonométrie rectiligne et sphérique et application de l'algèbre à la géométrie*. Although the phrase "analytic geometry" had appeared every now and then during the eighteenth century, it seems first to have been used as the title of a textbook by Lefrançois in an edition of his *Essais de Géométrie* of 1804, and by Biot in an 1805 edition of his *Essais de Géométrie Analytique*, the latter of which, translated into English as well as other languages was used for many years at West Point. We need not look in detail at the contents of the texts of Lacroix, Lefrançois, Biot, and others, for they resemble very closely the books you and I have used.

We have dwelt at such length on the work of Monge that one may get the impression that he was the outstanding figure of the Revolution. This is far from the truth, for the mathematician whose name was on the tongue of every Frenchman during the Revolution was not Monge but Carnot. It was Carnot, who, when the success of the Revolution was threatened by confusion within and invasion from without, organized the armies and led them to victory. As ardent a republican as Monge, he nevertheless shunned all political cliques. He had a high sense of intellectual honesty and tried to be impartial in reaching decisions. After investigation he absolved the loyalists of the charge that they had mixed powdered glass in flour for the army; but he felt bound by conscience to vote for the death of the king. (The American Tom Paine, sometimes regarded in this country as dangerously radical, voted *against* the execution of the king.) But reasoned impartiality is difficult to maintain in times of crisis, and Robespierre, whom Carnot had antagonized, threatened that Carnot would lose his head at the first military disaster. Had Carnot been merely a mathematician and a politician, like Monge and Condorcet, he might well have gone to the guillotine. But Carnot had won the admiration of his countrymen for his remarkable military successes; and when a voice in the Convention proposed his arrest, the deputies spontaneously rose to his defense, acclaiming him the "Organizer of Victory." Hence it was instead the head of Robespierre which fell, and Carnot survived to take an active part in the École Polytechnique. Carnot was greatly interested in education at all levels, even though he seems never to have taught a class.

Carnot led a charmed political life until 1797. He had gone from the National Assembly to the Legislative Assembly, to the National Convention, to the powerful Committee of Public Safety, to the Council of Five Hundred and the Directorate. In 1797, however, he refused to join a partisan coup d'etat and was promptly ordered deported. His name was stricken from the roles of the Institut and his chair of geometry was voted unanimously to General Bonaparte. Even Monge, fellow republican and mathematician, approved the intellectual outrage. About the only thing that can be said in extenuation of his action is that Monge seems to have been mesmerized by Napoleon. Monge followed his idol through thick and thin, his devotion being such that he became literally sick every time Napoleon lost a battle. This is in contrast to Carnot, who initially was responsible for Bonaparte's rise to power through his appointment to the Italian campaign. But Carnot did not hesitate to oppose the Frankenstein he had created, although it nearly cost him his life. Mathematically, his proscription turned out to be a good thing, for it gave him an opportunity, while in exile, to complete a work which had been on his mind for some time. One should expect that a man engaged, as was Carnot, in affairs of enormous practical exigency, would tend to think in terms of immediate practicality. Trajectories would appear to be a more likely subject for study than abstract ideas. But the work which Carnot had been planning during his politically busy days was, *mirabile dictu*, the *Réflexions sur la métaphysique du calcul infinitésimal* which appeared in 1797. This was not a work on applied mathematics; it came closer to philosophy than physics, and in this respect it adumbrated the period of rigor and concern for foundations so typical of the following century. Carnot's *Réflexions* became very popular and ran through a number of editions in several languages, proving that even in times that try men's souls pure mathematics finds many devotees.

Carnot was not the only one of our Revolutionary group who felt the need for greater rigor in mathematics. We have mentioned the lamentable state of geometry as portrayed by Bézout's *Cours de mathématiques*. This prompted Legendre, who was, after all, primarily an analyst, to revive some of the intellectual quality of Euclid. The result was the *Éléments de géométrie* which appeared in 1794, the very year of the Terror. Here, too, one sees the very antithesis of what generally is regarded as practical. As Lengendre says in the preface, his object is to present a geometry which shall satisfy the "spirit." It is indeed heartening to note that at least a number of mathematicians were able to resist the anti-intellectual forces of the day, and it should serve to inspire us to do likewise in times of crisis. The result of Legendre's efforts was a remarkable successful textbook—one of the mathematical products of the Revolution which had pervasive influence, for twenty editions appeared within the author's lifetime. Legendre wrote that his object was "to compose a very rigorous element" of geometry; but the author did not wax pedantic to the point of making a fetish of rigor at the expense of clarity.

Often we are inclined to think of American mathematics as influenced primarily by German scholarship, for a generation ago one went to Goettingen to be in touch with the foremost scholars in the field. We are prone to forget that during much of the nineteenth century it was French mathematics which dominated American teaching, and this was primarily through the work of the very men we have been considering. Textbooks by Lacroix, Biot, and Lagrange were published in America for use in the schools; but perhaps the most influential of all was the geometry of Legendre. *Davies' Legendre* became almost a synonym for geometry in America. As late as 1885 Dean Van Amringe of Columbia wrote in the preface of another edition, "It is believed that in clearness and precision of definition, in general simplicity and rigor of demonstration, in orderly and logical development of the subject, and in compactness of form, *Davies' Legendre* is superior to any work of its grade for the general training of the logical powers of pupils, and for their instruction in the great body of elementary geometric truth."

But if Carnot and Legendre were disciples of clear and rigorous thought, Lagrange was the high priest of the cult. At the height of the Terror Lagrange had thought seriously of leaving France; but just at this critical point the École Normale and the École Polytechnique were established, and Lagrange was invited to lecture on analysis. Lagrange seems to have welcomed the opportunity to teach, although it had been many years since he had done lecturing at Turin. In the interval he had been under the patronage of the nobility, but during the Revolution he did not take sides. Perhaps this was the result of political apathy, possibly it was due to Lagrange's mental depression at the time. At all events, his appointment to the newly established schools awoke him from his lethargy. The new curriculum called for new lecture notes, and these Lagrange supplied in the form of a classic in mathematics. His *Théorie des fonctions analytiques* appeared in the same year as Carnot's *Métaphysique,* and together they make 1797 a banner year for the rise of rigor. Lagrange's function theory, which developed some ideas he had presented in a paper about twenty-five years earlier, certainly was not useful in the narrower

sense, for the notation of the differential was far more expeditious and suggestive than the Lagrangian derived function, from which our name derivative comes. The whole motive of the work was not to try to make the calculus more utilitarian, but to make it more logically satisfying. The key idea is easy to describe. The function $f(x) = 1/(1 - x)$ when expanded by long division, yields the infinite series $1 + 1x + 1x^2 + 1x^3 + \ldots\ldots + 1x^n + \ldots$. If the coefficient of x^n is multiplied by $n!$, Lagrange called the result the value of the nth derived function of $f(x)$ for the point $x = 0$—with suitable modification for expansions of functions about other points. Lagrange thought that through this device he had eliminated the need for limits or infinitesimals; but alas, there are flaws in his fine scheme. Not every function can be so expanded; and, moreover, the question of the convergence of the infinite series brings back the need for the limit concept. But the work of Lagrange during the Revolution can be said to have had a broader influence through the initiation of a new subject which has ever since been the center of attention in mathematics—the theory of functions.

We have said little so far about Laplace, who in his day was regarded more highly as a mathematician than was Lagrange. There are two reasons for our relative neglect. In the first place, Laplace took virtually no part in revolutionary activities. He seems to have had a strong sense of intellectual honesty in science, but in politics he was completely without convictions. This does not mean that he was timid, for he seems to have associated freely with those of his scientific colleagues who were suspect during the period of crisis. It is said that he too would have been in danger of the guillotine except for his contributions to science, but this statement seems to be questionable. He played a role in the Committee on Weights and Measures, but this was not of great significance. He naturally was a professor at the École Normale and the École Polytechnique, but unlike Monge and Lagrange, he did not publish lecture notes. His publications were primarily on celestial mechanics, in which he stands pre-eminent in the period since Newton. Laplace did have one fling at political administration some years later, for Napoleon, a great admirer of men of science, appointed him Minister of the Interior—a post which Carnot likewise had held for a while under Napoleon. But it is well known that Laplace, unlike Carnot, showed no aptitude for the office. A second reason for our failure to emphasize the work of Laplace is that it did not have the immediate and persistent influence that can be traced to others in our group. He represents in a sense the end of an era rather than the beginning of a new period—although one must make some exception in the case of his work in probability.

We shall terminate our account with the date 1799, for at that time Napoleon seized power and one can regard the period of the Revolution as over. This was far from the end of the activity of the five survivors in our group, for every one of them continued to make contributions to mathematics, and some to politics as well. Honors came to all of them, Monge, Carnot, and Lagrange being named counts of the empire and Laplace achieving the title of marquis. Of our group of six, only Legendre seems never to have borne a title. Mathematically our story has a happy ending, for our scholars were able to continue their work until the end. Politically, however, at least two of our heroes were to suffer defeat. Carnot and Monge had strong political convictions, and both of them voted for the death of Louis XVI. Carnot, the more consistent of the two, ever was opposed to dictators, and in 1804 he was the only Tribune with sufficient courage and conviction to vote against naming Napoleon emperor. Yet later when he felt that the welfare of France demanded it, Carnot willingly served under Napoleon, both in the army and in governmental administration. Monge, on the other hand, supported his idol from the revolutionary corporal to the despotic emperor. He and Fourier accompanied Bonaparte on the Italian and Egyptian campaigns, and it was Monge who executed the distasteful task of determining what works of art were to be brought back to Paris as war booty. Following the restoration of the French monarchy, Carnot was forced to seek exile in Magdeburg, and Monge was banished and stripped of all his honors, including his place in the École Polytechnique and the Institut National. The turn in events was accepted courageously by Carnot, who continued his scholarly activities, but it broke the spirit of Monge, who died shortly afterward. Lagrange had died a few years before the Napoleonic crisis. Legendre seems to have remained politically neutral throughout the changes, for he was shy and retiring; but he produced a steady stream of publications on elliptic integrals and the theory of numbers, as well as contributions to other parts of mathematics. Toward the end of his life he, too, suffered politically. Because he resisted the move of the government to dictate to the Académie des Sciences, he was deprived of his pension. Laplace, on the other hand, made peace with each regime as it came along, including in editions of his works glowing tributes to whichever side happened to be in power. Posterity, as a result, has admired Laplace for his mathematics while deploring his political maneuvering.

It is now over a century and a half since the days of which we have been speaking, and we can look back on the period dispassionately. Are there any lessons to be drawn from our survey? One that can be drawn is that the things that really count in mathematics, and that have lasting influence, are not those which immediate practicality dictates. Even in times of crisis it is things of the "spirit" (in the French sense) which count most, and this spirit is perhaps best imparted by great teachers. But perhaps more important than this is the moral that, like Carnot, one should never lose heart, no matter how disillusioning the political or intellectual outlook may be.

The *Ladies' Diary* . . . Circa 1700

TERI PERL

RECENTLY I learned of the existence of a journal for women, published in England throughout the eighteenth century, which was largely devoted to problems and puzzles in mathematics. It was called the *Ladies' Diary*, and it was fascinating in that it revealed mathematics to be a proper, pleasant, and possible pastime for gentlewomen (fig. 1).

The *Ladies' Diary* was published annually from 1704 to 1841. Questions presented in one issue were answered in the following year's issue. The *Ladies' Diary* was an almanac(k) as well as a collection of practical exercises involving many branches of contemporary mathematics. During its day, it contained contributions by almost all mathematicians of eminence in England, among them Thomas Simpson (of Simpson's rule, 1710–1761), its editor from 1754 to 1760. In all, this journal may be said to have reflected the progress of mathematical science in England in its time. I. Grattan-Guinness, editor of *Annals of Science*, compares the *Ladies' Diary*, in level and purpose, to *Scientific American* today.*

According to Thomas Leybourn, writing about the *Ladies' Diary*,

> the diary was projected and begun in the year 1704 by Mr. John Tipper, who conducted it until 1713. It does not appear that the improvement of Mathematical Science was a particular object with the ingenious projector: indeed the law, which the first

FIGURE I

* Grattan-Guinness, 1974; personal communication.

Special thanks are due to Cecily Young Tanner, daughter of mathematicians William and Grace Chisholm Young, who told me about the existence of the diaries in London during the summer of 1974. Mrs. Tanner is herself a mathematician, now retired professor of mathematics at Imperial College, London.

Reprinted from *Mathematics Teacher* 70 (Apr., 1977): 354–58; with permission of the National Council of Teachers of Mathematics

contributors imposed on themselves, of not only proposing but also of answering all questions in rhyme, was not favourable to the development of Mathematical genius. [Leybourn 1817]

A look at specific issues of the diaries shows them to have been small—about ten centimeters by fifteen centimeters in size. In the preface to the 1727 issue, the editor

has the Happiness in a New Path to meeth with no competitors, for the Pleasure he takes in endeavouring to employ Some Spare Hours of the Fair-Sex in a Study Innocent, Useful, and Diverting: And is always willing to shew that Veneration for them whilst they are pleas'd to contribute their Performances, for Illustrating the DIARY, but publishing to the WORLD the HONOUR and Merit of the British FAIR.

The 1727 issue contained "an article by Mr. Lock on the Mathematicks." It also contained some letters to the "Author," some enigmas and answers (we could call these *word puzzles*), and some "Arithmetical Quests."

The first part of each issue was an almanac. In this section relevant items were collected, month by month, page by page. The layout of each page had several distinct features (fig. 2). First, the day and hour of the appearance of the new moon and full moon were given. This was followed by a listing of information associated with each day of that month; for example, the time of sunrise, important birthdates, the length of day and night in hours and minutes, the end of Oxford and Cambridge terms (March 25,F), moonrises, and holidays. At the bottom of each page was a listing of unusual weather for the same month the previous year. Dates of freezing days, windy days, stormy days, great rains as well as ordinary rains were to be found here.

In the earlier issues, many women's names appear as proposers, as well as answerers, of mathematics problems. One name appears frequently, referred to by Mr. Leybourn as "the ingenious" Mrs. Barbara Sidway. Here Mrs. Sidway solves a

FIGURE 2

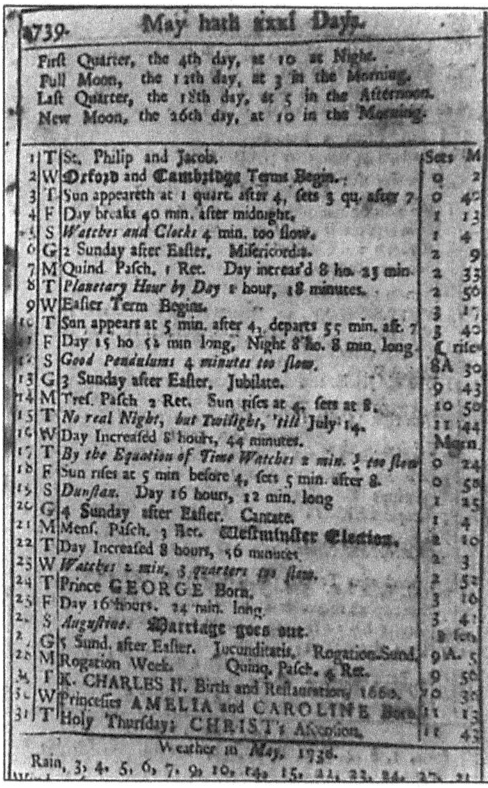

problem referred by a Mr. William Howney, and then goes on to propose one of her own. Mr. Howney's problem goes like this:

A General who had served the King successfully in his wars, asked as a reward for his services, a farthing for every different file of ten men each, which he could make with a body of 100 men. The King, thinking the request a very moderate one, readily assented. Pray what sum would it amount to?

Mrs. Sidway answered,

$$\frac{91}{1} \times \frac{92}{2} \times \frac{93}{3} \times \frac{94}{4} \times \frac{95}{5} \times \frac{96}{6} \times \frac{97}{7}$$

$$\times \frac{98}{8} \times \frac{99}{9} \times \frac{100}{10}$$

= 17310309456440 farthings
= 180315723501.9s.3d.

The following theorem is proposed to answer all questions of this nature: The number of combinations of m given things, all different from each other, taken by a given number n at a time, is equal to

$$\frac{m}{1} \times \frac{m-1}{2} \times \frac{m-2}{3} \times \frac{m-3}{4} \times \text{etc.}$$

to n terms.

Mrs. Sidway in turn asks readers to solve the question of how "from a given cone to cut the greatest cylinder possible."

Four men presented solutions in the following year's Diary. One of these was a straightforward geometric solution, which I have included here. The other is interesting because it involves *fluxions*, which was the British term for the derivative. Simpson, in his *New Treatise on Fluxions* (1737), defined a fluxion thus: "The magnitude by which any flowing quantity would be uniformly increased in a given portion of time with the generating celerity at any proposed position or instant (was [were]) it from thence to continue invariable) is the fluxion of the said quantity at that position or instant" (Kline 1972, p. 427).

The geometric solution was as follows:

Let ACB be the given cone [fig. 3], DE the diameter, and PQ the altitude of the greatest inscribed cylinder; then the solidity of the cylinder is $DE^2 \times .7854 \times PQ$, therefore $DE^2 \times PQ$ must be a maximum. But DE has to AP a given ratio, namely that which the diameter of the base has to the altitude,

and therefore when $DE^2 \times PQ$ is the greatest, $AP^2 \times PQ$ will also be the greatest: and this is the case when AP is double of PQ; hence the cone must be cut as a third of its altitude from the base.

The solution by fluxions is as follows:

Put $AQ = a$, $BC = b$, and $X = AP$; then by similar triangles $a : b :: X : \frac{b}{a}X = DE$; whence $DE^2 \times PQ = \frac{b^2}{a^2} \times X^2 \times (a - X)$; therefore, leaving out the constant factor $\frac{b^2}{a^2}$, $X^2(a - X)$ must be a maximum. Taking fluxions $2aX\dot{X} - 3X^2\dot{X} = 0$, and reducing, $X = \frac{2}{3}a$, whence $a - X = PQ = \frac{1}{3}a$, as before. [Leyborn 1817]

Before going on to other examples of problems that appeared in the diaries, I have a few comments on some different facets of the magazine. I found the advertisements in the *Ladies' Diary* particularly interesting. The following is a sample of advertisements that appeared in the first half of the eighteenth century.

Astronomy (new publication) by C. Leadbetter. Teacher of the Mathematicks. Printed by J. Wilcox in Little-Britain, and T. Heath near the Fountain-Tavern in the Strand.

Artificial Teeth set in so firm as to eat with them, and so exact as not to be distinguished from Natural; not to be taken out at night ... greatly helpful to Speech. Also Teeth clean'd or drawn by J. Watts and Sam Rutter, Operators, who apply themselves wholly to the said business, and live in Raquel-Court, Fleet Street, London.

One particular advertiser appears in the 1727 *and* 1728 issues:

Steel Spring and other sorts of Trusses for Ruptures at the Navel, Groin or elsewhere being very Easie and Effectual to Old or Young. Also Bag Trusses for fixed Tumours, and Instruments for the Lame or Crooked, by P. Bartlett at the Golden Ball in St. Paul's Churchyard near Cheapside, London. Persons in the Country, sending their bigness, and which side the Rupture is on, may be supply'd with the Trusses, and proper Directions. His Mother, Mrs. M. Bartlett, at the Golden Ball against St. Bride's Lane in Fleet Street, is Skilful in this Busi-

FIGURE 3

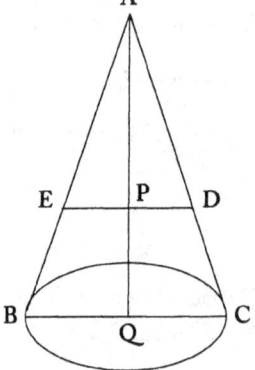

ness to her own Sex, who also makes Sturdy Stockings, Knee and Ankle-pieces very useful for swelt'd limbs.

A look at the diaries for 1776 and 1777 showed advertisements no longer appearing in the magazine. These later issues contain questions that seem different from those in women's magazines of today only in the style of language used. Here are a few sample "queries" I noted.

Ye learned, pray say, why people deceased are always interr'd with their heads to the west?

Whether there is anything that will prevent the cramp when people are swimming; and if there is, what?

On what principle is it that ladies' turn down an empty tea-cup in the bottom of a fruit pye, to prevent the syrrup from boiling out in baking?

A rebus from one of the later issues went like this:

If you a bullock set before
A shallow river's brink,
You may a city's name explore
I' the time you let him drink.

A problem proposed by a Miss Ann Nichols in 1761 is a good example of the rhyme format that was encouraged in the *Ladies' Diary*. Answered by several people in 1762, one writer pointed out an ambiguity in the problem as originally stated. Miss Nichols wrote—

Old John, who had in credit liv'd,
Tho' now reduc'd, a sum receiv'd:
This lucky hit's no sooner found,
Than clam'rous duns came swarming round:
To th' landlord, baker, many more,
John paid in all pounds ninety-four.
Half what remain'd, a friend he lent,
On Joan and self, one-fifth he spent;
And when of all these sums bereft,
One-tenth o' th' sum receiv'd had left.
Now shew your skill, ye learned fair,
And in your next that sum declare.

A Mr. Tho. Sadler replied—

If X be supposed equal the whole sum John received, then will

$$\frac{X-94}{2} + \frac{X-94}{5} + \frac{X}{10} + 94$$

or

$$\frac{8X+282}{10} = X,$$

per the conditions of the question: whence $X = 141$ £ the sum required. Mr. George Salmon, late of Mr. B. Donn's school, and several others, observe, that this question is ambiguous, it being doubtful whether $^1/_5$ of the whole sum received or only $^1/_5$ of what remained after the 94 £ was paid, was the part thereof spent on Joan and himself, and accordingly find the whole sum received to be either 141 or 235 pounds.

Here is an interesting puzzle problem from a later issue. Proposed in 1791 and answered in 1792, by which time the problems and solutions were all by men, this problem involves discovering the word described in the riddle by using a mapping from numbers to letters. Two solutions are included.

What's *high* in esteem,
As the ladies now deem,
To declare, from below, condescend:
In Diary next year
Pray let it appear,
'Twill oblige your poetical friend.
$XYZ = 160$; $X^2 + Y^2 + Z^2 = 465$; $X^3 + Y^3 + Z^3 = 8513$.

The answer by a Mr. Joseph Garnett was as follows:

By the question X, Y, and Z are to be whole numbers, and from the third equation no one can be above 20, and some one must be more than 14. Now from the first equation the three roots must evidently be some divisors of 160, which are 1, 2, 4, 5, 8, 10, 16, and 20, among which there are only two above 14, viz. 16 and 20, therefore the numbers are either

$$16 \begin{cases} 10 \text{ and } 1, \\ 5 \text{ and } 2 \end{cases} \quad \text{or } 20 \begin{cases} 8 \text{ and } 1 \\ 4 \text{ and } 2 \end{cases}$$

of which 20, 8, and 1 are the only ones that will answer the conditions, and the word is HAT.

Another solution to this problem came from a Mr. J. Holt of Manchester:

> Because no word can be formed without a vowel, the value of one of the unknown quantities must answer to a vowel, and be such that the product of the other two in the first equation be a composite number; but 1 and 5 only have these properties. Now if 5 be substituted for X in the second equation, then $Y^2 + Z^2 = 440$: but no two perfect squares whatever will make this number; therefore $X = 1$, consequently $Y^2 + Z^2 = 464$, $YZ = 160$; from this latter equation $Z = 160 + Y$, which value of Z substituted in the former, it becomes $Y^2 + 25600 + Y^2 = 464$, or $Y^4 - 464 Y^2 = -25600$, the roots of which quadratic equation are 8 and 20, which are the values of Y and Z. Hence the required word is HAT.

The problems I've included here are rather straightforward and relatively simple. I had very little time to explore the diaries this first time around, and I tended to skip problems that looked too complex. Of particular interest to me was that whereas the early issues of the *Ladies' Diary* contained many questions and solutions contributed by women, the later ones seem to have fewer and fewer contributions from women. I wonder why this was so. Perhaps this effect was a reflection of the attitude of the editor at a particular time. Perhaps it was a reflection of a larger change in the role of women in society in general. It might be interesting to explore this further. I found the existence of the diaries fascinating in itself and definitely worth a longer look. I hope to get back to them one day, and I hope that others will also consider them worth looking at more carefully.

REFERENCES

Kline, Morris. *Mathematical Thought from Ancient to Modern Times.* New York: Oxford University Press, 1972.

The Ladies' Diary: or, the Woman's Almanack. British Museum Library, Stanford University Library (fragments).

Leybourn, Thomas. *The Mathematical Questions, Proposed in the "Ladies' Diary," and Their Original Answers Together with Some New Solutions, 1704 to 1816.* 4 vols. London: J. Mawman, 1817.

HISTORICAL EXHIBIT 13

Women in Mathematics

In both ancient classical and traditional societies, evidence for the participation of women in mathematics is particularly scarce. It is known that the Pythagoreans (6th century B.C.) permitted women to attend some of their ceremonies; but to what extent these women were actually involved with mathematics remains unclear. History recognizes Hypatia (ca. A.D. 370–415) of Alexandria as the first, known, woman mathematician. She was the daughter of the Greek mathematician, Theon, acknowledged for his commentaries on Ptolemy's *Almagst* and Euclid's *Elements*. While Hypatia also distinguished herself as a commentator, writing on Diophantus's *Arithmetica* and Appollonius's *Conic Sections*, she was renowned for her knowledge of mathematics, medicine, and philosophy. Hypatia, as the leader of the Neo-Platonic school of philosophy in Alexandria, was a proponent of "pagan-science" and, as such, was singled out by a fanatical Christian mob and murdered.

With the Christian homogamy of Europe that followed the fall of the Roman Empire, mathematics became the prerogative of churchmen, academicians, and, later, merchants. Thus, by associating mathematics with male-dominated professions, Western traditions excluded women from major mathematical pursuits for many centuries. However, this has not necessarily been the case in non-Western societies. Recent research by Professor Michael Closs* of the University of Ottawa reveals that, during the classical period of their civilization (A.D. 300–900), the Mayan peoples of Mesoamerica employed female mathematician-scribes. An illustration from a painted pottery vessel of this period depicts such a female mathematician at work.

Women's involvement with mathematics is culturally bound and, in the contemporary world, is experiencing a rapid reorientation.

* Michael P. Closs, "I Am a Kahal; My parents were Scribes," *Research Reports on Ancient Maya Writing* (May, 1992), pp. 7–22.

Epilogue: The Process Continues

*T*he close of the 18th, and the beginnings of the 19th centuries witnessed the rise of European industrialization and the forces and movements associated with it. Agricultural societies now became involved in production of manufactured goods. Drawn by economic opportunities, rural workers migrated to the cities and towns that served as the centers for manufacturing activities. Toil in the fields was replaced by work in a factory or mill. Due to these migrations, urban populations greatly expanded. A new distribution of wealth spawned an enlarged middle-class, which soon began expressing itself as a political force. The quest for natural resources and economic markets to fuel the new industries led nations onto the paths of imperial expansion involving exploration, exploitation and the establishment overseas empires. It was a time of widespread political, economic and social unrest; a period of human action and reaction.

Unfolding industrial processes also gave rise to new scientific fields of concern: the physics of heat transfer; the patterns of fluid motion; the effects of vibrations; elasticity and plasticity of solid materials, and the dynamics of electromagnetic fields. An understanding and harnessing of such emerging phenomena depended on mathematics. New mathematical challenges were encountered. The concepts of Newtonian physics faltered and much of the "old mathematics" just didn't work. Mathematical relationships deviated from the expected norm; where mathematical continuities were sought they often did not exist. Contradictions and paradoxes began to appear, pathological functions were discovered, that is, mathematical functions which deviated from any previous known behavior. New and strange geometries were invented. Noncommutative mathematical systems were devised. It was a time of mathematical uncertainty and confusion. Many of the existing cornerstones of traditional mathematics were shaken, forcing a re-examination of the foundations of mathematics including even questioning the essence of mathematical proof. Analytical efforts moved from the application of fixed rules to attempting to understand the basic structures of mathematics itself. Questions arose as to the nature and properties of the real numbers, cardinality and the "sizes" of infinity.

But despite this climate of mathematical flux and uncertainty, some social and political forces acted to strengthen the place of mathematics in society. University curricula were liberalized to include more sciences and to support and nurture mathematical research. German universities led in this movement. Professional organizations devoted to the undertaking of mathematics and mathematical research were founded on a national basis. The London Mathematical Society originated in1865; in Paris, the Société Mathématique de France began operation in 1872 and, in 1884, the Circolo Matematico di Palermo appeared in Italy. In the following years, other major European countries would establish their own mathematical societies. Each of these organizations published transactions and journals which, in themselves, became vehicles for mathematical communication, information and cooperation. Centers of mathematical study and research were organized in France and Germany. Mathematical activities began in earnest in the United States. This enriched climate of inquiry and encouraged creativity produced new generations of mathematicians who readily took up the emerging mathematical challenges and continued the search for mathematical certainty.

Suggested Readings

For a broader perspective on European mathematics during the period 1000 to 1800 see:

Eves, Howard. *An Introduction to the History of Mathematics.* Philadelphia: Saunders College Publishing, 1990. Chapters 8 – 12.

Katz, Victor J. *The History of Mathematics: A Brief Version.* New York: Addison Wesley, 2003. Chapters 8 - 14.

More specific information can be found in the following:

Aiton, E. *Leibniz: A Biography.* Bristol: Higler, 1985.

Bardi, Jason S. *The Calculus Wars: Newton, Leibniz, and the Greatest Mathematical Clash of all Time.* New York: Thundermouth Press, 2006.

Baron, M. E. *The Origins of the Infinitesimal Calculus.* Oxford: Pergamon Press, 1969.

Barroclough, Geoffrey. *The Crucible of Europe.* Berkeley: University of California Press, 1976.

Berggren, Lennart J. "Medieval Arithmetic: Arab Texts and European Motivations." *Word, Image, Number: Communication in the Middle Ages.* pp 351 – 365.

Blackwell, Richard. *Behind the Scenes at Galileo's Trial.* Notre Dame: University of Notre Dame Press, 2006.

Bogolyubov, N.N. *Euler and Modern Science.* Washington, DC: The Mathematical Association of America, 2007.

Brady, Robert E. *Euler at 300: An Appreciation.* Washington, DC: The Mathematical Association Association of America, 2007.

Brown, Nancy M. *The Abacus and the Cross: The Story of the Pope who Brought the Light of Science to the Middle Ages.* New York: Basic Books, 2010.

Cardano, Girolamo. *The Great Art, or the Rule of Algebre.* Richard Winsted (tr.), Cambridge, MA: MIT Press, 1968.

Contreni, John and Casciani Santa (ed.). *Word, Image, Number: Communication in the Middle Ages.* Sismel: Edizoni Del Galluzzo, 2002.

Costa, Shelley. "The Lady's Diary: Gender, Mathematics and Civil Society in Eighteenth Century England", Osiris, 2002, 17: 49 – 73.

Crosby, Alfred W. *The Measure of Reality: Quantification and Western Society 1250 – 1600.* New York: Cambridge University Press, 1997.

Derbyshire, John. *Unknown Quantity: a Real and Imaginary History of Algebra.* Washington, DC: John Henry Press, 2006.

Devlin, Keith. *The Man of Numbers: Fibonacci's Arithmetic Revolution.* New York: Walker and Company, 2011.

Dolnick, Edward. *The Clockwork Universe: Isaac Newton, the Royal Society and the Birth of the Modern World.* New York: Harper Collins Books, 2011.

Drake, Stillman. *Galileo Pioneer Scientist.* Toronto: University of Toronto Press, 1990.

Dunham, William. *The Calculus Gallery from Newton to Lebesque.* Princeton: Princeton University Press, 2005.

Edwards, C. H. *The Historical Development of the Calculus.* New York: Springer, 1979.

Fauvel, John, et al., *Let Newton Be! A New Perspective on his Life and Works.* Oxford: Oxford University Press, 1989.

Ferguson, Kathy. *Tyco & Kepler: The Unlikely Partnership that Changed our Understanding of the Heavens.* New York: Walker Books, 2002.

Folkerts, Menso. *The Development of Mathematics in Medieval Europe: the Arabs, Euclid, Regionmontanus.* Aldershot, UK; Ashgate Publishing, 2006.

Gies, Joseph and Frances. *Leonardo of Pisa and the New Mathematics of the Middle Ages*. New York: Thomas Crowell, 1969.

Merchants and Moneymen: The Commercial Revolution. New York: Thomas Crowell, 1972.

Gimpel, Jean. *The Medieval Machine: The Industrial Revolution and the Middle Ages*. New York: Holt, Rinehart & Winston, 1976.

Glimp, D. and Warren, M. L (eds.). *The Arts of Calculation: Quantifying Thought in Early Modern Europe*. Basingstoke, UK; Palgrave McMillan, 2004.

Hall, Rupert A. *Philosophers at War: The Quarrel between Newton & Leibniz*. Cambridge: Cambridge University Press, 1980.

Hughes, Barnabas (ed). *Fibonacci's De Practica Geometrie*. New York: Springer, 2008.

Kearney, Hugh. *Science and Change 1500 – 1700*. New York: World Library Press, 1971.

Koestler, Arthur. *The Sleepwalkers: A History of Man's Changing View of the Universe*. New York: Perseus Books, 1959.

Kuhn, Thomas. *The Copernican Revolution*. New York: Modern Library Press, 1957.

Lunenburg, H. *Leonhardi Pisani Liber Abbaci oder Lesevergnüge eines Mathematikers*. Mannheim: Wissenschaftaverlag, 1993.

Mahoney, Michael. *The Mathematical Career of Pierre de Fermat*. Princeton: Princeton University Press, 1994.

McMullen, Ernan (ed.). *The Church and Galileo*. Notre Dame: Notre Dame University Press, 2005.

Murray, Alexander. *Reason and Society in the Middle Ages*. Oxford: Clarendon Press, 1978.

Pullan, J. M. *The History of Abacus*. New York: Frederick A. Prager, 1969

Sandifur, Edward. *The Early Mathematics of Leonhard Euler*. Washington, DC: The Mathematical Association of America, 2007.

How Euler Did It. Washington, DC: The Mathematical Association of America, 2007.

Sigler, Laurence (trans.). *Leonardo Pisano the Book of Squares; An Anointed Translation into Modern English*. Boston: Academic Press, 1987.

(trans.). *Fibonacci's Liber Abaci- A Translation into Modern English of Leonardo Pisano's Book of Calculation*. New York: Springer, 2002.

Stedall, Jacqueline A. *A Discourse Concerning Algebra: English Algebra to 1685*. New York: Oxford University Press, 2002.

Stevens, Wesley M. "Fields and Streams: Language and Printing of Arithmetic and Geometry in Early Medieval Schools", *Word, Image, Number: Communication in the Middle Ages*. pp. 113 – 205.

Swetz, Frank J. *Capitalism and Arithmetic: The New Math of the 15th Century*. Chicago: Open Court publishing, 1989.

"Figura Mercantesco: Merchants and the Evolution of a Number Concept in the Later Middle Ages", *Word, Image, Number: Communication in the Middle Ages*. pp. 391 – 412.

Vrooma, Jack. *René Descartes: A Biography*. New York: C. P. Putnam's Sons, 1970.

Westfal, Richard. *Never at Rest: A Biography of Isaac Newton*. Cambridge: Cambridge University Press, 1980.

Index